Prentice Hall Advanced Reference Series

Physical and Life Sciences

Equilibrium Statistical Physics

Michael Plischke
Simon Fraser University

Birger Bergersen
University of British Columbia

Prentice Hall, *Englewood Cliffs, New Jersey 07632*

Library of Congress Cataloging-in-Publication Data

PLISCHKE, MICHAEL.
  Equilibrium statistical physics.

  Bibliography: p.
  Includes index.
  1. Statistical physics. 2. Critical phenomena
(Physics) I. Bergersen, Birger. II. Title.
QC174.8.P55 1989   530.1'3   88-9883
ISBN 0-13-283276-3

Editorial/production supervision
  and interior design: BARBARA MARTTINE
Cover design: LUNDGREN GRAPHICS, LTD.
Manufacturing buyer: MARY ANN GLORIANDE

Prentice Hall Advanced Reference Series

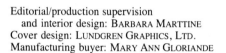
Printed in the United States of America

10 9 8 7 6 5 4 3 2 1

ISBN 0-13-283276-3

PRENTICE-HALL INTERNATIONAL (UK) LIMITED, *London*
PRENTICE-HALL OF AUSTRALIA PTY. LIMITED, *Sydney*
PRENTICE-HALL CANADA INC., *Toronto*
PRENTICE-HALL HISPANOAMERICANA, S.A., *Mexico*
PRENTICE-HALL OF INDIA PRIVATE LIMITED, *New Delhi*
PRENTICE-HALL OF JAPAN, INC., *Tokyo*
SIMON & SCHUSTER ASIA PTE. LTD., *Singapore*
EDITORA PRENTICE-HALL DO BRASIL, LTDA., *Rio de Janeiro*

Contents

3 Mean Field and Landau Theory   49

4 Dense Gases and Liquids   99

5 Critical Phenomena, Part 1   131

Preface

During the last decade each of the authors has regularly taught a graduate or senior undergraduate course in statistical mechanics. During this same period, the renormalization group approach to critical phenomena, pioneered by K. G. Wilson [1971], greatly altered our approach to condensed matter physics. Since its introduction in the context of phase transitions, the method has found application in many other areas of physics, such as many-body theory, chaos, the conductivity of disordered materials, and fractal structures. So pervasive is its influence that we feel that it now essential that graduate students be introduced at an early stage in their career to the concepts of scaling, universality, fixed points, and renormalization transformations, which were developed in the context of critical phenomena, but are relevant in many other situations.

In this book we describe both the traditional methods of statistical mechanics and the newer techniques of the last two decades. Most graduate students are exposed to only one course in statistical physics. We believe that this course should provide a bridge from the typical undergraduate course (usually concerned primarily with noninteracting systems such as ideal gases and paramagnets) to the sophisticated concepts necessary to a researcher.

We begin with a short chapter on thermodynamics and continue, in Chapter 2, with a review of the basics of statistical mechanics. We assume that the student has been exposed previously to the material of these two chapters and thus our treatment is rather concise. We have, however, included a substantial number of exercises that complement the review.

In Chapter 3 we begin our discussion of strongly interacting systems with a lengthy exposition of mean field theory. A number of examples are worked out in detail. The more general Landau theory of phase transitions is developed and used to discuss critical points, tricritical points, and first-order phase transitions. The limitations of mean field and Landau theory are described and the role of fluctuations is explored in the framework of the Landau-Ginzburg model.

Chapter 4 is concerned with the theory of dense gases and liquids. Many of the techniques commonly used in the theory of liquids have a long history and are well described in other texts. Nevertheless, we feel that they are sufficiently important that we could not omit them. The traditional method of virial expansions is presented and we emphasize the important role played in both theory and experiment by the pair correlation function. We briefly describe some of the useful and still popular integral equation methods based on the Ornstein–Zernike equation used to calculate this function as well as the modern perturbation theories of liquids. Simulation methods (Monte Carlo and molecular dynamics) are introduced. In the final section of the chapter we present an interesting application of mean field theory, namely the van der Waals theory of the liquid–vapor interface and a simple model of roughening of this interface due to capillary waves.

Chapters 5 and 6 are devoted to continuous phase transitions and critical phenomena. In Chapter 5 we review the Onsager solution of the two-dimensional Ising model on the square lattice and continue with a description of the series expansion methods, which were historically very important in the theory of critical phenomena. We formulate the scaling theory of phase transitions following the ideas of Kadanoff, introduce the concept of universality of critical behavior, and conclude with a mainly qualitative discussion of the Kosterlitz–Thouless theory of phase transitions in two-dimensional systems with continuous symmetry.

Chapter 6 is entirely concerned with the renormalization group approach to phase transitions. The ideas are introduced by means of technically straightforward calculations for the one- and two-dimensional Ising models. We discuss the role of the fixed points of renormalization transformations and show how the theory leads to universal critical behavior. The original ϵ-expansion of Wilson and Fisher is also discussed. This section is rather detailed, as we have attempted to make it accessible to students without a background in field theory.

In Chapter 7 we turn to quantum fluids and discuss the ideal Bose gas, the weakly interacting Bose gas, the BCS theory of superconductivity, and the phenomenological Landau–Ginzburg theory of superconductivity. Our treatment of these topics (except for the ideal Bose gas) is very much in the spirit of mean field theory and provides more challenging applications of the formalism developed in Chapter 3.

Chapter 8 is devoted to linear response theory. The fluctuation-dissipation theorem, the Kubo formalism, and the Onsager relations for transport coefficients are discussed. This chapter is consistent with our emphasis on equilibrium phenomena—in the linear response approximation the central role is played by equilibrium correlation functions. A number of applications of the formalism, such as the dielectric response of an electron gas, the elementary excitations of a Heisenberg ferromagnet, and the excitation spectrum of an interacting Bose fluid, are discussed in detail. The complementary approach to transport via the linearized Boltzmann equation is also presented.

Chapter 9 provides an introduction to the physics of disordered materials. We discuss the effect of disorder on the quantum states of a system and introduce (as an example) the notion of localization of electronic states by an explicit calculation for a one-dimensional model. Percolation theory is introduced and its analogy to thermal phase transitions is elucidated. The nature of phase transitions in disordered materials is discussed and we conclude with a very brief and qualitative description of the glass and spin glass transitions. These subjects are all very much at the forefront of current research and we do not claim to be at all comprehensive in our treatment. In compensation, we have provided a more extensive list of references to recent articles on these topics than elsewhere in the book.

We have found the material presented here suitable for an introductory graduate course, or with some selectivity, for a senior undergraduate course. A student with a previous course in statistical mechanics, some background in quantum mechanics, and preferably, some exposure to solid state physics should be adequately prepared. The notation of second quantization is used extensively in the latter part of the book and the formalism is developed in detail in the Appendix. The instructor should be forewarned that although some of the problems, particularly in the early chapters, are quite straightforward, those toward the end of the book can be rather challenging.

Much of this book deals with topics on which there is a great deal of recent research. For this reason we have found it necessary to give a large number of references to journal articles. Whenever possible, we have referred to recent review articles rather than to the original sources.

The writing of this book has been an ongoing (frequently interrupted) process for a number of years. We have benefited from discussion with, and critical comments from, a number of our colleagues. In particular, Ian Affleck, Leslie Ballentine, Robert Barrie, John Berlinsky, Peter Holdsworth, Zoltàn Ràcz, and Bill Unruh have been most helpful. Our students Dan Ciarniello, Victor Finberg, and Barbara Frisken have also helped to decrease the number of errors, ambiguities, and obscurities. The responsibility for the remaining faults rests entirely with the authors.

MICHAEL PLISCHKE
BIRGER BERGERSEN

1

Review of Thermodynamics

In this chapter we present a brief review of elementary thermodynamics. This chapter complements Chapter 2, in which the connection between thermodynamics and statistical mechanical ensembles is established. The reader may wish to use this chapter as a short refresher course and may wish to consult one of the many books on thermodynamics, such as that of Callen [1985] or Chapters 2 to 4 of the book by Reichl [1980], for a more complete discussion of the material. The outline of this chapter is as follows. In Section 1.A we introduce the notion of state variables and equations of state. Section 1.B contains a discussion of the laws of thermodynamics, definition of thermodynamic processes, and the introduction of the entropy. In Section 1.C we introduce the thermodynamic potentials that are most useful from a statistical point of view. The Gibbs–Duhem equation and a number of useful Maxwell relations are derived in Section 1.D. In Section 1.E we turn to the response functions, such as the specific heat, susceptibility, and compressibility, which provide the common experimental probes of macroscopic systems. Section 1.F contains a discussion of some general conditions of equilibrium and stability and we conclude, in Section 1.G, with a brief discussion of the thermodynamics of phase transitions and the Gibbs phase rule.

1.A STATE VARIABLES AND EQUATIONS OF STATE

A macroscopic system has many degrees of freedom, only a few of which are measurable. Thermodynamics thus concerns itself with the relation between a small number of variables which are sufficient to describe the bulk behavior of

1

the system in question. In the case of a gas or liquid the appropriate variables are the pressure P, volume V, and temperature T. In the case of a magnetic solid the appropriate variables are the magnetic field H, the magnetization M, and the temperature T. In more complicated situations, such as when a liquid is in contact with its vapor, more variables are needed: the pressure P, temperature T, volume of liquid and gas V_L, V_G, the interfacial area A, and surface tension σ. If the aforementioned thermodynamic variables are independent of time, the system is said to be in a *steady state*. If, moreover, there are no macroscopic currents in the system, such as a flow of heat or particles through the material, the system is in *equilibrium*. Any quantity which, in equilibrium, depends only on the thermodynamic variables, rather than on the history of the sample, is called a *state function*. In subsequent sections we shall meet a number of such quantities. For a large system the state variables can normally be taken to be either *extensive* (i.e., proportional to the size of the system) or *intensive* (i.e., independent of system size). Examples of extensive variables are the internal energy, the entropy, and the mass of the different constituents or their number, while the pressure, the temperature, and the chemical potentials are intensive. The postulate that quantities like the internal energy and entropy are extensive and independent of shape is equivalent to an assumption of additivity or, as we shall see in Section 2.A, of the existence of the *thermodynamic limit*. In the process of taking the thermodynamic limit, we let the size of the system become infinitely large, with the densities (of mass, energy, magnetic moment, polarization, etc.) remaining constant.

In equilibrium the state variables are not all independent and are connected by *equations of state*. The role of statistical mechanics is the derivation, from microscopic interactions, of such equations of state. Simple examples of equations of state are the ideal gas law,

$$PV - Nk_\mathrm{B}T = 0 \tag{1}$$

where N is the number of molecules in the system and k_B is Boltzmann's constant; the van der Waals equation,

$$\left(P + \frac{aN^2}{V^2}\right)(V - Nb) - Nk_\mathrm{B}T = 0 \tag{2}$$

where a, b are constants; the virial equation of state

$$P - \frac{Nk_\mathrm{B}T}{V}\left[1 + \frac{NB_2(T)}{V} + \frac{N^2B_3(T)}{V^2} + \cdots\right] = 0 \tag{3}$$

where the functions $B_2(T)$, $B_3(T)$ are called virial coefficients; and in the case of a paramagnet, the Curie law,

$$M - \frac{CH}{T} = 0 \tag{4}$$

where C is a constant called the Curie constant. Equations (1), (2), and (4) are approximations, and we shall use them primarily to illustrate various principles. Equation (3) is, in principle, exact, but as we shall see in Chapter 4, calculation of more than a few of the virial coefficients is very difficult.

1.B LAWS OF THERMODYNAMICS

In this section we explore the consequences of the zeroth, first, and second laws of thermodynamics. The zeroth law states that if system A is in equilibrium with system B and with system C, then system B is in equilibrium with system C. This statement can also be thought of as a statement that it is possible to define a universal standard for thermometers, pressure gauges, and other measuring devices.

(a) First Law

The first law of thermodynamics is a manifestation of the law of conservation of energy. However, it also partitions the change in energy of a system into two pieces, heat and work:

$$dE = đQ - đW \tag{1}$$

In (1) dE is the change in internal energy of the system, $đQ$ the amount of heat *added* to the system, and $đW$ the amount of work done *by* the system during an infinitesimal process. Aside from the partitioning of the energy into two parts, the formula distinguishes between the infinitesimals dE and $đQ$, $đW$. The difference beween the two measurable quantities $đQ$ and $đW$ is found to be the same for any process in which the system evolves between two given states, independent of the path. This indicates that dE is an exact differential or, equivalently, that the internal energy is a state function. The same is not true of the differentials $đQ$ and $đW$, hence the difference in notation.

Consider a system whose state can be specified by the values of a set of state variables x_j (e.g., the volume, the number of moles of the different constituents, the magnetization, the electric polarization, etc.) and the temperature. We write, for the work done during an infinitesimal process,

$$đW = - \sum_j X_j \, dx_j \tag{2}$$

where the X_j's can be thought of as generalized forces and the x_j's as generalized displacements.

Before going on to discuss the second law, we pause to introduce some terminology. A thermodynamic transformation or process is any change in the state variables of the system. A *spontaneous* process is one that takes place without any change in the external constraints on the system, due simply to the internal dynamics of the system. An *adiabatic* process is one in which no heat is exchanged between the system and its surroundings. A process is *isothermal* if the temperature is held fixed, *isobaric* if the pressure is fixed, *isochoric* if the density is constant, and *quasistatic* if the process is infinitely slow. A *reversible* process is by nature quasistatic and follows a path in thermodynamic space which can be exactly reversed. If this is not possible, the process is *irreversible*. An example of a reversible process is the slow adiabatic expansion of a gas against a piston on which a force is exerted externally which is infinitesimally less than PA, where P is the pressure of the gas and A the area of the piston. An example of an irreversible process is the free adiabatic expansion of a gas into a vacuum. In this case the initial state of the gas can be recovered if one compresses it and adds heat. This is, however, not the same thermodynamic path.

(b) Second Law

We present next two equivalent statements of the *second law of thermodynamics*. The *Kelvin* version is: *There exists no thermodynamic process whose sole effect is to extract a quantity of heat from a system and to convert it entirely to work.* The equivalent statement of *Clausius* is: *No process exists in which the sole effect is that heat flows from a reservoir at a given temperature to a reservoir at a higher temperature.*

A corollary of these statements is that the most efficient engine operating between two reservoirs at temperatures T_1 and T_2 is the Carnot engine. The Carnot engine is an idealized engine in which all the steps are reversible. We show the Carnot cycle for an ideal gas working substance in Figure 1.1. In step AB heat Q_1 is absorbed by the gas, which expands isothermally and does work in the process. The next step, BC, is adiabatic and further work is done. In step CD heat $(-Q_2)$ is given off to the low-temperature reservoir and work is done on the gas. Step DA returns the working substance to its original state adiabatically. The efficiency, η, of the engine is defined to be the ratio of work done in

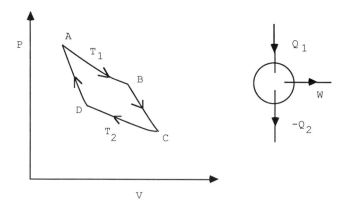

Figure 1.1 Carnot cycle for an ideal gas working substance.

the entire cycle to heat absorbed from the high-temperature reservoir:

$$\eta = \frac{W}{Q_1} = \frac{Q_1 + Q_2}{Q_1} \tag{3}$$

In (3) we have followed the convention of the first law that heat transfer is positive if added to the working system. Suppose now that a second more efficient engine operates between the same two temperatures. We can use this engine to drive the Carnot engine backwards—since it is reversible, Q_1, Q_2, and W will simply change sign and η will remain the same.

In Figure 1.2 the Carnot engine is denoted by C, the other hypothetical engine, with efficiency $\eta_o > \eta_c$ is denoted by O. We use all the work done by engine O to drive engine C. Let the heat absorbed from the reservoirs be Q_{1c}, Q_{1o}, Q_{2c}, Q_{2o}. By assumption we have

$$\eta_o = \frac{W}{Q_{1o}} > \frac{-W}{Q_{1c}} = \eta_c \tag{4}$$

These equations imply that $|Q_{1c}| > Q_{1o}$ and the net effect of the entire process is to transfer heat from the low-temperature reservoir to the high-temperature reservoir. This violates the Clausius statement of the second law. Similarly, if we take only part of the work output of engine O, and adjust it so that there is no net heat transfer to the low-temperature reservoir, a contradiction of the Kelvin statement of the second law results. We conclude that no engine operating between two reservoirs at fixed temperatures is more efficient than a Carnot engine. Equivalently, all reversible engines operating between fixed temperatures have the same efficiency and are Carnot engines.

The result that all Carnot engines operating between two temperatures have the same efficiency can be used to define a temperature scale. One possible definition is

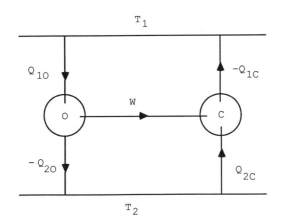

Figure 1.2   Carnot engine (C) driven by an irreversible engine (O).

$$\frac{T_2}{T_1} = 1 - \eta_c(T_1, T_2) \tag{5}$$

where $\eta_c(T_1, T_2)$ is the Carnot efficiency. Using an ideal gas as a working substance, one can easily show (Problem 1.1) that this temperature scale is identical with the ideal gas (or absolute) temperature scale. Substituting for η in equation (3), we have, for a Carnot cycle,

$$\frac{Q_1}{T_1} + \frac{Q_2}{T_2} = 0 \tag{6}$$

With this equation we are in a position to define the entropy. Consider an arbitrary reversible cyclic process such as the one drawn in Figure 1.3. We can cover the region of the P–V plane, enclosed by the reversible cycle (R in Figure 1.3), with a set of Carnot cycles operating between temperatures arbitrarily close to each other. For each Carnot cycle we have, from (6),

$$\sum_j \frac{\Delta Q_j}{T_j} = 0 \tag{7}$$

As the number of Carnot cycles goes to infinity, the integral of dQ/T over the uncompensated segments of these cycles approaches

$$\int_R \frac{dQ}{T} = 0 \tag{8}$$

Thus the expression dQ/T is an exact differential for reversible processes and we define the state function whose differential it is to be the entropy S. For reversible processes the first law can therefore be written in the form

$$dE = T\,dS - dW = T\,dS + \sum_j X_j\,dx_j \tag{9}$$

The fact that the Carnot cycle is the most efficient cycle between two tempera-

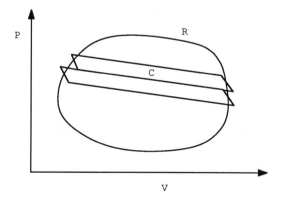

Figure 1.3   Arbitrary reversible process covered by infinitesimal Carnot cycles.

tures allows us to obtain an inequality for arbitrary processes. Consider a possibly irreversible cycle between two reservoirs at temperatures T_1 and T_2.

$$\frac{\Delta Q_1 + \Delta Q_2}{\Delta Q_1} \leq \frac{\Delta Q_{1C} + \Delta Q_{2C}}{\Delta Q_{1C}} = \eta_C \tag{10}$$

or

$$\frac{\Delta Q_2}{\Delta Q_1} \leq -\frac{T_2}{T_1} \tag{11}$$

$$\frac{\Delta Q_1}{T_1} + \frac{\Delta Q_2}{T_2} \leq 0 \tag{12}$$

Generalizing to a finite process, we obtain

$$\int \frac{dQ}{T} \leq 0 \tag{13}$$

where the equality holds for reversible processes. Since the entropy is a state function $\int dS = 0$ for any reversible closed cycle and we therefore obtain $T\Delta S \geq \Delta Q$. Combining this with the first law we have, for arbitrary processes,

$$T\Delta S \geq \Delta E + \Delta W \tag{14}$$

where, once again, the equality holds for reversible processes.

A further consequence of the foregoing discussion is that the entropy of an isolated system cannot decrease in any spontaneous process. Imagine a spontaneous process in which the system evolves from point A to point B (Figure 1.4) in the thermodynamic space. (Note that the irreversible path cannot be represented as a curve in the P–T plane. The dotted line represents a reversible path connecting the same endpoints.) Since the system is isolated $\Delta Q = 0$ and

$$\int_A^B dS \geq \int_A^B \frac{dQ}{T} = 0 \tag{15a}$$

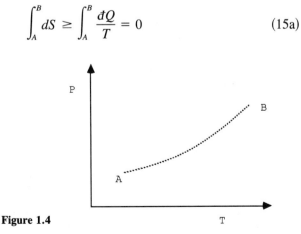

Figure 1.4

or

$$S(B) - S(A) \geq 0 \qquad (15b)$$

Since spontaneous processes tend to drive a system toward equilibrium, we conclude that the equilibrium state of an isolated system is the state of maximum entropy.

1.C THERMODYNAMIC POTENTIALS

The term *thermodynamic potential* derives from an analogy with mechanical potential energy. In certain circumstances the work obtainable from a macroscopic system is related to the change in the appropriately defined thermodynamic potential. The simplest example is the internal energy $E(S, V)$ for a PVT system. The second law for reversible processes reads

$$dE = T dS - P dV = đQ - đW \qquad (1)$$

In a reversible *adiabatic* transformation the decrease in internal energy is equal to the amount of work done by the expanding system. If the transformation is adiabatic but not reversible, $\Delta Q = 0$ and the first law yields

$$\Delta E = -(\Delta W)_{\text{irrev}} \qquad (2)$$

with the same change in E as in a reversible transformation connecting the same endpoints in the thermodynamic space. However, the change in entropy is not necessarily zero and must be calculated along a reversible path:

$$\Delta E = (\Delta Q)_{\text{rev}} - (\Delta W)_{\text{rev}} \qquad (3)$$

Subtracting and using $(dQ)_{\text{rev}} = T dS$, we find that

$$(\Delta W)_{\text{rev}} - (\Delta W)_{\text{irrev}} = \int T \, dS \geq 0 \qquad (4)$$

Thus the decrease in internal energy is equal to the maximum amount of work obtainable through an adiabatic process, and this maximum is achieved if the process is reversible.

We now generalize the formulation to allow other forms of work, as well as the exchange of particles between the system under consideration and its surroundings. This more general internal energy is a function of the entropy, the volume, the other extensive generalized displacements, and the number of particles of each species: $E = E(S, V, \ldots, x_i, \ldots, \{N_j\})$ with a differential (for reversible processes)

$$dE = T \, dS + \sum_i X_i \, dx_i + \sum_j \mu_j \, dN_j \qquad (5)$$

Here N_j is the number of molecules of type j and the "chemical potential" μ_j is defined by (5). We are now in a position to introduce a number of other useful thermodynamic potentials. The *Helmholtz free energy, A,* is related to the internal energy through a Legendre transformation:

$$A = E - TS \tag{6}$$

The quantity A is a state function with differential

$$dA = dE - T\, dS - S\, dT$$

$$= -S\, dT - \sum_i X_i\, dx_i + \sum_j \mu_j\, dN_j \tag{7}$$

As in the case of the internal energy, the change in Helmholtz free energy may be related to the amount of work obtainable from the system. In a general infinitesimal process

$$dA = dE - d(TS)$$

$$= đQ - T\, dS - S\, dT - đW \tag{8}$$

Thus

$$đW = (đQ - T\, dS) - S\, dT - dA \tag{9}$$

In a reversible transformation $đQ = TdS$. If the process is isothermal as well as reversible, we have

$$đW = -dA$$

and the Helmholtz free energy plays the role of a potential energy for reversible isothermal processes. If the process in question is isothermal but not reversible, we have $đQ - T\, dS \le 0$ and

$$(đW)_{\text{irrev}} = đQ - TdS - dA \le -dA \tag{10}$$

which shows that $-dA$ is the maximum amount of work that can be extracted, at constant temperature, from the system. We also see, from (10), that if the temperature and generalized displacements are fixed ($đW = 0$), a spontaneous process can only decrease the Helmholtz free energy and conclude that the equilibrium state of a system at fixed $(T, V, \ldots, x_j, \ldots, \{N\})$ is the state of minimum Helmholtz free energy.

Another thermodynamic potential which is often useful is the *Gibbs potential G.* For a *PVT* system we write

$$G = A + PV \tag{11}$$

This function is again a state function with a differential

$$dG = dA + P\, dV + V\, dP = -S\, dT + V\, dP \tag{12}$$

In a general process

$$dG = dE - d(TS) + d(PV)$$

$$= (đQ - T\, dS) - (đW - P\, dV) + V\, dP - S\, dT$$

We see that the relations

$$dW - P \, dV = 0$$

$$dQ - T \, dS \leq 0$$

imply that the Gibbs potential can only decrease in a spontaneous process at fixed T and P. It is also easy to show that the change in Gibbs potential is the maximum amount of work that can be extracted from a system through a process at fixed T and P. It follows that this occurs when the process is reversible.

In our microscopic statistical treatment of magnetic materials we will model these materials by systems of magnetic moments (or "spins") which can orient themselves in an applied magnetic field H. The work done by the system, if the applied field is held constant, is then [1]

$$dW = -\mathbf{H} \cdot d\mathbf{M} \tag{13}$$

Instead of defining yet another potential, we modify the definition of the Gibbs potential for a purely magnetic system to read

$$G(\mathbf{H}, T) = E(S, \mathbf{M}) - TS - \mathbf{M} \cdot \mathbf{H} \tag{14}$$

$$dG = -S \, dT - \mathbf{M} \cdot d\mathbf{H} \tag{15}$$

It should be noted that (14) is not a universally accepted convention. Some authors refer to (14) as the Helmholtz free energy.

One further potential that is very useful in statistical physics is the *grand potential* $\Omega_G(T, V, \{\mu\})$. This potential is obtained from the internal energy through the transformation

$$\Omega_G(T, V, \{\mu\}) = E - TS - \sum_i \mu_i \, N_i \tag{16}$$

and has the differential

$$d\Omega_G = -S \, dT - P \, dV - \sum_i N_i \, d\mu_i \tag{17}$$

1.D GIBBS–DUHEM AND MAXWELL RELATIONS

The internal energy E has as its natural independent variables the entropy S, the volume V, and other generalized displacements which are all *extensive* variables. If these quantities are rescaled by a factor λ, the internal energy must itself change by the same factor:

$$E(\lambda S, \{\lambda x_i\}, \{\lambda N_j\}) = \lambda E(S, \{x_i\}, \{N_j\}) \tag{1}$$

---

[1] For a discussion of work done by a magnet during a process in which its magnetization is changed, see Appendix B in Callen [1985].

Differentiating both sides with respect to λ using (1.C.5) on the right-hand side, we obtain the *Gibbs–Duhem equation:*

$$E(S, \{x_i\}, \{N_j\}) = TS + \sum_i X_i x_i + \sum_j \mu_j N_j \tag{2}$$

For a single-component *PVT* system, (2) reduces to

$$E = TS - PV + \mu N \tag{3}$$

or

$$G(P, T, N) = \mu N \tag{4}$$

Taking the differential of (2) and using (1.C.5), we find that

$$0 = S\,dT + \sum_i x_i\,dX_i + \sum_j N_j\,d\mu_j \tag{5}$$

which illustrates the fact that the intensive variables T, $\{X_i\}$, $\{\mu_j\}$ are not all independent. An r-component *PVT* system thus has $r + 1$ independent intensive thermodynamic variables. Another consequence is that at least one extensive variable is needed to specify completely the state of the system.

It follows from the differential form (1.C.7) for a single-component *PVT* system that

$$\left(\frac{\partial A}{\partial T}\right)_{N,V} = -S \qquad \left(\frac{\partial A}{\partial V}\right)_{T,N} = -P \qquad \left(\frac{\partial A}{\partial N}\right)_{T,V} = \mu \tag{6}$$

It is a well-known result from the theory of partial differentiation that higher-order derivatives are independent of the order in which the differentiation is carried out; that is, if ϕ is a single-valued function of the independent variables $x_1, x_2, \ldots, x_n$, then

$$\frac{\partial}{\partial x_i}\left(\frac{\partial \phi}{\partial x_j}\right) = \frac{\partial}{\partial x_j}\left(\frac{\partial \phi}{\partial x_i}\right) \tag{7}$$

By applying this result to (6) we immediately obtain the *Maxwell relations*

$$\left(\frac{\partial S}{\partial V}\right)_{T,N} = \left(\frac{\partial P}{\partial T}\right)_{V,N} \tag{8}$$

$$\left(\frac{\partial S}{\partial N}\right)_{V,T} = -\left(\frac{\partial \mu}{\partial T}\right)_{V,N} \tag{9}$$

$$\left(\frac{\partial P}{\partial N}\right)_{V,T} = -\left(\frac{\partial \mu}{\partial V}\right)_{T,N} \tag{10}$$

Similarly, in the case of the Gibbs potential we find from (1.C.12)

$$\left(\frac{\partial G}{\partial T}\right)_{P,N} = -S \qquad \left(\frac{\partial G}{\partial P}\right)_{T,N} = V \qquad \left(\frac{\partial G}{\partial N}\right)_{T,P} = \mu \tag{11}$$

from which we have the additional Maxwell relations:

$$\left(\frac{\partial S}{\partial P}\right)_{T,N} = -\left(\frac{\partial V}{\partial T}\right)_{P,N} \tag{12}$$

$$\left(\frac{\partial V}{\partial N}\right)_{T,P} = \left(\frac{\partial \mu}{\partial P}\right)_{T,N} \tag{13}$$

$$\left(\frac{\partial S}{\partial N}\right)_{T,P} = -\left(\frac{\partial \mu}{\partial T}\right)_{P,N} \tag{14}$$

Further equations of this type can be found for magnetic systems, using (1.C.15), or, in the case of PVT systems, by using the internal energy or the grand potential. The usefulness of these relations is demonstrated in the next section, in which we derive relations between some of the most commonly measured response functions.

1.E RESPONSE FUNCTIONS

A great deal can be learned about a macroscopic system through its response to various changes in externally controlled parameters. Important response functions for a PVT system are the specific heats at constant volume and pressure,

$$C_V = \left(\frac{dQ}{dT}\right)_V = T\left(\frac{\partial S}{\partial T}\right)_V$$
$$C_P = \left(\frac{dQ}{dT}\right)_P = T\left(\frac{\partial S}{\partial T}\right)_P \tag{1}$$

the isothermal and adiabatic compressibilities,

$$K_T = -\frac{1}{V}\left(\frac{\partial V}{\partial P}\right)_T$$
$$K_S = -\frac{1}{V}\left(\frac{\partial V}{\partial P}\right)_S \tag{2}$$

and the coefficient of thermal expansion

$$\alpha = \frac{1}{V}\left(\frac{\partial V}{\partial T}\right)_P \tag{3}$$

Intuitively, we expect the specific heats and compressibilities to be positive and $C_P > C_V$, $K_T > K_S$. In this section we derive relations between these response functions. The intuition that the response functions are positive will be justified in the following section in which we discuss thermodynamic stability. We begin with the assumption that the entropy has been expressed in terms of T and V and that the number of particles is kept fixed. Then

$$dS = \left(\frac{\partial S}{\partial T}\right)_V dT + \left(\frac{\partial S}{\partial V}\right)_T dV \tag{4}$$

and

$$T\left(\frac{\partial S}{\partial T}\right)_P = T\left(\frac{\partial S}{\partial T}\right)_V + T\left(\frac{\partial S}{\partial V}\right)_T\left(\frac{\partial V}{\partial T}\right)_P \tag{5}$$

or

$$C_P - C_V = T\left(\frac{\partial S}{\partial V}\right)_T\left(\frac{\partial V}{\partial T}\right)_P \tag{6}$$

We now use the Maxwell relation (1.D.8) and the chain rule

$$\left(\frac{\partial z}{\partial x}\right)_y\left(\frac{\partial y}{\partial z}\right)_x\left(\frac{\partial x}{\partial y}\right)_z = -1 \tag{7}$$

which is valid for any three variables obeying an equation of state of the form $f(x, y, z) = 0$ to obtain

$$\left(\frac{\partial S}{\partial V}\right)_T = \left(\frac{\partial P}{\partial T}\right)_V = -\left(\frac{\partial P}{\partial V}\right)_T\left(\frac{\partial V}{\partial T}\right)_P \tag{8}$$

and

$$C_P - C_V = T\left[-\left(\frac{\partial P}{\partial V}\right)_T\right]\left(\frac{\partial V}{\partial T}\right)_P^2 = \frac{TV}{K_T}\alpha^2 \tag{9}$$

In a similar way we obtain a relation between the compressibilities K_T and K_S. Assume that the volume V has been obtained as function of S and P. Then

$$dV = \left(\frac{\partial V}{\partial P}\right)_S dP + \left(\frac{\partial V}{\partial S}\right)_P dS \tag{10}$$

and

$$-\frac{1}{V}\left(\frac{\partial V}{\partial P}\right)_T = -\frac{1}{V}\left(\frac{\partial V}{\partial P}\right)_S - \frac{1}{V}\left(\frac{\partial V}{\partial S}\right)_P\left(\frac{\partial S}{\partial P}\right)_T \tag{11}$$

or

$$K_T - K_S = -\frac{1}{V}\left(\frac{\partial V}{\partial S}\right)_P\left(\frac{\partial S}{\partial P}\right)_T \tag{12}$$

The Maxwell relation (1.D.12) and the equation

$$\left(\frac{\partial V}{\partial S}\right)_P = \frac{(\partial V/\partial T)_P}{(\partial S/\partial T)_P} \tag{13}$$

yield

$$K_T - K_S = \frac{TV}{C_P}\alpha^2 \tag{14}$$

Thus (9) and (14) together produce the interesting and useful results

$$C_P(K_T - K_S) = K_T(C_P - C_V) = TV\alpha^2 \tag{15}$$

and

$$\frac{C_P}{C_V} = \frac{K_T}{K_S} \tag{16}$$

An analogous derivation, in the case of magnetic systems produces the equations

$$C_H(\chi_T - \chi_S) = T\left(\frac{\partial M}{\partial T}\right)_H^2$$

$$\chi_T(C_H - C_M) = T\left(\frac{\partial M}{\partial T}\right)_H^2 \tag{17}$$

where χ_T and χ_S are the isothermal and adiabatic susceptibilities, $\chi_T = (\partial M/\partial H)_T$, $\chi_S = (\partial M/\partial H)_S$, and C_H and C_M are the specific heats at constant applied field and constant magnetization, respectively.

1.F GENERAL CONDITIONS FOR EQUILIBRIUM AND STABILITY

We consider two systems in contact with each other. It is intuitively clear that if heat can flow freely between the two systems and if the volumes of the two systems are not separately fixed, the parameters will evolve so as to equalize the pressure and temperature of the two systems. These conclusions can easily be obtained from the principle of maximum entropy. Suppose that the two systems have volumes V_1 and V_2, energies E_1 and E_2, and that the number of particles in each as well as the combined energy and total volume are fixed. In equilibrium, the total entropy

$$S = S_1(E_1, V_1) + S_2(E_2, V_2) \tag{1}$$

must be a maximum. Thus

$$dS = \left(\frac{\partial S_1}{\partial E_1}\right)_{V_1} dE_1 + \left(\frac{\partial S_2}{\partial E_2}\right)_{V_2} dE_2 + \left(\frac{\partial S_1}{\partial V_1}\right)_{E_1} dV_1 + \left(\frac{\partial S_2}{\partial V_2}\right)_{E_2} dV_2$$

$$= \left[\left(\frac{\partial S_1}{\partial E_1}\right)_{V_1} - \left(\frac{\partial S_2}{\partial E_2}\right)_{V_2}\right] dE_1 + \left[\left(\frac{\partial S_1}{\partial V_1}\right)_{E_1} - \left(\frac{\partial S_2}{\partial V_2}\right)_{E_2}\right] dV_1 = 0 \tag{2}$$

where we have used the constraint $E_1 + E_1 = $ const., $V_1 + V_2 = $ const. We have

$$\left(\frac{\partial S_j}{\partial E_j}\right)_{V_j} = \left(\frac{\partial E_j}{\partial S_j}\right)_{V_j}^{-1} = \frac{1}{T_j} \tag{3}$$

and

$$\left(\frac{\partial S_j}{\partial V_j}\right)_{E_j} = \frac{P_j}{T_j} \tag{4}$$

which together with (2) yield

$$\frac{1}{T_1} = \frac{1}{T_2} \tag{5}$$

$$\frac{P_1}{T_1} = \frac{P_2}{T_2} \tag{6}$$

or $T_1 = T_2$, $P_1 = P_2$, which is the expected result. More generally, one finds that when the conjugate displacements are unconstrained all generalized forces of two systems in equilibrium must be equal.

To this point we have required only that the equilibrium state correspond to a *stationary* state of the entropy. Requiring this stationary state to be a maximum will provide conditions on the second derivatives of the entropy. These conditions are local in nature. A stronger (global) condition is that the entropy be a *concave* function of the generalized displacements (see Problem 1.10).

Some of the most useful *stability criteria* are obtained from the Gibbs potential rather than from the entropy and we proceed to consider a small (but macroscopic) system in contact with a much larger reservoir. This reservoir is assumed to be so large that the fluctuations or spontaneous processes in the small system do not change the temperature or pressure of the reservoir, which we denote by T_0 and P_0. The Gibbs potential, as we have seen in Section 1.C, is a minimum in equilibrium and for the small system we have

$$G_1(P_0, T_0) = E_1 - T_0 S_1 + P_0 V_1 \tag{7}$$

Suppose now that there is a fluctuation in the entropy and volume of this system. To second order in the fluctuating quantities,

$$\delta G_1 = \delta S_1 \left(\frac{\partial E_1}{\partial S_1} - T_0\right) + \delta V_1 \left(\frac{\partial E_1}{\partial V_1} + P_0\right)$$
$$+ \frac{1}{2}\left[(\delta S_1)^2 \left(\frac{\partial^2 E_1}{\partial S_1^2}\right) + 2(\delta S_1)(\delta V_1)\left(\frac{\partial^2 E_1}{\partial S_1 \partial V_1}\right) + (\delta V_1)^2 \left(\frac{\partial^2 E_1}{\partial V_1^2}\right)\right] \tag{8}$$

which must be greater than zero if the state specified by P_0, T_0 is the state of minimum Gibbs potential. Since $(\partial E_1/\partial S_1) = T_0$ and $(\partial E_1/\partial V_1) = -P_0$ we obtain the condition

$$(\delta S)^2 \frac{\partial^2 E}{\partial S^2} + 2(\delta S)(\delta V)\frac{\partial^2 E}{\partial S \partial V} + (\delta V)^2 \frac{\partial^2 E}{\partial V^2} > 0 \tag{9}$$

where we have dropped the subscripts. The fluctuations in the entropy and volume are independent of each other and we can guarantee that the expression (9) is positive if we require that $E(S, V)$ satisfies the conditions

$$\frac{\partial^2 E}{\partial S^2} > 0 \qquad (10)$$

$$\frac{\partial^2 E}{\partial V^2} > 0 \qquad (11)$$

$$\frac{\partial^2 E}{\partial S^2}\frac{\partial^2 E}{\partial V^2} - \left(\frac{\partial^2 E}{\partial S\,\partial V}\right)^2 > 0 \qquad (12)$$

The inequality (10) reduces to

$$\left(\frac{\partial T}{\partial S}\right)_V = \frac{T}{C_V} > 0 \quad \text{or} \quad C_V > 0 \qquad (13)$$

while (11) implies

$$-\left(\frac{\partial P}{\partial V}\right)_S = \frac{1}{VK_S} > 0 \quad \text{or} \quad K_S > 0 \qquad (14)$$

and (12) yields

$$\frac{T}{VK_S C_V} > \left(\frac{\partial T}{\partial V}\right)_S^2 \qquad (15)$$

Equations (13)–(15) are special cases of Le Châtelier's principle, which states that if a system is in equilibrium, any spontaneous changes in its parameters will bring about processes that tend to restore the system to equilibrium. In our situation such spontaneous processes raise the Gibbs potential. Other stability criteria can be obtained by using one of the other thermodynamic potentials of Section 1.C.

1.G THERMODYNAMICS OF PHASE TRANSITIONS

A typical phase diagram for a one-component PVT system looks like Figure 1.5. The solid lines separate the P–T plane into regions in which a unique phase is the stable thermodynamic state. As the system passes through one of these lines, called *coexistence curves,* a *phase transition* occurs, generally accompanied by the absorption or liberation of latent heat. In Figure 1.5 there are two special points, the *triple point* P_t, T_t and the *critical point*, P_c, T_c. At the critical point the properties of the fluid and vapor phase become identical and much of our study of phase transitions in later chapters will focus on the region of the phase diagram around this point. We note that the properties of the system vary smoothly along any curve which does not cross a coexistence curve. Thus it is possible to pass continuously from the vapor to the liquid phase by taking the system to high enough temperature, increasing the pressure, and then lowering the temperature again. It is not possible to avoid the liquid–solid coexistence curve—this curve extends to $P = \infty$, $T = \infty$ (as far as

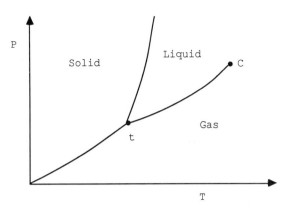

Figure 1.5   Schematic phase diagram of a simple one-component *PVT* system.

we know). The analogous phase diagram for a ferromagnetic substance is shown in Figure 1.6 in the *H–T* plane with a critical point at $H_c = 0$, T_c.

The phase diagrams of Figures 1.5 and 1.6 are particularly simple, partly because of the choice of variables. The fields *H* and *P* as well as the temperature *T* are required (1.F) to be equal in the coexisting phases. Conversely, the densities conjugate to these fields (i.e., the magnetization *M*, density ρ, specific volume *v*, or the entropy per particle *s* can take on different values in the two phases. Thus the phase diagram of the ferromagnet of Figure 1.6 drawn in the *T–M* plane takes the form shown in Figure 1.7. The points *A*, *B* on the bell-shaped curve represent the stable states which the system approaches as the coexistence curve of Figure 1.6 is approached from one phase or the other. A state inside the bell-shaped region, such as the one marked × on the vertical line, is not a stable single phase state—the system separates into two regions, one with the magnetization of point *A*, the other with the magnetization of point *B*.

Similarly, the phase diagram of the simple fluid of Figure 1.5, drawn in the ρ–*T* plane, is schematically shown in Figure 1.8. The liquid-gas coexistence curve in Figure 1.8, while lacking the symmetry of the bell-shaped curve

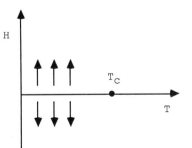

Figure 1.6 Phase diagram of a ferromagnet in the *H–T* plane.

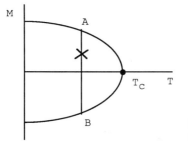

Figure 1.7   Phase diagram of the ferromagnet of Figure 1.6 in the *M–T* plane.

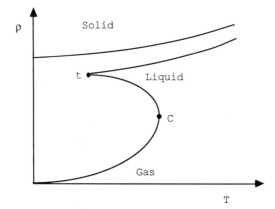

Figure 1.8   Phase diagram of Figure 1.5 in the *ρ–T* plane.

of Figure 1.7, has many of the same properties. Indeed, as the critical point is
approached, we will find, in later chapters, that certain magnetic systems behave in a similar way as liquid–gas systems. We now discuss some of the properties of coexistence curves.

We consider a single-component *PVT* system on either side of the liquid
–gas or liquid–solid coexistence curve. The two coexisting phases may be
thought of as systems in contact and in equilibrium with each other. We therefore have

$$T_1 = T_2$$
$$P_1 = P_2 \tag{1}$$
$$\mu_1 = \mu_2$$

where the subscripts 1 and 2 refer to the two phases. From the Gibbs–Duhem
equation (1.D.4) we therefore obtain

$$g_1(T, P) = g_2(T, P) \tag{2}$$

where g_1 and g_2 are the Gibbs potential per particle in phases 1 and 2, respectively. This equality (2) must hold along the entire coexistence curve and
hence

$$dg_1 = -s_1\,dT + v_1\,dP = dg_2 = -s_2\,dT + v_2\,dP \tag{3}$$

for differentials (dT, dP) along the coexistence curve. Thus

$$\left(\frac{dP}{dT}\right)_{\text{coex}} = \frac{s_1 - s_2}{v_1 - v_2} = \frac{L_{12}}{T(v_1 - v_2)} \tag{4}$$

where L_{12} is the latent heat per particle needed to transform the system from phase 2 to phase 1. Equation (4) is known as the Clausius–Clapeyron equation. As a simple example, consider a transition from liquid to vapor with $v_1 \gg v_2$ and with $v_1 = k_B T/P$. Then

$$\frac{dP}{dT} \approx \frac{PL_{12}}{k_B T^2}$$

and if L_{12} is roughly constant along the coexistence curve, we have

$$P(T) \approx P_0 \exp\{-\frac{L_{12}}{k_B}\left(\frac{1}{T} - \frac{1}{T_0}\right)\} \tag{5}$$

where P_0, T_0 is a reference point on the coexistence curve. Using approximate equations of state for the solid and liquid phases, one can derive an equation similar to (5) for the solid-liquid coexistence curve.

As a final topic we now briefly discuss the Gibbs phase rule. This rule allows one to limit the topology of a phase diagram on the basis of some very general considerations. Consider first a single-component PVT system with a phase diagram as shown in Figure 1.5. For two-phase coexistence the chemical potential $\mu(P, T)$ must be the same in the two phases, yielding a *curve* in the P–T plane. Similarly, three-phase coexistence implies that

$$\mu_1(P, T) = \mu_2(P, T) = \mu_3(P, T)$$

which, in general, will have a solution only at an *isolated point,* the triple point. Four-phase coexistence is ruled out unless there are hidden fields separate from the temperature and pressure.

One can also see that the critical point P_c, T_c will be an isolated point for a PVT system. At the critical point the liquid and vapor densities, or specific volumes, are equal. This condition yields a second equation,

$$v_1(P_c, T_c) = \left.\frac{\partial g_1}{\partial P}\right|_{P_c, T_c} = v_2(P_c, T_c) = \left.\frac{\partial g_2}{\partial P}\right|_{P_c, T_c} \tag{6}$$

which together with $\mu_1(P_c, T_c) = \mu_2(P_c, T_c)$ determines a unique point in the P–T plane.

In a multicomponent system the situation is more complicated. We take as thermodynamic variables P, T, and c_{ij}, $i = 1, 2, \ldots, r$, where c_{ij} is the mole fraction of constituent i in phase j of an r-component system. Suppose that there are s coexisting phases. Since

$$\sum_{i=1}^{r} c_{ij} = 1 \tag{7}$$

there are $s(r - 1) + 2$ remaining independent variables. Equating the chemical potentials for the r components gives $r(s - 1)$ equations for these variables. If a solution is to exist, we must have at least as many variables as equations, that is,

$$s(r - 1) + 2 \geq r(s - 1)$$

or

$$s \leq r + 2 \tag{8}$$

Therefore, at most, $r + 2$ phases can coexist in a mixture of r constituents.

PROBLEMS

1.1. *Equivalence of Carnot and Ideal Gas Temperature Scale.* Consider a Carnot engine working between reservoirs at temperatures T_1 and T_2 with an ideal gas working substance obeying the equation of state

$$PV = Nk_B T$$

which may be taken to be a definition of a temperature scale. Show that the efficiency of the cycle is given by

$$\eta = 1 - \frac{T_2}{T_1}$$

where $T_1 > T_2$.

1.2. *Adiabatic Processes in Paramagnets.*

(a) A Curie paramagnet $(M = CH/T)$ is magnetized adiabatically and reversibly. Find a relation between M and H and between M and T during such a process.

(b) Use the results of part (a) to verify the formula for the efficiency of a Carnot cycle with a magnet as the working substance between temperatures T_1 and T_2.

1.3. *Stability Analysis for an Open System.* Carry out the stability analysis of Section 1.F for a system that is kept at a fixed volume but is free to exchange energy and particles with a reservoir. The temperature and chemical potential of the reservoir are not affected by the fluctuations. In particular, show that

$$C_{V,N} > 0 \qquad \left(\frac{\partial N}{\partial \mu}\right)_{V,S} > 0 \qquad C_{V,N}\left(\frac{\partial N}{\partial \mu}\right)_{V,S} < \left(\frac{\partial S}{\partial \mu}\right)^2_{V,N}$$

1.4. *Specific Heats of Magnets.* Derive the relations

$$C_H - C_M = \frac{T}{\chi_T}\left(\frac{\partial M}{\partial T}\right)^2_H$$

$$C_H(\chi_T - \chi_S) = T\left(\frac{\partial M}{\partial T}\right)^2_H$$

for a magnet.

1.5. *Second Law for the Entropy Density.* The entropy density $s = S/N$ can be expressed in terms of the intensive variables $e = E/N$ and $v = V/N$. The second law then takes the form

$$ds = \left(\frac{\partial s}{\partial e}\right)_v de + \left(\frac{\partial s}{\partial v}\right)_e dv = \frac{1}{T}de + \frac{P}{T}dv$$

Generalize this expression to a multicomponent system by writing

$$s = \frac{S(E, V, N_1, N_2, \ldots, N_r)}{N} = s(e, v, x_1, x_2, \ldots, x_r)$$

where $x_j = N_j/N$ is the mole fraction of constituent j and N is the total number of particles. Show that

$$ds = \frac{1}{T}de + \frac{P}{T}dv - \sum_{j=1}^{r-1} (\mu_j - \mu_r)dx_j$$

1.6. *Brayton Cycle.* The Joule or Brayton cycle is shown in the *P-S* plane in Figure 1.9. Assuming that the working substance is an ideal gas, show that the efficiency of the cycle is

$$\eta = 1 - \left(\frac{P_A}{P_B}\right)^{\frac{C_P - C_V}{C_P}}$$

where C_P and C_V are the heat capacities at constant pressure and volume, respectively. You may assume that C_P and C_V are constant.

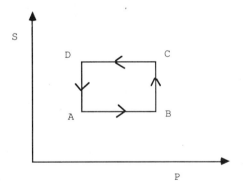

Figure 1.9 Brayton cycle.

1.7. *Enthalpy Change in a Spontaneous Process.* Show that the enthalpy, $H = E + PV$, of an isolated system can only decrease due to spontaneous processes that take place at fixed pressure. It may be helpful to visualize this result by considering the following example. A mixture of 1 mol of O_2 and 2 mol of H_2 is kept at pressure P in a cylindrical vessel by a piston and is thermally insulated. The energy released if the mixture is ignited will produce heating of the H_2O vapor and also kinetic energy of the piston. Only if all the kinetic energy of the piston is dissipated as heat in the vapor as the system comes to rest will the final enthalpy of the vapor be the same as the initial enthalpy of the mixture.

1.8. *Effect of an Inert Gas on Vapor Pressure.* A liquid is in equilibrium with its vapor at temperature T and pressure P. Suppose that an inert ideal gas, which is insoluble in the liquid, is introduced into the container and has partial pressure P_i. The temperature is kept fixed. Show that if the volume per particle, v_L, in the liquid is much smaller than the specific volume, v_G, in the gas the vapor pressure will *increase* by an amount δP given by

$$\frac{\delta P}{P} = \frac{P_i v_L}{k_B T}$$

if P_i is small enough.

1.9. *Entropy of Mixing.* Calculate the entropy of mixing of two volumes V_A and V_B of an ideal gas of species A and B, respectively, both initially at the same temperature T and pressure P and, with a final volume $V = V_A + V_B$

1.10. *Convexity and Local Stability.* The equilibrium state of a system is the state of maximum entropy. It is easily seen that this statement implies the relation

$$S(E + \Delta E, V, N) + S(E - \Delta E, V, N) - 2S(E, V, N) \leq 0$$

which can be derived by considering an energy fluctuation between two subsystems of the same size in thermal contact with each other. More generally,

$$S(E + \Delta E, V + \Delta V, N) + S(E - \Delta E, V - \Delta V, N) - 2S(E, V, N) \leq 0$$

which is the mathematical statement that S is a *concave* function of its extensive variables. The *local* stability requirements

$$\left(\frac{\partial^2 S}{\partial E^2}\right)_{V,N} \leq 0, \qquad \left(\frac{\partial^2 S}{\partial V^2}\right)_{E,N} \leq 0, \qquad \left(\frac{\partial^2 S}{\partial E^2}\right)\left(\frac{\partial^2 S}{\partial V^2}\right) - \left(\frac{\partial^2 S}{\partial E \partial V}\right)^2 \leq 0$$

are special cases of the concavity condition.

(a) Show that at fixed S and V the equilibrium state of a system is the state of minimum internal energy and that this implies that E is a *convex* function of S and V.

(b) Show that the Gibbs potential is a convex function of P and T.

1.11. *Derivation of Equation of State.* In a certain system the internal energy E is related to the entropy S, particle number N, and volume V through

$$E = \text{const. } N\left(\frac{N}{V}\right)^d \exp\left\{\frac{dS}{Nk_B}\right\}$$

(a) Show that the system satisfies the ideal gas law independent of the value of the constant d.

(b) Find the coefficient γ in the adiabatic equation of state $Pv^\gamma = \text{const}$ and the molar specific heats C_P and C_V of the system.

2

Statistical Ensembles

In this chapter we develop the foundations of *equilibrium* statistical mechanics. No attempt will be made to derive the equilibrium theory from the detailed microscopic dynamics. Instead, we will content ourselves with postulating the statistical laws. In Section 2.A we consider a closed classical system and develop the concept of the microcanonical ensemble from the ergodic hypothesis. We define the entropy, the thermodynamic limit, and make contact with thermodynamics. In Section 2.B we extend the formalism to systems in thermal contact with the surroundings and introduce the canonical ensemble. We also discuss fluctuations of thermodynamic variables. The grand canonical ensemble is discussed in Section 2.C for systems that are free to exchange particles with the outside and we demonstrate the general relation between fluctuations in particle number and the compressibility. In Section 2.D we modify the formalism to include quantum systems and introduce the density matrix. An alternative, information-theoretic approach is discussed in Section 2.E and a number of useful thermodynamic variational principles are introduced in Section 2.F.

We assume that the reader has had some previous exposure to the material under discussion. Our treatment is therefore fairly condensed. Some general references that we have found particularly useful are Huang [1987], Kubo et al. [1965], Ma [1985], Reichl [1980], Toda et al. [1983], and Wannier [1966].

2.A ISOLATED SYSTEMS: MICROCANONICAL ENSEMBLE

We consider first a system of $3N$ degrees of freedom described by the canonical variables $q_1, \ldots, q_j, \ldots, q_{3N}, p_1, \ldots, p_j, \ldots, p_{3N}$. These variables are assumed to obey classical Hamiltonian dynamics:

$$\begin{aligned}
\dot{p}_j &= -\frac{\partial H}{\partial q_j} \\[2mm]
\dot{q}_j &= \frac{\partial H}{\partial p_j}
\end{aligned} \qquad j = 1, 2, \ldots, 3N \qquad (1)$$

A possible example of such a system is that of N particles confined to a volume V by an idealized surface which does not permit the flow of heat or particles.

In general, a Hamiltonian system will have a number of conserved quantities, the most obvious and universal being the total energy E. If the system is fully integrable,[1] the motion can be completely expressed in terms of $6N$ constants such as the action variables and the initial values of the $3N$ angles. Most systems are not completely integrable, but there may still be a few first integrals expressable as conservation laws, such as those of total linear and angular momentum. If we restrict our attention to a particular frame of reference (e.g., the one in which there is no net rotational or translational motion), the energy will typically be the only remaining easily identifiable conserved quantity.

We define a $6N$-dimensional phase space and represent the state of the system, at a particular instant, by a $6N$-dimensional vector $\mathbf{x}$ in this space, with components given by the generalized coordinates and momenta q_j, p_j. If we specify the energy E of the system, the motion of the N-particle system will be confined to a $(6N - 1)$-dimensional surface given by

$$H(\mathbf{x}) = E \qquad (2)$$

We denote this surface by $\Gamma(E)$. A measurement of some macroscopic property of the system (e.g., the pressure of particles in a box) will, in general, be carried out over a period of time $t_0 < t < t_0 + \tau$ during which the phase point covers some part of $\Gamma(E)$. If the instantaneous value of the property in question is $\phi(\mathbf{x}(t))$, the measured value will be

$$\langle \phi \rangle = \frac{1}{\tau} \int_{t_0}^{t_0 + \tau} dt\, \phi(\mathbf{x}(t)) \qquad (3)$$

Since we are unable to work out the detailed dynamics $\mathbf{x}(t)$, we are forced to make some assumptions at this stage. The simplest assumption is referred to as the *ergodic hypothesis: During any significant time interval τ the phase point $\mathbf{x}(t)$ will spend equal time intervals in all regions of $\Gamma(E)$.*

There is nothing obvious about this hypothesis. In Problem 2.1 at the end

[1] See, for example, Goldstein [1980], in particular for a discussion of action angle variables.

of the chapter the reader will find one example of a system that satisfies this hypothesis and one that does not. Both examples are trivial in the sense that the equations of motion of the dynamical variables are integrable, whereas in real systems, the equations of motion are invariably nonintegrable and relatively little is known in general about the behavior of large (10^{23} particles) systems of equations. Suffice to say that a proof of the ergodic hypothesis under fairly general conditions is lacking. In any case, while the ergodic hypothesis is sufficient to allow us to replace a time average by an unweighted average over the surface $\Gamma(E)$, it is not necessary.

A hypothesis that is weaker than the ergodic hypothesis but nevertheless leads to statistical mechanics, as we know it, is the mixing hypothesis. Briefly stated, *mixing* refers to the notion that an initial (compact) distribution of points on $\Gamma(E)$ very quickly distorts into a convoluted object which permeates the entire surface while still occupying the same volume (as required by Liouville's theorem). Mixing can be proven to occur, for example, for systems of hard spheres (Sinai, 1966, 1970). When one talks about a distribution of points on $\Gamma(E)$, one is no longer discussing a single system, and for a truly isolated system, mixing is of course irrelevant. However, the same instability of the trajectory $\mathbf{x}(t)$, with respect to small changes in initial conditions that produces mixing will also allow small amounts of noise, due to interaction with the container by which an "isolated" system is confined, to distort the trajectory sufficiently that in effect one obtains ergodicity.

The foregoing discussion is quite incomplete and was included only to highlight some of the difficulties in the foundation of statistical mechanics. An excellent discussion of these topics can be found in Rasetti [1986] and in the Appendix of the book by Balescu [1975].

We shall henceforth simply assume that we can replace the time average (3) by an average over the surface $\Gamma(E)$. This average is a special case of an *ensemble average*. An ensemble can be visualized as a collection of snapshots of the system at different times. We specify an ensemble through the probability density $\rho(\mathbf{x})$ for a selected member of the ensemble to occupy the phase space point $\mathbf{x}$. The ensemble average of an observable $\phi(\mathbf{x}(t))$ is then

$$\langle \phi \rangle = \int d^{6N}x \rho(\mathbf{x})\phi(\mathbf{x}) \tag{4}$$

where the integral should be carried out over the entire $6N$-dimensional phase space. In the special case that we are considering—a closed classical system of energy E,

$$\rho(\mathbf{x}) = C\delta(H(\mathbf{x}) - E) \tag{5}$$

where C is a normalizing factor. The ensemble corresponding to (5) is called the *microcanonical ensemble*.

As we shall see, the entropy S plays a particularly fundamental role when the microcanonical ensemble is used. To arrive at an acceptable definition of the entropy, which can easily be extended to quantum systems and which can be used in semiclassical arguments, we make a slight generalization of the microcanonical ensemble.

Instead of working with the constant energy surface $\Gamma(E)$ we assume that all **x** such that

$$E < H(\mathbf{x}) < E + \delta E \tag{6}$$

occur with equal probability in the ensemble. We discuss the tolerance δE in greater detail later in this section and in Section 2.D. Consider first a system of distinguishable particles, for example, a perfect crystal or a polymer chain where the particles can be identified by their positions in the lattice or in the chain. We define

$$\Omega(E) = \frac{1}{h^{3N}} \int_{E \leq H(\mathbf{x}) \leq E + \delta E} d^{6N}x \tag{7}$$

Ω is proportional to the volume of phase space which satisfies (6). The entropy $S(E)$ is then defined as

$$S(E) = k_B \ln \Omega(E) \tag{8}$$

where k_B is Boltzmann's constant.

The factor in front of the integral in (7) requires some comment. To make $\Omega(E)$ dimensionless a constant of dimension $[\text{action}]^{-3N}$ is required. The particular value chosen is arbitrary in a purely classical context. We have chosen the value h^{-3N} to make contact with the Bohr-Sommerfeld semiclassical theory. According to this theory the quantization rule for a single degree of freedom is $\int p\,dq = rh$, where r is an integer. A volume of phase space $\int_W dp\,dq$ which is large enough that the discreteness of the levels is unimportant will then on the average contain W/h quantum states. It is intuitively straightforward[2] to make the generalization that a volume

$$\int_W dp_1\, dp_2 \cdots dp_{3N}\, dq_1\, dq_2 \cdots dq_{3N}$$

will contain on the average W/h^{3N} states. When writing (7) we assumed that the system was composed of distinguishable particles. If we are dealing with a system of identical particles, we do not distinguish between states which differ only by the labeling of the particles occupying the various regions of phase space. In this case our measure of the number of available states in the ensem-

---

[2] The result is obvious when the action angle variables of an integrable system are used. Since phase space volumes are invariant under canonical transformations, the result will hold for any choice of generalized coordinates of an integrable system. In the case of nonintegrable systems (which are the ones we are mainly interested in since we are assuming the ergodic hypothesis), the program of semiclassical quantization is remarkably difficult to carry out (see, e.g., Berry, 1981).

ble becomes[3]

$$\Omega(E) = \frac{1}{h^{3N} N!} \int_{E \leq H(\mathbf{x}) \leq E + \delta E} d^{6N} x \tag{9}$$

The entropy, S, is again given by (8).

In a purely classical treatment the tolerance δE is, strictly speaking, not necessary. The energy levels of a large but finite quantum system are discrete but closely spaced. To obtain a continuous function $\Omega(E)$ in the quantum case, we have introduced this tolerance δE. However, we must require that δE play no role in the *thermodynamic limit* $N \rightarrow \infty$. This requires that $E/N < \delta E << E$. We show this explicitly for a monatomic ideal gas of N particles in a volume V.

$$H(\mathbf{x}) = \sum_{j=1}^{N} \frac{\mathbf{p}_j^2}{2m} = E \tag{10}$$

The equation $\Sigma \mathbf{p}_j^2 = 2mE$ defines the surface of a $3N$-dimensional sphere of radius $(2mE)^{1/2}$. If $\delta E << E$, we have

$$\Omega(E, V, N) = \frac{V^N}{h^{3N} N!} A_{3N}[(2mE)^{1/2}] \frac{m \, \delta E}{(2mE)^{1/2}} \tag{11}$$

where

$$A_n(r) = \frac{2\pi^{n/2} r^{n-1}}{\Gamma(n/2)} \tag{12}$$

is the surface area of a n-dimensional sphere of radius r and, for integer n, $\Gamma(n) = (n-1)!$. We obtain

$$\Omega(E, V, N) = \frac{V^N}{h^{3N} N!} \frac{(2m\pi E)^{3N/2}}{(3N/2 - 1)!} \frac{\delta E}{E} \tag{13}$$

If $|\ln(\delta E/E)| << N$, we find, using Stirling's formula: $\ln N! = N \ln N - N$,

$$S(E. V, N) = Nk_B \ln \frac{V}{N} + \frac{3N}{2} k_B \ln \frac{4\pi mE}{3Nh^2} + \frac{5}{2} Nk_B \tag{14}$$

This expression is independent of δE.

We conclude that if δE is chosen as specified, the entropy is an extensive variable; that is, $S \propto N$ in the limit $N \rightarrow \infty$ if the number of particles per unit volume is kept constant and the energy is proportional to N. The expression (14) for the entropy of an ideal gas is referred to as the Sackur–Tetrode formula.

---

[3] Historically, the $N!$ in (9) was introduced by Gibbs to remove a spurious entropy of mixing of two volumes of the same ideal gas (see Problem 2.2). If we are dealing with a mixture consisting of N_1 particles of type 1, N_2 particles of type 2, and so on, we must replace $N!$ by $N_1! \, N_2! \, \dots \, .$

We are now in a position to make contact with thermodynamics. The differential of the entropy for a system of N particles in a volume V is (1.C.5)

$$dS(E, N, V) = \frac{1}{T}dE + \frac{P}{T}dV - \frac{\mu}{T}dN \tag{15}$$

Once we have calculated the microcanonical entropy, (15) provides the statistical definition of the temperature, pressure, and chemical potential:

$$T = \left(\frac{\partial S}{\partial E}\right)_{N,V}^{-1} \qquad \mu = -T\left(\frac{\partial S}{\partial N}\right)_{E,V} \qquad P = T\left(\frac{\partial S}{\partial V}\right)_{E,N} \tag{16}$$

Returning to our example of the classical ideal gas, we find the expected results:

$$\frac{1}{T} = \frac{3Nk_B}{2E} \quad \text{or} \quad E = \frac{3}{2}Nk_B T \tag{17}$$

$$P = \frac{Nk_B T}{V} \quad \text{or} \quad PV = Nk_B T \tag{18}$$

$$\mu = k_B T \ln\left[\frac{P}{k_B T}\left(\frac{h^2}{2\pi m k_B T}\right)^{3/2}\right] \tag{19}$$

The variables T, P, and μ are independent of the size of the system in the limit $N \to \infty$ and are thus *intensive* (Section 1.A). The ideal gas is an example of a *normal* system. We will consider a system to be normal if in the thermodynamic limit, $N \to \infty$ with N/V = constant, the energy and entropy are *extensive* (i.e., proportional to N), and T, P, and μ are intensive. Not all systems are normal in this sense. Examples of systems which are not are (a) a system that is self-bound by gravitational forces, and (b) a system with a net macroscopic charge. In both cases there will be contributions to the energy which increase more rapidly than linearly with the number of particles (charges). Neutral systems are, however, normal (Lebowitz and Lieb 1969). An important property of normal thermodynamic systems is the Gibbs–Duhem relation (see Section 1.D).

2.B SYSTEMS AT FIXED TEMPERATURE: CANONICAL ENSEMBLE

We now consider two systems which are free to exchange energy but which are isolated from the rest of the universe by an ideal insulating surface. The numbers N_1, N_2 and volumes V_1, V_2 are fixed for each subsystem. The total energy E will be constant under our assumptions and we assume further that the two subsystems are sufficiently weakly interacting that we can write

$$E_T = E_1 + E_2 \tag{1}$$

where E_1 and E_2 are the energies of the subsystems. We again assume a tolerance δE, chosen suitably so that the statistical weights Ω, Ω_1, Ω_2 are proportional to δE. We then have

$$\Omega(E) = \int_{E \,<\, E_T \,<\, E \,+\, \delta E} \frac{dE_T}{\delta E} \int \frac{dE_1}{\delta E} \Omega_2(E_T - E_1)\Omega_1(E_1) \qquad (2)$$

If the subsystems are sufficiently large, the product $\Omega_2(E_T - E_1)\Omega_1(E_1)$ will be a sharply peaked function of E_1. The reason for this is that Ω_1 and Ω_2 are rapidly increasing functions of E_1 and $(E_T - E_1)$, respectively.[4] From the definition (2.A.8) we note that the entropy is a monotonically increasing function of Ω and that the product $\Omega_1 \Omega_2$ will be at a maximum when the total entropy

$$S(E, E_1) = S_1(E_1) + S_2(E - E_1) \qquad (3)$$

is at a maximum. We can now make an argument, similar to Section 1.F, that the most likely value $\langle E \rangle$, of E, is the one for which

$$\frac{\partial S_1}{\partial E_1} + \frac{\partial S_2}{\partial E_2}\frac{\partial E_2}{\partial E_1} = 0 \qquad (4)$$

Since $\partial E_2/\partial E_1 = -1$ we find, using (2.A.16), that

$$\frac{1}{T_1} - \frac{1}{T_2} = 0$$

or $T_1 = T_2 = T$. The most probable partition of energy between the two systems is the one for which the two temperatures are the same. This is the basis of the zeroth law of thermodynamics, which we see follows naturally from the ensemble concept. In the preceding section we gave a statistical mechanical definition of the temperature, pressure, and chemical potential and we showed that these definitions produced the correct thermodynamic results in the special case of a classical ideal gas. We can now appeal to the zeroth law of thermodynamics and imagine that an arbitrary system is in thermal contact with an ideal gas reservoir. If the system is sufficiently large, it is overwhelmingly probable that the partition of energy between the system and the reservoir will be such as to leave the temperatures effectively the same and we conclude that our definition of temperature is in agreement with thermodynamics. In Section 2.C we make a similar argument with regard to the chemical potential, while we have left the question of the pressure to Problem 2.3.

To establish the *canonical ensemble*, we again consider two systems in thermal contact in such a way that the volume and particle number in each subsystem are held fixed. We now assume that subsystem 2 is very much larger than subsystem 1. The probability $p(E_1)\,dE_1$ that subsystem 1 has energy be-

---

[4] The reader is encouraged to verify this statement explicitly for the case of two ideal gases in contact using (2.A.13).

tween E_1 and $E_1 + dE_1$ is[5]

$$p(E_1)\,dE_1 = \frac{\Omega_1(E_1)\Omega_2(E - E_1)\,dE_1}{\int dE_1\,\Omega_1(E_1)\Omega_2(E - E_1)} \tag{5}$$

We have

$$\Omega_2(E - E_1) = \exp\left\{\frac{S_2(E - E_1)}{k_B}\right\} \tag{6}$$

Since $E_1 << E$, we may expand S_2 in a Taylor series:

$$S_2(E - E_1) = S_2(E) - E_1\frac{\partial S_2}{\partial E} + \frac{1}{2}E_1^2\frac{\partial^2 S_2}{\partial E^2} + \cdots \tag{7}$$

The temperature of the large system is T and we have

$$\frac{\partial S_2}{\partial E} = \frac{1}{T}$$

$$\frac{\partial^2 S_2}{\partial E^2} = \frac{\partial(1/T)}{\partial E} = -\frac{1}{T^2}\left(\frac{\partial T}{\partial E}\right)_{V_2, N_2} = -\frac{1}{T^2 C_2} \tag{8}$$

where C_2 is the heat capacity of system 2 at constant V and N. Since the second system is very much larger than the first, we have $E_1 << C_2 T$ and

$$\Omega_2(E - E_1) = \text{const. } \exp\left\{-\frac{E_1}{k_B T}\right\} \tag{9}$$

With the notation $\beta = 1/(k_B T)$ we thus obtain

$$p_C(E_1) = \frac{1}{\delta E Z_C}\Omega_1(E_1)\exp\{-\beta E_1\} \tag{10}$$

The probability density $p_C(E_1)$ is called the *canonical distribution,* and the normalizing term

$$Z_C = \int\frac{dE_1}{\delta E}\Omega_1(E_1)\exp\{-\beta E_1\} \tag{11}$$

is the *canonical partition function.* The canonical distribution is often a more convenient tool than the microcanonical distribution. In the microcanonical ensemble we were dealing with states with a specified energy $E(S, V, N)$ and T, P, and μ were derived quantities. In the canonical ensemble the system is kept at fixed temperature. We have already seen in Section 1.C that the change of independent variable from S to T is achieved by replacing the energy E, as dependent variable, by the Helmholtz free energy A:

$$A = E - TS \tag{12}$$

---

[5] It is now understood that dE_1 is much larger than the tolerance δE.

Using (2.A.15), we find

$$dA = dE - TdS - S\,dT$$

$$= \mu\,dN - P\,dV - S\,dT \tag{13}$$

The statistical mechanical definition of the free energy is

$$A = -k_B T \ln Z_C \tag{14}$$

We will now show that the two definitions (12) and (14) agree in the sense that for a large system the free energy defined by (14) is given by

$$A = \langle E \rangle - T\langle S \rangle \tag{15}$$

where $\langle E \rangle$ and $\langle S \rangle$ are the most probable values of E and S in the canonical ensemble. To see this, let us rewrite the partition function

$$Z_C = \int \frac{dE}{\delta E} \Omega(E, V, N) \exp\left\{-\frac{E}{k_B T}\right\} = \int \frac{dE}{\delta E} \exp\left\{-\frac{1}{k_B}\left[\frac{E}{T} - S(E, V, N)\right]\right\} \tag{16}$$

If the system is large, it is overwhelmingly probable that the energy will be close to the most likely value $\langle E \rangle$ given by

$$-\frac{1}{k_B}\left[\frac{\langle E \rangle}{T} - S(\langle E \rangle, V, N)\right] = \text{max.} \tag{17}$$

We expand the exponent around its maximum value at $\langle E \rangle$:

$$-\frac{1}{k_B}\left(\frac{E}{T} - S\right) = -\frac{1}{k_B}\left(\frac{\langle E \rangle}{T} - \langle S \rangle\right) + \frac{1}{2k_B}(E - \langle E \rangle)^2 \frac{\partial^2 S}{\partial E^2}\bigg|_{E = \langle E \rangle} + \cdots \tag{18}$$

Using (11), we find that

$$Z_C \approx \exp\left\{-\frac{1}{k_B}\left(\frac{\langle E \rangle}{T} - \langle S \rangle\right)\right\} \int \frac{dE}{\delta E} \exp\left\{-\frac{(E - \langle E \rangle)^2}{2Ck_B T^2}\right\}$$

$$\approx \frac{(2\pi k_B T^2 C)^{1/2}}{\delta E} \exp\left\{-\beta(\langle E \rangle - T\langle S \rangle)\right\}$$

or

$$-\frac{1}{\beta}\ln Z_C = \langle E \rangle - T\langle S \rangle - \frac{1}{\beta}\ln\frac{(2\pi k_B T^2 C)^{1/2}}{\delta E} \tag{19}$$

In the limit that the system is very large, the logarithmic term is small compared to the other terms because both $\langle E \rangle$ and $\langle S \rangle$ are extensive, and we have the desired result.

We have argued that in the canonical ensemble the energy and entropy fluctuate about their mean values. Instead of using condition (17), we can calculate the mean energy by taking an *ensemble average* over all possible values of the energy:

$$\langle E \rangle = \frac{\int dE\, E\Omega(E) \exp\{-\beta E\}}{\int dE\, \Omega(E) \exp\{-\beta E\}}$$

$$= -\frac{\partial}{\partial\beta} \ln Z_C = \frac{\partial}{\partial\beta}(\beta A) \qquad (20)$$

The mean-square fluctuation in the energy is given by

$$\langle (E - \langle E \rangle)^2 \rangle = \langle E^2 \rangle - \langle E \rangle^2$$

$$= -\frac{\partial \langle E \rangle}{\partial\beta} = k_B T^2 \frac{\partial \langle E \rangle}{\partial T} = k_B T^2 C_{V,N} \qquad (21)$$

where $C_{V,N}$ is the heat capacity at constant N and V. The heat capacity is an extensive variable, as is the energy, and the root-mean-square (rms) fluctuation in the energy will thus be proportional to $N^{1/2}$. The mean fluctuation is therefore large for a large system but is a small fraction, which vanishes in the thermodynamic limit, of the total energy:

$$\frac{\sqrt{\langle (E - \langle E \rangle)^2 \rangle}}{\langle E \rangle} \sim \frac{1}{N^{1/2}}$$

The relationship (21) between the response function $C_{V,N}$ and the mean-square fluctuation of the energy is a special case of a very general result known as the fluctuation-dissipation theorem (see Chapter 8). We shall encounter a number of such relations in the following sections.

2.C SYSTEMS WITH FIXED CHEMICAL POTENTIAL: GRAND CANONICAL ENSEMBLE

In a number of instances it is not convenient or possible to fix the number of particles in a system. An example is the case of chemical equilibrium between a number of different species. As an external parameter, such as the temperature, is varied, the concentration of the various constituents will change and we must formulate the statistical treatment in terms of a partition function which allows concentrations to adjust. Moreover, we shall see in Section 2.D that in quantum statistical mechanics a description in terms of a variable number of particles is usually more practical.

We proceed, as in Sections 2.B and 1.F, by considering two systems in contact. They are free to exchange energy and particles, but the volumes of the two subsystems are held fixed. We first derive an equilibrium condition and then let one system be extremely large compared to the other to obtain the probability density in phase space for the smaller system.

Let systems 1 and 2 have energy E_1, E_2 and particle numbers N_1, N_2 with

$$E_1 + E_2 = E_T$$
$$N_1 + N_2 = N_T$$

(1)

The microcanonical partition function for the composite system is given by

$$\Omega(E, N_T) = \sum_{N_1=0}^{N_T} \int \frac{dE_T}{\delta E} \int \frac{dE_1}{\delta E} \Omega_1(E_1, N_1)\Omega_2(E_T - E_1, N_T - N_1)$$

(2)

The product $\Omega_1(E_1 N_1)\Omega_2(E_T - E_1, N_T - N_1)$ will be sharply peaked near the values of E_1 and N_1 that maximize it. This occurs for values of E_1 and N_1 near the ones that maximize the total entropy

$$S_1(E_1, N_1) + S_2(E - E_1, N_T - N_1)$$

(3)

Differentiation using (2.A.16) yields

$$\frac{1}{T_1} = \frac{1}{T_2}$$

$$\frac{\mu_1}{T_1} = \frac{\mu_2}{T_2}$$

(4)

In addition to the previous condition, that at equilibrium two systems in thermal contact have the same temperature, we now have a second condition, namely that the chemical potentials μ_1, μ_2 of the two systems must be equal.

We can construct the *grand canonical* ensemble and partition function by using arguments similar to those of the preceding section. We let subsystem 2 be very much larger than subsystem 1. This allows us to expand

$$\Omega_2(E - E_1, N_T - N_1) = \exp \left\{ \frac{S(E - E_1, N_T - N_1)}{k_B} \right\}$$

$$= \Omega_2(E, N_T) \exp \left\{ -\frac{1}{k_B} \left(E_1 \frac{\partial S}{\partial E} + N_1 \frac{\partial S}{\partial N} \right) \right\}$$

$$= \text{const. } \exp \{ -\beta(E_1 - \mu N_1) \}$$

(5)

We can thus write for the grand canonical probability density,

$$p_G(E_1, N_1) = \frac{1}{\delta E\, Z_G} \Omega_1(E_1, N_1) \exp \{ -\beta(E_1 - \mu N_1) \}$$

(6)

where the normalizing factor, the *grand partition function*, is

$$Z_G(\mu, T, V) = \sum_{N_1=0}^{\infty} \exp \{ \beta \mu N_1 \} Z_C(T, N_1)$$

(7)

In (7) we have, for convenience, taken the upper limit of the summation to be ∞ rather than N_T; the summand becomes negligibly small for values of N_1 comparable to or larger than N_T. In the grand canonical ensemble the system is kept at constant volume, temperature, and chemical potential, in distinction from the canonical ensemble, where V, T, N are kept fixed. In thermodynamics the change from N to μ as an independent variable is accomplished by a Legendre transformation in which the dependent variable A, the Helmholtz free energy, is replaced by the grand potential, $\Omega_G = A - \mu N$. For a normal thermodynamic system we have, from the Gibbs–Duhem relation (1.D.3),

$$\Omega_G = -PV \tag{8}$$

Using (2.B.13), we obtain

$$d\Omega_G = -S\,dT - P\,dV - N\,d\mu \tag{9}$$

and

$$N = -\left(\frac{\partial \Omega_G}{\partial \mu}\right)_{V,T} \quad S = -\left(\frac{\partial \Omega_G}{\partial T}\right)_{V,\mu} \quad P = -\left(\frac{\partial \Omega_G}{\partial V}\right)_{\mu,T} \tag{10}$$

In analogy with the treatment of the preceding section, we postulate the *statistical mechanical* definition of the grand potential:

$$\Omega_G = -k_B T \ln Z_G \tag{11}$$

It is a straightforward matter to show that the definitions (8) and (11) are equivalent in the sense that for a large system,

$$k_B T \ln Z_G = \mu \langle N \rangle - \langle A \rangle \tag{12}$$

The proof is left as an exercise for the reader (Problem 2.6). In the grand canonical ensemble the number of particles fluctuates about the mean particle number

$$\langle N \rangle = \frac{1}{Z_G} \sum_N N \exp\left\{\beta \mu N\right\} Z_C(N) = \frac{1}{\beta} \frac{\partial}{\partial \mu} \ln Z_G \tag{13}$$

A convenient measure of the expected magnitude of the fluctuations is

$$(\Delta N)^2 = \langle (N - \langle N \rangle)^2 \rangle = \langle N^2 \rangle - \langle N \rangle^2$$

$$= \frac{1}{\beta} \frac{\partial \langle N \rangle}{\partial \mu} \tag{14}$$

It is clear from this that since $\partial \langle N \rangle / \partial \mu \sim \langle N \rangle$, we must have

$$\frac{\Delta N}{\langle N \rangle} \approx \langle N \rangle^{-1/2} \tag{15}$$

A useful equation is obtained if we rewrite (14) in terms of the isothermal compressibility

$$K_T = -\frac{1}{V}\left(\frac{\partial V}{\partial P}\right)_{N,T} \tag{16}$$

To do this we note that from the Gibbs–Duhem equation,

$$d\Omega_G = -S\,dT - P\,dV - N\,d\mu = -P\,dV - V\,dP \tag{17}$$

and have

$$d\mu = \frac{V}{N}dP - \frac{S}{N}dT \tag{18}$$

Let $v = V/N$ be the specific volume. Since μ is intensive we can express it as $\mu(v, T)$ and obtain

$$\left(\frac{\partial\mu}{\partial v}\right)_T = v\left(\frac{\partial P}{\partial v}\right)_T \tag{19}$$

We can change v by changing either V or N:

$$\left(\frac{\partial}{\partial v}\right)_{V,T} = \left(\frac{\partial N}{\partial v}\right)_{V,T}\left(\frac{\partial}{\partial N}\right)_{V,T} = -\frac{N^2}{V}\left(\frac{\partial}{\partial N}\right)_{V,T}$$

$$\left(\frac{\partial}{\partial v}\right)_{N,T} = \left(\frac{\partial V}{\partial v}\right)_{N,T}\left(\frac{\partial}{\partial V}\right)_{N,T} = N\left(\frac{\partial}{\partial V}\right)_{N,T} \tag{20}$$

but the way in which v is changed cannot affect (19). Therefore,

$$-\frac{N^2}{V}\left(\frac{\partial\mu}{\partial N}\right)_{V,T} = V\left(\frac{\partial P}{\partial V}\right)_{N,T} \tag{21}$$

Substitution of this result into (14) and (16) finally yields

$$\frac{(\Delta N)^2}{\langle N \rangle} = \frac{K_T}{\beta v} \tag{22}$$

2.D QUANTUM STATISTICS

The simplest consequence of a quantum-mechanical treatment is that the energy levels are discrete. A second consequence originates in the symmetry requirements on the many particle wave functions imposed by the Pauli principle for fermions and by the corresponding requirement on integer spin particles. As we shall see, this leads to combinatorical factors which are different from the simple $N!$ with which we avoided the Gibbs paradox in classical statistics.

The modifications of our statistical formulation due to the discreteness of the energy spectrum are rather simple when the canonical and grand canonical ensembles are used. If E_γ is the energy of the γ th quantum state and its degeneracy is g_γ, we write the canonical partition function as

$$Z_C = \sum_\gamma g_\gamma \exp\{-\beta E_\gamma\} \tag{1}$$

where the sum extends over all energy levels of the system. The correct combinatorical factors for particles obeying Bose–Einstein or Fermi-Dirac statistics are most easily obtained if we consider our system to be quantum states in contact with a heat bath. In a noninteracting many-particle system, for example, the solutions of the Schrödinger equation, consistent with the requirements of quantum statistics, are Slater determinants of single-particle wave functions in the case of fermions or symmetrized linear combinations of product states in the case of bosons (see the Appendix). A unique many-particle state can thus be completely specified through the occupation numbers of the single-particle basis states. In the canonical ensemble the sum of these occupation numbers must be equal to the number of particles, N, whereas in the grand canonical ensemble we do not have this restriction and we may freely sum over all allowed values of the occupation number for each single-particle state. We will illustrate these principles by a number of examples (see also Problems 2.8 and 2.9).

Example 1: Harmonic Oscillator

The energy levels of a single oscillator are $(n + 1/2)h\nu$, where $n = 0, 1, 2, \ldots$ and the states are nondegenerate. The partition function is

$$Z = \sum_{n=0}^{\infty} e^{-\beta(n + 1/2)h\nu} = \frac{e^{-\beta h\nu/2}}{1 - e^{-\beta h\nu}} \tag{2}$$

The ensemble for this partition function can be thought of as grand canonical in the sense that the number of quanta is variable, or canonical in the sense that there is only one oscillator. The average energy is

$$\langle E \rangle = \frac{1}{Z} \sum_{n=0}^{\infty} \left(n + \frac{1}{2} \right) h\nu e^{-\beta(n + 1/2)h\nu}$$
$$= -\frac{\partial}{\partial \beta} \ln Z = \frac{1}{2} h\nu + \frac{h\nu}{e^{\beta h\nu} - 1} \tag{3}$$

Example 2: Noninteracting Fermions

We suppose that the single-particle states are labeled by a wave vector $\mathbf{k}$ and a spin index σ. The Pauli principle requires the occupation number of each such state to be 0 or 1 and the contribution of this state to the grand partition function is

$$\sum_{n=0}^{1} \exp \{-n\beta (E_{\mathbf{k}, \sigma} - \mu)\} = 1 + \exp \{-\beta (E_{\mathbf{k}, \sigma} - \mu)\}$$

where the energy $E_{\mathbf{k}, \sigma} = \hbar^2 k^2 / 2m$ in the case of free particles. The grand partition function is the product of such terms,

$$Z_G = \prod_{\mathbf{k}, \sigma} [1 + \exp \{-\beta (E_{\mathbf{k}, \sigma} - \mu)\}] \tag{4}$$

and the mean number of particles $\langle N \rangle$ is (2.C.13)

$$\langle N \rangle = \sum_{\mathbf{k},\sigma} \frac{1}{\exp\{\beta(E_{\mathbf{k},\sigma} - \mu)\} + 1} \tag{5}$$

Thus the term

$$\langle n_{\mathbf{k},\sigma} \rangle = \frac{1}{\exp\{\beta(E_{\mathbf{k},\sigma} - \mu)\} + 1} \tag{6}$$

is the mean occupation number of the state $(\mathbf{k}, \sigma)$ or, equivalently, the probability that it is occupied, while the probability that it is unoccupied is

$$1 - \langle n_{\mathbf{k},\sigma} \rangle = \frac{1}{1 + \exp\{-\beta(E_{\mathbf{k},\sigma} - \mu)\}} \tag{7}$$

We note that as $T \rightarrow 0$ at fixed density $\langle N \rangle/V$ the chemical potential μ must approach a finite positive limit called the Fermi energy ϵ_F. In this limit the occupation number (6) is 1 for energies less than ϵ_F and 0 otherwise. Conversely, from (7) we see that if $e^{-\beta\mu} >> 1$, the probability of occupation of any state is small (note that E is positive). In this case the system of fermions is *nondegenerate* and approximately satisfies Boltzmann statistics:

$$\langle n_{\mathbf{k},\sigma} \rangle \approx \exp\{-\beta(E_{\mathbf{k},\sigma} - \mu)\}$$

From (2.A.18–2.A.19) we find the condition for degeneracy:

$$v = \frac{V}{N} << \left(\frac{h^2}{2\pi m k_B T}\right)^{3/2} \tag{8}$$

We refer to the quantity

$$\lambda = \left(\frac{h^2}{2\pi m k_B T}\right)^{1/2} \tag{9}$$

as the thermal wavelength and note that if this quantity is smaller than typical interparticle separations, the system will be nondegenerate.

Example 3: Noninteracting Bosons

In the case of bosons (integer spin, which we take to be zero) a single particle state can have an arbitrary occupation number. The contribution of the state labeled by $\mathbf{k}$ (energy $\hbar^2 k^2/2m$) to the grand partition function is

$$\sum_{n=0}^{\infty} \exp\{-n\beta(E_{\mathbf{k}} - \mu)\} = \frac{1}{1 - \exp\{-\beta(E_{\mathbf{k}} - \mu)\}}$$

and, as in the case of fermions, the grand partition function is the product of such terms:

$$Z_G = \prod_{\mathbf{k}} [1 - \exp\{-\beta(E_{\mathbf{k}} - \mu)\}]^{-1} \tag{10}$$

The average number of particles occupying the state $(\mathbf{k})$ is

$$\langle n_{\mathbf{k}} \rangle = -k_B T \frac{\partial}{\partial \mu} \ln\left[1 - \exp\left\{-\beta(E_{\mathbf{k}} - \mu)\right\}\right] = \frac{1}{\exp\left\{\beta(E_{\mathbf{k}} - \mu)\right\} - 1} \tag{11}$$

A Bose system, like a Fermi system, obeys Maxwell–Boltzmann statistics to a good approximation if the thermal wavelength (9) is small compared to typical interparticle separations $(V/N)^{1/3}$. The statistical mechanics of a noninteracting Bose system is discussed in more detail in Section 7.A.

We conclude this section by introducing the concept of the density matrix. Let us first assume that we know all the quantum states $|n\rangle$ of a given system. These states are eigenstates of the Hamiltonian

$$H|n\rangle = E_n|n\rangle \tag{12}$$

and we assume that they are *orthonormal:*

$$\langle n|n'\rangle = \delta_{nn'} \tag{13}$$

We also assume that these states are *complete,* that is, all accessible states are expressible as a linear combination of this basis set. If $|a\rangle$ is an arbitrary state, we can thus write

$$|a\rangle = \sum_n a_n |n\rangle \tag{14}$$

where, from (13), we have

$$a_n = \langle n|a\rangle \tag{15}$$

and

$$|a\rangle = \sum_n |n\rangle\langle n|a\rangle \tag{16}$$

Formally, this allows us to express the completeness requirement as

$$\bar{1} = \sum_n |n\rangle\langle n| \tag{17}$$

where $\bar{1}$ is the unit operator. In the canonical ensemble the probability that the system is in state $|n\rangle$ is given by

$$p_n = \frac{\exp\left\{-\beta E_n\right\}}{\sum_n \exp\left\{-\beta E_n\right\}} \tag{18}$$

and the thermal average of an arbitrary operator $\tilde{A}$ is

$$\langle \tilde{A} \rangle = \sum_n \langle n|\tilde{A}|n\rangle \frac{\exp\left\{-\beta E_n\right\}}{\sum_n \exp\left\{-\beta E_n\right\}} = \frac{1}{Z_C} \sum_n \langle n|\tilde{A}|n\rangle \exp\left\{-\beta E_n\right\} \tag{19}$$

The operator

$$\tilde{\rho} = \frac{1}{Z_C} \sum_n |n\rangle\langle n| \exp\{-\beta E_n\} \tag{20}$$

is commonly referred to as the *density operator* and its matrix representation as the *density matrix*. This formalism can easily be generalized to the grand canonical ensemble. We define the grand Hamiltonian ($\tilde{N}$ is the number operator):

$$K = H - \mu\tilde{N} \tag{21}$$

A complete set of states then contains states with any number of particles. If $\{|n\rangle\}$ is a complete set of eigenstates of K with

$$K|n\rangle = K_n|n\rangle \tag{22}$$

we may define the grand canonical density operator to be

$$\tilde{\rho} = \frac{1}{Z_G} \sum_n |n\rangle\langle n| \exp\{-\beta K_n\} \tag{23}$$

Frequently, the complete set of eigenstates of the Hamiltonian is not known, but it may be possible to find a set of states $|v\rangle$ which are not eigenstates of the Hamiltonian, but are complete. The density matrix is not necessarily diagonal in this representation:

$$\rho_{vv'} = \langle v|\tilde{\rho}|v'\rangle = \frac{1}{Z_G}\langle v|e^{-\beta K}|v'\rangle \tag{24}$$

We see that the density matrix can be written, independently of representation, as

$$\tilde{\rho} = \frac{1}{Z_G} e^{-\beta K} \tag{25}$$

It is easy to see that the thermal average of an operator $\tilde{A}$ can be expressed, in an arbitrary basis, as

$$\langle\tilde{A}\rangle = \frac{1}{Z_G} \sum_n \langle n|\tilde{A}|n\rangle \exp\{-\beta K_n\}$$

$$= \frac{1}{Z_G} \sum_{v,v'} \langle v|\tilde{A}|v'\rangle\langle v'|e^{-\beta K}|v\rangle \tag{26}$$

$$= \frac{1}{Z_G} \mathrm{Tr}\, \tilde{A}e^{-\beta K} = \mathrm{Tr}\, \tilde{A}\tilde{\rho}$$

In this general notation, the partition function is given by

$$Z_G = \mathrm{Tr}\, e^{-\beta K} \tag{27}$$

and the normalization condition on the density matrix is

$$\mathrm{Tr}\, \tilde{\rho} = 1 \tag{28}$$

2.E MAXIMUM ENTROPY PRINCIPLE

The link provided by the ergodic hypothesis between the time evolution of a dynamical system and the ensemble concept is a weak one. It is therefore of interest to show that it is possible to formulate the foundation of equilibrium statistical mechanics in terms of information theory (Jaynes, 1957), in which the thermodynamic variables are inferred from a least possible bias estimate on the basis of available information. In this way the link with dynamics is severed altogether and the approach leads to a maximum entropy principle, which yields results equivalent to the ones found earlier in this chapter.

Consider first the situation in which a state variable x is capable of attaining any one of Ω discrete values x_i ($i = 1, 2, \ldots, \Omega$). Let p_i be the associated probability distribution. This distribution must be normalized:

$$\sum_{i=1}^{\Omega} p_i = 1 \tag{1}$$

and there may be a number of other constraints of the form

$$\bar{f} = \langle f(x) \rangle = \sum_{i=1}^{\Omega} p_i f(x_i) \tag{2}$$

For example, the energy, pressure, or concentrations of different constituents may have a known mean value established by the appropriate contact with the surroundings. In general, nothing else is known about the p_i's. In the case of the microcanonical ensemble the only constraints were that the total energy should lie within a tolerance δE of a prescribed value and that the particle number N and volume V be fixed. We then have no reason to prefer one allowed state over another and have no other choice but to assume that all such states are equally likely or $p_i = 1/\Omega$. Indeed, it is intuitively obvious that any other choice would represent a bias, or additional information about the system.

From the above it is clear that we require an unambiguous measure of "the amount of uncertainty" represented by a probability distribution. It is natural to require of such a measure that it should be positive, with zero corresponding to absolute certainty ($p_i = 1$ for some i), and that it should increase monotonically with increasing uncertainty. Shannon [1948] has shown that if, in addition, we require that our measure be additive for independent sources of uncertainty (extensivity), the choice is essentially unique:

$$S = -K \sum_{i=1}^{\Omega} p_i \ln p_i \tag{3}$$

where K is an arbitrary positive constant. If we choose $K = k_B$, equation (3) gives the *information-theoretic entropy*. It is easy to generalize this definition to a continuous probability distribution $\rho(\mathbf{x})$:

$$\int dx \, \rho(\mathbf{x}) = 1 \tag{4}$$

Consider the case of N identical particles in a $6N$-dimensional phase space. We have seen that according to the semiclassical rules there are

$$\frac{d^{6N}x}{h^{3N}N!} \tag{5}$$

states in the phase space volume $d^{6N}x$. We find, for the entropy,

$$S = -k_B \int d^{6N}x \, \rho(\mathbf{x}) \ln [h^{3N}N! \, \rho(\mathbf{x})] \tag{6}$$

It is easy to see that if we let

$$\rho(\mathbf{x}) = \frac{1}{h^{3N}N! \, \Omega(E)} \tag{7}$$

where $\Omega(E)$ is given by (2.A.9), we get back the expression (2.A.8) for the entropy of the microcanonical ensemble. To see that the probability density (7) maximizes the expression (6) subject to the constraint (4), we use the method of Lagrange multipliers. Requiring the functional derivative

$$\frac{\delta}{\delta\rho(x)}\{-k_B \int d^{6N}x \, [\rho \ln (h^{3N}N! \, \rho) - \lambda\rho]\} = 0$$

gives

$$\rho = \frac{1}{h^{3N}N!} e^{\lambda - 1}$$

The multiplier λ is determined by the normalization condition (4) from which (7) follows. The extension to the case in which there are one or more constraints of the form (2) is straightforward. If we wish to constrain the expectation value of the function $f(x)$, we maximize the expression

$$-k_B \sum_i [p_i \ln p_i - \lambda p_i + \beta p_i f(x_i)]$$

where β and λ are Lagrange multipliers. This procedure yields

$$p_i = \exp \{\lambda - 1 - \beta f(x_i)\} \tag{8}$$

With the definition

$$Z(\beta) = \sum_i \exp \{-\beta f(x_i)\}$$

we find from the normalizing constraint (1),

$$e^{\lambda - 1} = \frac{1}{Z(\beta)}$$

The remaining Lagrange multiplier β is determined from the condition

$$\langle f(x) \rangle = \sum_i p_i f(x_i) = -\frac{\partial}{\partial \beta} \ln Z(\beta)$$

If we let $f(x_i)$ be the energy of the state x_i, we see that we recover the canonical distribution with the identification

$$\beta = \frac{1}{k_{\mathrm{B}} T} \quad \text{and} \quad \lambda = \beta A + 1$$

where A is the Helmholtz free energy. It is easy to see that the second variation of S is negative, indicating that we have found a true maximum of the entropy subject to the constraint. The extension of the method to the grand canonical ensemble is left as an exercise (Problem 2.10).

It should be noted that while, as we have seen, the information-theoretic approach and the approach taken in Sections 2.A to 2.D lead to similar results, there is a subtle difference in the concept of entropy. In Section 2.A we defined the entropy for the microcanonical ensemble. The definition was unambiguous in the thermodynamic limit but depended on the tolerence δE for a small system. When we proceeded to the other ensembles in which the energy and particle number could fluctuate, the entropy too became a fluctuating state variable.

In the information-theoretic context, the entropy is a function of the probability distribution in the ensemble and is not fluctuating since it has nothing to do with the state in which the system happens to be. On the other hand, the definition is perfectly unambiguous for systems of any size. Also, there is no restriction to equilibrium situations. The probabilities p_i will be time dependent if the system evolves dynamically.

2.F THERMODYNAMIC VARIATIONAL PRINCIPLES

We saw in Section 1.C that the thermodynamic equilibrium state of a system at fixed temperature and volume is the one that minimizes the Helmholtz free energy. The same result followed from maximum-likelihood arguments in Section 2.B and from the information-theoretic argument of Section 2.E. Similarly, at equilibrium at constant pressure and temperature the Gibbs free energy is at a minimum. These results can be used in statistical calculations. A common approach is to find an approximate expression for the appropriate free energy in terms of certain parameters, and then to determine the equilibrium values of these parameters by minimizing the free energy. This method will be used frequently in our treatment of phase transitions in Chapter 3. At this point we present a simple example that makes use of this method.

Example 1: Schottky Defects in a Crystal

Consider a solid in which the atoms are located on a regular lattice. In equilibrium, at nonzero temperature, the crystal will contain a certain number of Schottky defects or *vacancies*. These can be thought of as atoms which have migrated from interior (bulk) positions to the surface (see Figure 2.1). The problem is to find the equilibrium concentration of vacancies at a given temperature T and pressure P. Let N be the total number of atoms, n the number of vacancies, v_c the volume per unit cell, S_c the configurational entropy of the vacancies, ΔA the change in the free energy of vibration of the crystal due to the presence of a single vacancy, and ϵ the energy cost of creating a vacancy by transferring an atom to the surface.

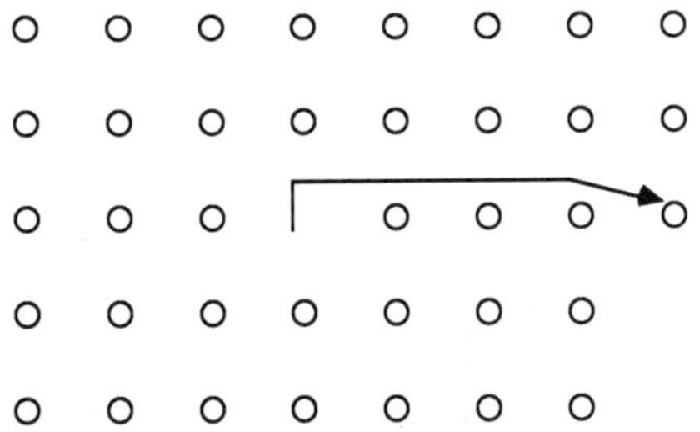

Figure 2.1 An atom migrates to the surface of a crystal, leaving behind a Schottky defect.

The configurational entropy can be obtained by noting that the number of sites is $N + n$. The number of ways of placing n vacancies on these sites is

$$\frac{(N + n)!}{N! \, n!}$$

giving

$$S_c = k_B \ln \frac{(N + n)!}{N! \, n!} = N k_B \ln \frac{N + n}{N} + n k_B \ln \frac{N + n}{n} \tag{1}$$

Since the vacancies are in equilibrium at constant pressure, the correct free energy to use is the Gibbs free energy $G = E - TS + PV$. The part of the free energy that depends on n is

$$G_v = n(\epsilon + \Delta A + P v_c) - T S_c$$

The requirement that G_v be at a minimum yields

$$\epsilon + \Delta A + P v_C = T \frac{\partial S_c}{\partial n} = k_B T \ln \frac{N + n}{n} \approx k_B T \ln \frac{N}{n}$$

and solving for n we find

$$n = N \exp \{ -\beta (\Delta A + P v_c + \epsilon) \}$$

It is shown in Problem 2.11 that for high temperatures, $e^{-\beta \Delta A}$ approaches a constant typically of the order of 10^1, while for low T, ΔA approaches a constant. At atmospheric pressures Pv_c will be in the range of 10^{-4} to 10^{-5} eV, which is negligible compared with a typical value of $\epsilon \sim 1$ eV. In the laboratory, pressures in the range 10 to 100 kbar are fairly standard and Pv_c then becomes comparable with the vacancy formation energy.

PROBLEMS

2.1. *Ergodic Hypothesis.*

(a) Consider a harmonic oscillator with Hamiltonian

$$H = \tfrac{1}{2}p^2 + \tfrac{1}{2}q^2 = \tfrac{1}{2}(\mathbf{x}\cdot\mathbf{x})$$

Show that any phase space trajectory $x(t)$ with energy E will, on the average, spend equal time in all regions of the constant energy surface $\Gamma(E)$.

(b) Consider two linearly coupled harmonic oscillators

$$H = \tfrac{1}{2}(p_1^2 + p_2^2 + q_1^2 + q_2^2 - 2q_1q_2)$$

Express the phase space trajectory $x = (p_1, p_2, q_1, q_2)$ in terms of the initial values of the amplitude and phase of the normal coordinates. Show that there are regions of the constant energy surface which are not visited for any particular trajectory $x(t)$. (If you are not familiar with normal coordinates, a good place to look is Goldstein [1980].)

2.2. *Gibbs Paradox.*

(a) Suppose that (2.A.7) and (2.A.8) were the correct expressions for the entropy of an ideal gas. Consider two volumes $V_A = V_B = V$, each containing N identical particles with the same mean energy. Show that if the two systems are joined together,

$$S_{A+B} = S_A + S_B + 2Nk_B \ln 2$$

that is, there is an entropy of mixing $2Nk_B \ln 2$.

(b) Show that if the correct expression (2.A.9) is used,

$$S_{A+B} = S_A + S_B$$

(c) Estimate the entropy of mixing of 1 mol of Ar and 1 mol of Kr at 1 atm and 300 K.

2.3. *Pressure of Two Systems Separated by a Movable Wall.* Consider two large systems which are isolated from the rest of the universe by an ideal insulating wall. The two subsystems are free to exchange energy, but the number of particles in each subsystem is held fixed, as is the total volume

$$V = V_1 + V_2$$

The two subsystems are separated by a freely movable wall. Show that the most probable partition of the two volumes is the one in which the pressure, as defined in (2.A.16), is equal in the two subsystems.

2.4. *Equipartition.*

(a) A classical harmonic oscillator

$$H = \frac{p^2}{2m} + \frac{Kq^2}{2}$$

is in thermal contact with a heat bath at temperature T. Calculate the partition function for the oscillator in the canonical ensemble and show explicitly that

$$\langle E \rangle = k_B T \qquad \langle (E - \langle E \rangle)^2 \rangle = k_B^2 T^2$$

(b) Consider a system of particles in which the force between the particles is derivable from a potential which is a generalized homogeneous function of degree γ, that is,

$$U(\lambda \mathbf{r}_1, \lambda \mathbf{r}_2, \ldots, \lambda \mathbf{r}_N) = \lambda^\gamma U(\mathbf{r}_1, \mathbf{r}_2, \ldots, \mathbf{r}_N)$$

Show that the equation of state for this system is of the form

$$PT^{-1+3/\gamma} = f(VT^{-3/\gamma})$$

where $f(x)$ can be calculated (at least in principle) once U is specified.

2.5. *Dielectric Function of a Diatomic Gas.* Consider a dilute gas made up of molecules which have a permanent electric dipole moment μ (Figure 2.2). The energy of a molecule in an electric field E pointing in the z direction can be written

$$E = T_{\text{transl}} + T_{\text{rot}} - \mu E \cos \theta$$

Treat the dipoles as classical rods with moment of inertia I. Then

$$T_{\text{rot}} = \tfrac{1}{2} I (\dot{\theta}^2 + \dot{\phi}^2 \sin^2 \theta)$$

Assume that the canonical partition function for the gas can be written

$$Z = \frac{Z_1^N}{N!}$$

where the partition function for a single molecule is $Z_1 = Z_{\text{transl}} \cdot Z_{\text{rot}}$.
(a) Show that

$$Z_{\text{rot}} = \frac{2I}{\hbar^2 \beta^2 E \mu} \sinh \beta E \mu$$

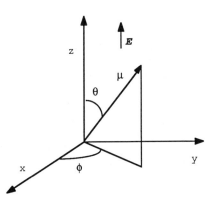

Figure 2.2 Polar molecule with dipole moment μ in an electric field E.

(b) Show that the polarization, P, satisfies

$$P = \frac{N}{V}\langle\mu\cos\theta\rangle = \frac{N}{V}\left(\mu\coth\beta\mu E - \frac{k_B T}{E}\right)$$

(c) Expand the hyperbolic cotangent, for small $\beta\mu E$, and show that in the weak-field limit the dielectric constant ϵ, given by

$$\epsilon E \equiv E + 4\pi P$$

satisfies

$$\epsilon = 1 + \frac{4\pi N}{3} \frac{1}{V}\beta\mu^2$$

2.6. *Equivalence of Thermodynamic and Statistical Mechanical Definitions.* Show that in the limit that we are dealing with a very large system

$$k_B T \ln Z_G = \mu\langle N\rangle - \langle A\rangle$$

Find an estimate for the correction term in analogy with equation (2.B.19).

2.7. *Classical Ideal Gas in the Grand Canonical Ensemble.*
   (a) Show that for a classical ideal gas of particles of mass m,

$$Z_G = \exp\left\{e^{\beta\mu}V\left(\frac{2m\pi k_B T}{h^2}\right)^{3/2}\right\}$$

   (b) Use the expression for Z_G to calculate the mean values of N, P, and S. Show that $PV = Nk_B T$ and that the Gibbs–Duhem relation gives $E = \frac{3}{2}Nk_B T$.

2.8. *Blackbody Radiation.*
   (a) The frequency of an electromagnetic mode of wave vector $\mathbf{k}$ is $\omega_\mathbf{k} = ck$. In a box of volume V there will be

$$\frac{2V\,d^3k}{(2\pi)^3}$$

   such modes in a region d^3k surrounding a given wave vector. The Hamiltonian of the system is $\Sigma\,\hbar\omega_\mathbf{k}n_\mathbf{k}$, where $n_\mathbf{k}$ is the number of excitations in the mode with wave vector $\mathbf{k}$. Show that the Helmholtz free energy of the electromagnetic radiation in the cavity is

$$A = \frac{Vk_B T}{\pi^2 c^3}\int_0^\infty \omega^2\,d\omega\,\ln\left(1 - e^{-\beta\hbar\omega}\right)$$

   (b) By comparing the expressions $P = -\partial A/\partial V$ and $E = \partial(\beta A)/\partial\beta$ for the pressure and internal energy, respectively, show that

$$PV = \frac{E}{3}$$

2.9. *Diatomic Molecule.* If the energy stored in the rotational and vibrational modes is not too large, we may approximate the Hamiltonian of a diatomic molecule by

$$H = T_{\text{transl}} + T_{\text{rot}} + T_{\text{vib}}$$

neglecting any effect of centrifugal forces on the vibrational modes and the effect of the vibrational distortions on the moment of inertia I. In a dilute gas the density will be low enough that the translational motion can be treated classically. The energy of rotation is

$$T_{\text{rot}} = \frac{\hbar^2 j(j + 1)}{2I}$$

where j is the rotational quantum number. The state with quantum number j has degeneracy $g_j = 2j + 1$ (we have, for simplicity, assumed that the two atoms in the diatomic molecule are different). The vibrational degree of freedom may be taken to be a harmonic oscillator with frequency ω_{vib}.

(a) Find an expression for the specific heat.

(b) Discuss the three cases

    (i) $T \ll \theta_{\text{rot}} = \hbar^2/(2Ik_B) \ll \hbar\omega_{\text{vib}}/k_B = \theta_{\text{vib}}$

    (ii) $\theta_{\text{rot}} \ll T \ll \theta_{\text{vib}}$

    (iii) $\theta_{\text{rot}} \ll \theta_{\text{vib}} \ll T$

The Euler summation formula

$$\sum_{n=0}^{\infty} f(n) = \int_0^{\infty} dx f(x) + \tfrac{1}{2}f(0) - \tfrac{1}{12}f'(0) + \tfrac{1}{720}f^{(3)}(0) + \cdots$$

may be useful in the solution of part (ii).

2.10. *Maximum Entropy Principle for the Grand Canonical Ensemble.* Construct the grand canonical ensemble from the maximum entropy principle by maximizing the entropy subject to the constraints that the mean energy and particle number are fixed.

2.11. *Effect of Lattice Vibrations on Vacancy Formation.* To establish the effect qualitatively, consider the following crude model. Each atom vibrates as an independent three-dimensional Einstein oscillator of frequency ω_0. Assume further that if a nearest-neighbor site is vacant, the frequency of the mode corresponding to vibration in the direction of the vacancy changes from ω_0 to ω. Let q be the number of nearest neighbors.

(a) Show that in this simple model,

$$\Delta A = nqk_B T \ln \frac{\sinh(\beta\hbar\omega/2)}{\sinh(\beta\hbar\omega_0/2)}$$

where n is the total number of vacancies.

(b) Consider as an example a simple cubic lattice. Each mode then corresponds to the vibration of two springs. If one of them is cut, the simplest assumption one can make is

$$\omega = \frac{\omega_0}{\sqrt{2}}$$

Show that for high temperatures, $\beta\hbar\omega \ll 1$,

$$e^{-\beta\Delta A/n} \approx 8$$

while for low temperatures, $\beta \hbar \omega \gg 1$,

$$\Delta A \approx -\tfrac{3}{2} n \hbar \omega_0 (2 - \sqrt{2})$$

2.12. *Partition Function at Fixed Pressure.* Consider a system of N noninteracting molecules in a container of cross-sectional area A. The bottom of the container (at $z = 0$) is rigid. The top consists of an airtight piston of mass M which slides without friction.

(a) Construct the partition function Z of the $(N + 1)$-particle system (N molecules of mass m, one piston of mass M). You may neglect the effect of gravity on the gas molecules.

(b) Show that the thermodynamic potential $-k_B T \ln Z$ is, in the thermodynamic limit, identical to the Gibbs potential of an ideal gas of N molecules.

3

Mean Field and Landau Theory

In this chapter we begin our discussion of the statistical mechanics of systems that display a change in phase as a function of an intensive variable such as the temperature or the pressure. In recent years a great deal of progress has been achieved in our understanding of phase transitions, notably through the development of the renormalization group approach of Wilson, Fisher, Kadanoff, and others. We postpone a discussion of this theory until Chapter 6. In this chapter we discuss an older approach known as mean field theory, which generally gives us a qualitative description of the phenomena of interest. A common feature of mean field theories is the identification of an *order parameter*. One approach is to express an approximate free energy in terms of this parameter and minimize the free energy with respect to the order parameter (we have used this approach in Section 2.F in connection with our discussion of Schottky defects in a crystal). Another, often equivalent approach is to approximate an interacting system by an equivalent noninteracting system in a *self-consistent* external *field* expressed in terms of the order parameter.

To understand the phenomena associated with the sudden changes in the material properties which take place during a phase transition, it has proven most useful to work with simplified models that single out the essential aspects of the problem. One important such model, the Ising model, is introduced in Section 3.A and discussed in the Weiss molecular field approximation—an example of the self-consistent field approach mentioned earlier. In Section 3.B we discuss the same model in the Bragg–Williams approximation, which is a free-energy minimization approach. In Section 3.C we show that the usefulness of the Ising model is not limited to the magnetic problems for which it was originally intended and apply it to a problem involving order–disorder transitions in alloys. Further applications are given as problems.

In Section 3.D we discuss an improved version of mean field theory, the Bethe approximation. This method gives better numerical values for the critical temperature and other parameters of the system, although we show in Section 3.E that certain properties associated with the critical behavior of mean field theories are always the same.

The most serious fault of mean field theories lies in the neglect of fluctuations of the order parameter that are correlated over large distances. As we shall see, the importance of this omission depends very much on the dimensionality of the problem, and in problems involving one- and two-dimensional systems the results predicted by mean field theory are frequently qualitatively wrong. In Section 3.F we illustrate this by discussing properties of the exact solution to the Ising model in one dimension.

Because of its close relation to mean field theory we discuss in Section 3.G the Landau theory of phase transitions. Symmetry considerations are in general important in determining the order of a transition. We illustrate this in Section 3.H by considering the Maier–Saupe model for liquid crystals.

In Section 3.I we extend the Landau theory to the case where more than one thermodynamic quantity can be varied independently. We use as an example the concentration and temperature dependence of ^{3}He–^{4}He mixtures and discuss the occurrence of tricritical points. In Section 3.J we discuss the limitations of mean field theory and derive the Ginzburg criterion for the relevance of fluctuations. We conclude our discussion of Landau theory in Section 3.K by considering multicomponent order parameters which are needed for a discussion of the Heisenberg ferromagnet and other systems.

In Section 3.L we consider the van der Waals theory of fluids as an example of a mean field theory for a continuous system and show that a generalization of the approach can be used to discuss mixtures.

An important reference for much of the material in this chapter is Landau and Lifshitz [1980]. Many examples are discussed in Kubo et al. [1965].

3.A ISING MODEL IN THE MEAN FIELD APPROXIMATION

We consider here a simple model, known as the Ising model, for a magnetic material. N magnetic atoms are assumed to be located on a regular lattice, with the magnetic moments interacting with each other through an exchange interaction of the form

$$H_{\text{Ising}} = -J_0 \sum_{\langle ij \rangle} S_i^z S_j^z \tag{1}$$

where J_0 is a constant and the symbol $\langle ij \rangle$ denotes that the sum is to be carried out over nearest-neighbor pairs of lattice sites. The exchange interaction is assumed to have a preferred direction so that the Ising Hamiltonian is diagonal in the representation in which each spin S_i^z is diagonal. The z component of the spins then takes on the discrete values $-S, -S + \hbar, \ldots, S$. The eigenstates of (1) are labeled by the values of S_i^z on each site, and the model has no dynamics. This makes the model easier to work with than the Heisenberg model,

$$H = -J_0 \sum_{\langle ij \rangle} S_i \cdot S_j \qquad (2)$$

where the local spin operators S_i^α do not commute with H. We specialize to the case $S = \frac{1}{2}\hbar$, and also add a Zeeman term for the energy in a magnetic field directed along the z direction to obtain the final version of the Ising Hamiltonian (which in accordance with the definitions in Chapter 1 should be considered as an enthalpy, but in conformity with common usage will be referred to as an energy),

$$H = -J \sum_{\langle ij \rangle} \sigma_i \sigma_j - h \sum_i \sigma_i \qquad (3)$$

where $\sigma_i = \pm 1$ and h is proportional to the magnetic field, but has the unit of energy. To obtain an intuitive feeling for the behavior of such a system, consider the limits $T \to 0$ and $T \to \infty$ for the temperature in the case $J > 0$. At $T = 0$ the system will be in its ground state, with all spins pointing in the direction of the applied field. At $T = \infty$ the entropy dominates and the spins will be randomly oriented. In certain cases the two régimes will be separated by a *phase transition,* that is, there will be a temperature T_c at which there is a sudden change from an ordered phase to a disordered phase as the temperature is increased. Suppose that at a certain temperature the expectation value of the magnetization is m, that is,

$$\langle \sigma_i \rangle = m \qquad (4)$$

for all i. We refer to m as the *order parameter* of the system. Consider the terms in (3) which contain a central spin σ_0. These terms are, with j restricted to nearest-neighbor sites of site 0,

$$H(\sigma_0) = -\sigma_0 \left(J \sum_j \sigma_j + h \right) \qquad (5a)$$

$$= -\sigma_0 (qJm + h) - J\sigma_0 \sum_j (\sigma_j - m) \qquad (5b)$$

where q is the number of nearest neighbors of site 0. If we disregard the second term in (5b), we are left with a noninteracting system; that is, each spin is in an effective magnetic field composed of the applied field and an average exchange field due to the neighbors. The magnetization has to be determined self-consistently from the condition

$$m = \langle \sigma_0 \rangle = \langle \sigma_j \rangle \tag{6}$$

This approximation constitutes a form of mean field theory—the fluctuating values of the exchange field are replaced by an effective average field—and is commonly referred to as the Weiss molecular field theory. We obtain the constitutive equation for m:

$$m = \langle \sigma_0 \rangle = \frac{\text{Tr } \sigma_0 \exp \{-\beta E(\sigma_0)\}}{\text{Tr } \exp \{-\beta E(\sigma_0)\}}$$
$$= \tanh \left[\beta (qJm + h) \right] \tag{7}$$

To find $m(h, T)$ we must solve (7) numerically. However, it is easy to see that $m(h) = -m(-h)$ and that for each $h \neq 0$ there is at least one solution, and sometimes three. For $h = 0$ there is always one solution $m = 0$, and if $\beta qJ > 1$, two further solutions at $\pm m_0$. We will show in Section 3.B that the *equilibrium* state for $T < T_c = qJ/k_B$ is either of the *broken symmetry* states with spontaneous magnetization $\pm m_0(T)$. As $T \to 0$, $\tanh[\beta qJm] \to \pm 1$ for $m \neq 0$, and $m_0 \to \pm 1$. As $T \to T_c$ from below, $|m_0(T)|$ decreases and we may obtain its asymptotic dependence by making a low-order Taylor expansion of the hyperbolic tangent, that is,

$$m_0 = \beta qJm_0 - \tfrac{1}{3}(\beta qJ)^3 m_0^3 + \cdots \tag{8}$$

or

$$m_0^2 = 3 \left(\frac{k_B T}{qJ} \right)^3 \left(\frac{qJ}{k_B T} - 1 \right) \tag{9}$$

$$m_0(T) = \pm 3^{1/2} \left(\frac{T}{T_c} \right)^{3/2} \left(\frac{T_c}{T} - 1 \right)^{1/2} \tag{10}$$

The order parameter m_0 approaches zero in a singular fashion as T approaches T_c from below, vanishing asymptotically as

$$\left(\frac{T_c}{T} - 1 \right)^{1/2}$$

The exponent for the power law behavior of the order parameter is in general given the symbol β and in more sophisticated theories, as well as in real ferromagnets, is not the simple fraction $\tfrac{1}{2}$ found here.

3.B BRAGG–WILLIAMS APPROXIMATION

In this section we apply the density matrix formulation of Section 2.D and the thermodynamic variational principle of Section 2.F for the Gibbs free energy to the Ising model. Our version of mean field theory corresponds to a particularly simple choice of trial density matrix, namely

$$\rho = \prod_i \rho_i \tag{1}$$

where the product over i is over all lattice sites and the symbol $\prod$ indicates a direct product.[1] The free energy, denoted by G, associated with the trial density matrix ρ is

$$G(h, T) = \text{Tr} \left[\rho H + k_B T \rho \ln \rho\right] \tag{2}$$

Since any combination of the site spin orientations is an eigenstate of the Hamiltonian, the density matrix will not couple different spin states. The most general normalized density matrix of the type (1) is then

$$\rho = \prod_i \begin{bmatrix} \frac{1}{2}(1 + m_i) & 0 \\ 0 & \frac{1}{2}(1 - m_i) \end{bmatrix} \tag{3}$$

Translational invariance requires that $m_i = m = $ constant. With this density matrix the trial free energy takes the form

$$G(h, T) = -\frac{qJN}{2}m^2 - Nhm$$

$$+ Nk_B T\left(\frac{1 + m}{2}\ln\frac{1 + m}{2} + \frac{1 - m}{2}\ln\frac{1 - m}{2}\right) \tag{4}$$

Minimizing with respect to the variational parameter m, we obtain

$$0 = -qJm - h + \tfrac{1}{2}k_B T \ln\frac{1 + m}{1 - m} \tag{5}$$

or

$$m = \tanh\left[\beta(qJm + h)\right] \tag{6}$$

as in the previous approach. Now let $h = 0$. The function $G(0, T)$ is then of the form

$$G(0, T)/N = \tfrac{1}{2}qJm^2$$

$$+ \tfrac{1}{2}k_B T[(1 + m) \ln(1 + m) + (1 - m) \ln(1 - m) - 2\ln(2)] \tag{7}$$

For small values of m we may expand in a power series to obtain

$$G(0, T) = \frac{m^2}{2}(k_B T - qJ) + \frac{k_B T}{12}m^4 + \cdots - k_B T \ln 2 \tag{8}$$

---

[1] The matrix ρ_i is 2×2, while ρ will be a $2^N \times 2^N$ matrix. Physically, the approximation (3.B.1) that ρ is a direct product means that the orientation of any spin is statistically independent of the orientation of any other spin. When this approximation does not hold we say that the spins are correlated.

with all higher order terms positive and of even power in m. The form of $G(0, T)$ is shown for T above and below $T_c = qJ/k_B$ in Figure 3.1.

It is clear that the ordered phase is the state of lower free energy when $T < T_c$. The type of transition seen in this system is known as *continuous*, or second order, since the order parameter increases continuously from zero as a function of $(T_c - T)$ below the transition (Figure 3.2). We shall encounter examples of discontinuous, or first-order, transitions later in this chapter.

The theory presented above is remarkably general. Note that neither the type of lattice nor the spatial dimensionality plays a role in the transition—the sole parameter characterizing the system is the number, q, of nearest neighbors. Therefore, in this approximation, the Ising models on the two-dimensional triangular lattice and the three-dimensional simple cubic lattice have identical properties. This result is quite incorrect and we demonstrate below how one can be misled by mean field arguments.

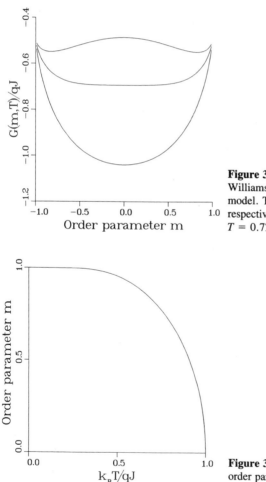

Figure 3.1 Free energy in the Bragg-Williams approximation for the Ising model. The three curves correspond to, respectively, $T = 1.5T_c$, $T = T_c$, and $T = 0.7T_c$.

Figure 3.2 Temperature dependence of order parameter below T_c.

Consider a one-dimensional chain with free ends. The Hamiltonian in zero field is

$$H = -J \sum_{i=1}^{N-1} \sigma_i \sigma_{i+1} \tag{9}$$

with ground-state energy $E_0 = -(N-1)J$. Suppose that the system is at a very low temperature and consider the class of excitations defined by $\sigma_l = 1$, $l \le i$ and $\sigma_l = -1$, $l > i$:

$$\uparrow \quad \uparrow \quad \dots \quad \uparrow \quad \downarrow \quad \downarrow \quad \dots \quad \downarrow$$
$$1 \quad 2 \qquad i \quad i+1 \qquad \quad N$$

There are $N - 1$ such states, all with the same energy $E = E_0 + 2J$. At temperature T the free-energy change due to these excitations is

$$\Delta G = 2J - k_B T \ln (N - 1)$$

which as $N \to \infty$ is less than zero for all $T > 0$. These excitations disorder the system; the expectation value of the magnetization is zero. Therefore, there cannot be a phase transition to a ferromagnetic state in the one-dimensional Ising model with nearest-neighbor (or indeed with finite range) interactions.

A similar argument can be used to give a crude estimate of the transition temperature for the two-dimensional Ising model. Consider a two-dimensional $N \times N$ square lattice with free surfaces. We wish to study the set of excitations of the type shown in the figure below, that is, excitations which divide the lattice into two large domains separated by a wall that extends from end to end without loops.

The energy of the domain wall is

$$\Delta E = 2LJ$$

where L is the number of segments in the wall. If we start the wall from the left there are at least two, sometimes three choices for the direction of the next step if we neglect the possibility of reaching the upper and lower boundaries. The entropy associated with nonlooping chains of length L is then at least $k_B \ln 2^L$. There are N possible starting points for the chain. If we assume two choices per site, one of which takes us to the right, the average length of the chain will be $2N$ with a standard deviation $\propto N^{1/2}$, which is small compared to

N if N is large enough. The free energy associated with dividing the lattice into two domains is thus approximately

$$\Delta G \approx 4NJ - k_B T \ln (N \times 2^{2N})$$

The system is therefore stable against domain formation if

$$T < T_c \approx \frac{2J}{k_B \ln 2} = \frac{2.885 J}{k_B}$$

This estimate is surprisingly close to the exact result (see Section 5.A) $T_c = 2.269185 \ldots J/k_B$.

A more sophisticated version of this type of argument was first devised by Peierls[2] to prove that in two dimensions a phase transition *does* occur. The one-dimensional argument was presented here to raise a warning flag—mean field arguments, although useful, are not invariably correct. We return to this topic in Section 3.F, where we discuss some exact properties of the Ising model in one dimension.

3.C BINARY ALLOYS IN THE BRAGG–WILLIAMS APPROXIMATION

As a second example of a system undergoing a phase transition we consider a binary alloy of Cu and Zn (brass). At compositions in a narrow range around 50 at % Cu, 50 at % Zn, the atoms occupy the sites of a body-centered cubic (bcc) lattice forming β-brass. The distribution of atoms on these sites is disordered above a temperature T_c which is close to 740 K. Below T_c there is ordering with atoms of each kind preferentially distributed on one of the two simple cubic sublattices of the bcc lattice (β' phase, see Figure 3.3).

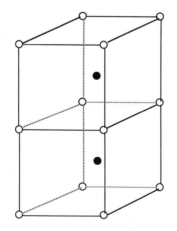

Figure 3.3 Ordered structure for β'-brass: open circles, Zn; filled circles, Cu.

[2] See Peierls [1936]. A clear exposition of the arguments is also given in Chapter 15.4 of Wannier [1966].

The simplest model that can account for the low-temperature structure is one in which the energy of nearest-neighbor pairs depends on what kind of pair it is. Define the quantities N_{AA}, N_{BB}, N_{AB} to be the number of nearest-neighbor pairs of the Cu–Cu, Zn–Zn, and Cu–Zn type, respectively. We take the energy of the configuration to be

$$E = N_{AA}e_{AA} + N_{AB}e_{AB} + N_{BB}e_{BB} \tag{1}$$

where e_{AA}, e_{AB}, and e_{BB} are, respectively, the energies of an AA, AB, and BB bond.

Let N be the numbers of lattice sites and N_A, N_B the number of Cu and Zn atoms, respectively. Referring to Figure 3.3, we introduce the occupation numbers for each of the two simple cubic sublattices: N_{A1} and N_{B1} are the number of atoms of each type on sublattice 1, N_{A2} and N_{B2} the number of atoms of each type on sublattice 2. We have

$$N_{A1} + N_{A2} = N_A = c_A N \qquad N_{B1} + N_{B2} = N_B = c_B N$$

$$N_{A1} + N_{B1} = \frac{N}{2} \qquad\qquad N_{A2} + N_{B2} = \frac{N}{2} \tag{2}$$

For the sake of definiteness we let $N_A \leq N_B$ and define the order parameter

$$m = \frac{N_{A1} - N_{A2}}{N_A} \tag{3}$$

With this definition $-1 \leq m \leq 1$ and

$$N_{A1} = \tfrac{1}{2}N_A(1 + m) \qquad N_{B1} = \tfrac{1}{2}(N_B - N_A m)$$

$$N_{A2} = \tfrac{1}{2}N_A(1 - m) \qquad N_{B2} = \tfrac{1}{2}(N_B + N_A m) \tag{4}$$

Up to this point our treatment is exact. We now make the crucial approximation

$$N_{AA} = q\frac{N_{A1}N_{A2}}{\tfrac{1}{2}N} \qquad N_{BB} = q\frac{N_{B1}N_{B2}}{\tfrac{1}{2}N}$$

$$N_{AB} = q\left(\frac{N_{A1}N_{B2}}{\tfrac{1}{2}N} + \frac{N_{A2}N_{B1}}{\tfrac{1}{2}N}\right) \tag{5}$$

The mean energy is obtained substituting (5) into (1), while the entropy can be evaluated using the method of Section 2.F. The appropriate free energy is

$$A = E - TS \tag{6}$$

with

$$E = \tfrac{1}{2}qN(e_{AA}c_A^2 + 2e_{AB}c_A c_B + e_{BB}c_B^2) - qNc_A^2 \epsilon m^2 \tag{7}$$

where

$$\epsilon = \tfrac{1}{2}e_{AA} + \tfrac{1}{2}e_{BB} - e_{AB} \tag{8}$$

and

$$S = -k_B \left(N_{A1} \ln \frac{N_{A1}}{\frac{1}{2}N} + N_{A2} \ln \frac{N_{A2}}{\frac{1}{2}N} + N_{B1} \ln \frac{N_{B1}}{\frac{1}{2}N} + N_{B2} \ln \frac{N_{B2}}{\frac{1}{2}N} \right)$$

$$= -\tfrac{1}{2} N k_B \Big(c_A (1 + m) \ln [c_A (1 + m)] + c_A (1 - m) \ln [c_A (1 - m)]$$

$$+ (c_B - mc_A) \ln [c_B - mc_A] + (c_B + mc_A) \ln [c_B + mc_A] \Big) \qquad (9)$$

The quantities c_A and c_B are fixed while m must be adjusted to minimize the free energy. Differentiation of A with respect to m gives

$$0 = -2qc_A \epsilon m + \tfrac{1}{2} k_B T \ln \frac{(1 + m)(c_B + c_A m)}{(1 - m)(c_B - c_A m)} \qquad (10)$$

For low temperature and $\epsilon > 0$ there are three solutions of (10): a trivial solution $m = 0$, and two nonzero solutions symmetric about $m = 0$. At high temperatures only the trivial solution exists. As can easily be verified by differentiating (10) with respect to m, the trivial solution yields a minimum of the free energy for

$$T > T_c = \frac{2q\epsilon c_A c_B}{k_B} \qquad (11)$$

This situation is very similar to that of Section 3.B; if we were to plot the free energy (6) as a function of the order parameter m, the resulting figure would look similar to Figure 3.1. In the special case $c_A = c_B = \frac{1}{2}$, (10) can be written

$$0 = -q\epsilon m + k_B T \ln \frac{1 + m}{1 - m} \qquad (12)$$

which is equivalent to (3.B.5) if we put $h = 0$ and $\epsilon = 2J$. The nature of the phase transition in the alloy system is thus identical to the phase transition of the Ising ferromagnet. This last conclusion is independent of the mean field approximation, as we now show.

We introduce the variables n_{iA}, n_{iB}, where $n_{iA} = 1$ if an atom of type A occupies site i and $n_{iA} = 0$ otherwise. Similarly, $n_{iB} = 1 - n_{iA}$. These variables can be expressed in terms of Ising spin variables

$$n_{iA} = \tfrac{1}{2}(1 + \sigma_i)$$
$$n_{iB} = \tfrac{1}{2}(1 - \sigma_i) \qquad (13)$$

with $\sigma_i = \pm 1$. With $\epsilon = 2J$, the energy (1) becomes

$$H = J \sum_{\langle ij \rangle} \sigma_i \sigma_j + \frac{q}{4}(e_{AA} - e_{BB}) \sum_i \sigma_i + \frac{q}{4} N (e_{AA} + e_{BB} + 2e_{AB}) \qquad (14)$$

Remembering that we have a system with a fixed total number of particles of each kind, we see that the last two terms are constant and therefore irrelevant. We obtain the equivalent Hamiltonian

$$H = J \sum_{\langle ij \rangle} \sigma_i \sigma_j \tag{15}$$

If $J > 0$, that is, if it is energetically favorable for atoms of opposite kind (spin) to be nearest neighbors, this Hamiltonian represents an antiferromagnet. The difference in concentration of each species is analogous to the magnetization in the Ising model, and since we are dealing with a system with fixed magnetization (rather than external field), we use the symbol A in preference to G for the free energy.

We now show that under certain circumstances the ferromagnetic and antiferromagnetic Ising models are equivalent. Consider a crystal structure that can be divided into two sublattices, so that the nearest neighbors to the sites on one of the sublattices belong to the other (e.g., the square, honeycomb, simple cubic, and body-centered lattices, but not the triangular, face-centered cubic, or hexagonal close-packed lattices). We may then make the transformation $\sigma_i = -\tau_i$ for i on one of the sublattices and $\sigma_i = +\tau_i$ for i on the other sublattice and have

$$H = -J \sum \tau_i \tau_j \tag{16}$$

Simply renaming τ_j to σ_j completes the transformation of the antiferromagnet into a ferromagnet. Thus the two systems have identical thermodynamic properties at all temperatures.

In our derivations we allowed the concentrations of the two components of the alloy to vary freely. In practice, β-brass occurs only in a fairly narrow concentration range[3] around 50% Cu, 50% Zn. At other stoichiometries the face-centered cubic, more complex cubic structures, or the hexagonal closed packed structure may be thermodynamically stable, or the system may be in a mixture of different phases. In general, there is no guarantee that a particular choice of lattice structure, division into sublattices, or selection of order parameter is the correct one. One should therefore be guided by physical intuition in trying out a number of different alternatives, selecting the one with lowest free energy.

The homogeneous phase with the lowest free energy may be unstable against *phase separation*. Consider a sample with concentration $c_A = c_0$ and let the minimum single-phase free energy be $A(c_0)$. If the sample were to split up into two phases, one with a fraction y of the total number of sites, the constraint that the overall concentration of A atoms is c_0 is expressed through the *lever rule*,

$$yc_1 + (1 - y)c_2 = c_0 \tag{17}$$

where c_1 and c_2 are the concentrations of A atoms in the two phases. The

---

[3] See, for example, Kittel [1976].

homogeneous phase is stable against phase separation if for all c_1 and c_2,

$$yA(c_1) + (1 - y)A(c_2) > A(c_0) \tag{18}$$

Geometrically, (18) corresponds to requiring that $A(x)$ be a *convex* function. If $A(x)$ can be differentiated twice, (18) is equivalent to the condition that

$$\frac{\partial^2 A}{\partial c^2} > 0 \qquad \text{for all } c \tag{19}$$

When the convexity requirement is violated, phase separation will occur. The resulting free energy lies on the convex envelope of $A(c)$. The equilibrium concentrations c_1 and c_2 are given by the lever rule (17) and by the condition $\partial A/\partial c|_{c_1} = \partial A/\partial c|_{c_2}$. This constitutes the double tangent construction of Figure 3.4. A simple model for phase separation is found in Problem 3.1.

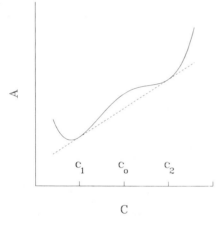

Figure 3.4 Phase separation. At equilibrium, material of initial concentration c_o will split up into two phases with concentrations c_1 and c_2, with the amount of material in each phase given by the lever rule (17).

It is interesting to carry out the equivalent variational treatment of Section 3.B for the alloy system considered here. In the binary alloy problem the simplest choice of density matrix is

$$\rho = \prod_{i \in 1} \rho_{i1} \prod_{j \in 2} \rho_{j2} \tag{20}$$

where the subscripts 1 and 2 refer to the sublattices and where, using (4) we have

$$
\rho_{i1} = \begin{bmatrix} c_A(1 + m) & 0 \\ 0 & c_B - c_A m \end{bmatrix}
$$

$$
\rho_{i2} = \begin{bmatrix} c_A(1 - m) & 0 \\ 0 & c_B + c_A m \end{bmatrix}
\tag{21}
$$

This form takes into account the normalization of the density matrix and gives the correct number of atoms of each species. The completion of this treatment is left as an exercise (Problem 3.2).

3.D IMPROVED MEAN FIELD THEORY: BETHE APPROXIMATION

In this section we wish to consider an approximation scheme due to Bethe [1935]. An extension of the approach, which provides the same results for the order parameter but also yields an expression for the free energy, is due to Fowler and Guggenheim [1940].

Consider again the simple Ising model of Section 3.A:

$$H = -J \sum_{\langle ij \rangle} \sigma_i \sigma_j - h \sum_i \sigma_i \tag{1}$$

In our mean field approximation we ignored all correlations between spins and, even for nearest neighbors made the approximation $\langle \sigma_i \sigma_j \rangle = \langle \sigma_i \rangle \langle \sigma_j \rangle = m^2$. It is possible to improve this approximation in a systematic fashion. We suppose that the lattice has coordination number q and now retain as variables a central spin and its shell of nearest neighbors. The remainder of the lattice is assumed to act on the nearest-neighbor shell through an effective exchange field which we will calculate self-consistently. The energy of the central cluster can be written as

$$H_c = -J\sigma_0 \sum_{j=1}^{q} \sigma_j - h \sum_{j=0}^{q} \sigma_j - h' \sum_{j=1}^{q} \sigma_j \tag{2}$$

The situation is depicted in Figure 3.5 for the square lattice. The fluctuating field acting on the peripheral spins $\sigma_1 \ldots \sigma_4$ has been replaced by an effective field h', just as we previously replaced the interaction of σ_0 with its first neighbor shell by a mean energy.

The partition function of the cluster is given by

$$Z_c = \sum_{\sigma_i = \pm 1} e^{-\beta H_c} = e^{\beta h}(2 \cosh [\beta(J + h + h')])^q$$

$$+ e^{-\beta h}(2 \cosh [\beta(J - h - h')])^q \tag{3}$$

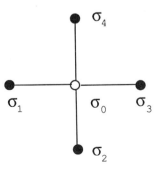

Figure 3.5 Spin cluster used in the Bethe approximation calculation.

The expectation value $\langle\sigma_0\rangle$ is given by

$$\langle\sigma_0\rangle = \frac{1}{Z_c}(e^{\beta h}\{2 \cosh[\beta(J + h + h')]\}^q - e^{-\beta h}\{2 \cosh[\beta(J - h - h')]\}^q)$$

(4)

while for $j = 1, \ldots, q$,

$$\langle\sigma_j\rangle = \frac{1}{Z_c}(2e^{\beta h} \sinh[\beta(J + h + h')]\{2 \cosh[\beta(J + h + h')]\}^{q-1}$$

$$-2e^{-\beta h} \sinh[\beta(J - h - h')]\{2 \cosh[\beta(J - h - h')]\}^{q-1})$$

(5)

For simplicity we now set $h = 0$. Since the ferromagnet is translationally invariant we must require $\langle\sigma_j\rangle = \langle\sigma_0\rangle$. This yields the equation

$$\cosh^q[\beta(J + h')] - \cosh^q[\beta(J - h')]$$

$$= \sinh[\beta(J + h')]\cosh^{q-1}[\beta(J + h')]$$

(6)

$$-\sinh[\beta(J - h')]\cosh^{q-1}[\beta(J - h')]$$

or

$$\frac{\cosh^{q-1}[\beta(J + h')]}{\cosh^{q-1}[\beta(J - h')]} = e^{2\beta h'}$$

(7)

This equation always has a solution $h' = 0$ corresponding to the disordered high-temperature phase. As $h' \to \infty$ the left side of (7) approaches unity while the right side diverges. Two further solutions [$-h'$ satisfies (7) if h' does] will therefore exist if the slope at $h' = 0$ of the left side is greater than 2β. This condition can be written

$$\coth \beta_c J = q - 1$$

(8)

On the square lattice this yields $k_B T_c/J = 2.885$, which may be compared with the exact result $k_B T_c/J = 2.269$ of Onsager [1944] and the prediction $k_B T_c/J = 4$ of the simple mean field theory of the preceding sections. We see that we have achieved a substantial improvement in the prediction of the critical temperature. It is interesting to note that for the one-dimensional Ising model ($q = 2$) the Bethe approximation does not predict a phase transition. This is in agreement with the exact results of Section 3.F.

     It is often important to have expressions for the free energy as well as the order parameter. We write the Hamiltonian (1) in the form

$$H = H_0 + \lambda V$$

(9)

where

$$H_0 = -h \sum_i \sigma_i \qquad V = -\sum_{\langle ij \rangle} \sigma_i \sigma_j$$

(10)

The physical system we wish to consider has $\lambda = J$ but one can also imagine systems with different values of the exchange coupling strength. The free energy associated with (9) is

$$G = -\beta^{-1} \ln \text{Tr } e^{-\beta H_0 - \beta \lambda V} \tag{11}$$

We define

$$G_0 = -\beta^{-1} \ln \text{Tr } e^{-\beta H_0} \tag{12}$$

and

$$\langle O \rangle_\lambda = \frac{\text{Tr } O e^{-\beta H_0 - \beta \lambda V}}{\text{Tr } e^{-\beta H_0 - \beta \lambda V}} \tag{13}$$

and see that

$$\frac{\partial G}{\partial \lambda} = \frac{1}{\lambda} \langle \lambda V \rangle \tag{14}$$

Finally,

$$G = G_0 + \int_0^J \frac{d\lambda}{\lambda} \langle \lambda V \rangle_\lambda \tag{15}$$

Equation (15) is exact if the expectation value $\langle \lambda V \rangle_\lambda$ can be computed exactly. Useful approximations to the free energy can frequently be obtained by substituting approximate expressions into (15). As an example, consider the one-dimensional Ising model ($q = 2$) in zero field. Then

$$G_0 = -Nk_B T \ln 2 \tag{16}$$

There is no phase transition in this case and $h' = 0$. The Bethe approximation then corresponds to writing

$$\langle \sigma_i \sigma_j \rangle_\lambda = \frac{e^{\beta \lambda} - e^{-\beta \lambda}}{e^{\beta \lambda} + e^{-\beta \lambda}} = \tanh \beta \lambda \tag{17}$$

when i, j are nearest neighbors. We obtain

$$G = -Nk_B T \ln 2 - \tfrac{1}{2} N q \int_0^J d\lambda \tanh \beta \lambda = -Nk_B T \ln (2 \cosh \beta J) \tag{18}$$

As we shall see in Section 3.F, this happens to be exact, whereas in general the Bethe approach produces only approximate free energies.

It should be obvious that still better results can be obtained by considering larger clusters. However, all approximations that depend in an essential way on truncation of correlations beyond a certain distance will break down in the vicinity of a critical point. To show this explicitly, in the next section we discuss the critical properties of mean field theories.

3.E CRITICAL BEHAVIOR OF MEAN FIELD THEORIES

In Section 3.A we showed that as $T \to T_c$, the order parameter (magnetization) of our Ising model had the asymptotic form [see (3.A.10)]

$$m(T) \propto (T_c - T)^{1/2} \tag{1}$$

as $T \to T_c$ from below. We now calculate several other thermodynamic functions as $T \to T_c$. Consider first the susceptibility per spin,

$$\chi(h, T) = \left(\frac{\partial m}{\partial h}\right)_T \tag{2}$$

From (3.A.7) we obtain

$$\chi(0, T) = \frac{\beta \, \text{sech}^2 \, \beta q J m}{1 - \beta q J \, \text{sech}^2 \, \beta q J m} \approx \frac{1}{k_B T_c (1 - T_c/T)} \tag{3}$$

as $T \to T_c^+$. For $T < T_c$ we use the asymptotic expansion for m to obtain

$$\chi(0, T) \approx \frac{1}{2 k_B T_c (1 - T/T_c)} \tag{4}$$

and we see that the susceptibility diverges as the critical point is approached from either the low- or high-temperature side. It is conventional to write, for T near T_c,

$$\chi(0, T) \approx A_\pm |T - T_c|^{-\gamma}$$

and we conclude that in our mean field theory, $\gamma = 1$. The exact solution of the two-dimensional Ising model (Section 5.A) yields $\gamma = \frac{7}{4}$; for the three-dimensional Ising model, γ is not known exactly, but is approximately 1.25. This breakdown of mean field theory can be understood in terms of the following rigorous expression for the susceptibility:

$$\chi = \left(\frac{\partial m}{\partial h}\right)_T = \frac{\partial}{\partial h} \left(\frac{\text{Tr} \, \sigma_0 e^{-\beta H}}{\text{Tr} \, e^{-\beta H}}\right)_T$$
$$= \frac{1}{k_B T} \sum_j \left(\langle \sigma_0 \sigma_j \rangle - \langle \sigma_0 \rangle \langle \sigma_j \rangle\right) \tag{5}$$

It is clear that χ can diverge only if the pair distribution function $\langle \sigma_0 \sigma_j \rangle$ is of long range; for example, in three dimensions it must not decay faster than

$$\frac{1}{|\mathbf{r}_j - \mathbf{r}_0|^3}$$

for large separations at $T = T_c$. In our simple mean field approximation, and also in the more sophisticated Bethe approximation, we clearly discarded long-range correlations, and it is therefore not surprising that finite cluster approximations will break down as $T \to T_c$.

Let us next examine the specific heat in both the simple mean field and the Bethe approximations. In zero magnetic field the internal energy in the mean field approximation of Section 3.A and 3.B is given by

$$E = \langle H \rangle = -J \sum_{\langle ij \rangle} \langle \sigma_i \rangle \langle \sigma_j \rangle$$

$$= -\tfrac{1}{2} J q N m^2 \tag{6}$$

giving

$$C_h = \left(\frac{\partial E}{\partial T} \right)_h = -\tfrac{1}{2} N J q \left(\frac{\partial m^2}{\partial T} \right) \to \tfrac{3}{2} N k_B \qquad \text{as } T \to T_c^-$$

$$= 0 \qquad \qquad \text{for } T > T_c \tag{7}$$

that is, the mean field theory produces a discontinuity at the transition. This behavior is in contrast to more correct theories and experimental results which yield a power law of the form

$$C_h = E_{\pm} |T - T_c|^{-\alpha} \tag{8}$$

where α is a conventional notation for the specific heat exponent. The determination of the specific heat singularity in the Bethe approximation is somewhat more tedious. It is easy to show that the correlation function $\langle \sigma_0 \sigma_j \rangle$ which determines the internal energy is given by

$$\langle \sigma_0 \sigma_j \rangle =$$
$$\frac{\sinh \beta (J + h') \cosh^{q-1} \beta (J + h') + \sinh \beta (J - h') \cosh^{q-1} \beta (J + h')}{\cosh^q \beta (J + h') + \cosh^q \beta (J - h')}$$

$$\tag{9}$$

if j is a nearest neighbor to site 0. For $T > T_c$, $h' = 0$ and

$$E = -\tfrac{1}{2} q J N \langle \sigma_0 \sigma_j \rangle = -\tfrac{1}{2} q J N \tanh \beta J$$

For $h' \neq 0$ we note that $\langle \sigma_0 \sigma_j \rangle_{h'} = \langle \sigma_0 \sigma_j \rangle_{-h'}$ and we must therefore have

$$\langle \sigma_0 \sigma_j \rangle_{h'} = \langle \sigma_0 \sigma_j \rangle |_{h'=0} + a(T) h'^2 + \cdots$$

The first piece of this expansion joins continuously with the high-temperature form of the internal energy. The second term will yield a discontinuity at T_c if $\partial h'^2 / \partial T$ approaches a constant as $T \to T_c$. We leave the explicit demonstration of this as an exercise (Problem 3.4). In a similar way it is possible to show that $m(T) = \langle \sigma_0 \rangle \propto |T - T_c|^{1/2}$ in the Bethe approximation. The critical properties of cluster theories thus seem to be in a sense universal and not dependent on the level of sophistication of the approximation.

Another quantity that shows similar behavior in all mean field theories is the critical isotherm $m(T_c, h)$. In the simplest mean field theory we have (3.A.7)

$$m = \tanh\left[\beta(qJm + h)\right]$$

at $T_c = qJ/k_B$ we have, on expanding the hyperbolic tangent,

$$m \approx m + \beta h - \tfrac{1}{3}(m + \beta h)^3$$

which gives, near $h = 0$,

$$h \propto m^\delta \qquad (10)$$

with $\delta = 3$. We again leave it as an exercise for the reader to show that $\delta = 3$ also in the Bethe approximation. In Section 3.G we discuss a general theory of phase transitions due to Landau which exhibits the same behavior as mean field and cluster theories near the critical point. Later in this chapter and in other sections we return to specific examples of mean field theories.

3.F ISING CHAIN: EXACT SOLUTION

The one-dimensional Ising model is one of very few models in statistical mechanics which are exactly solvable. Moreover, the solution is sufficiently simple that the thermodynamic properties can be evaluated without too much difficulty. Let us first consider a chain of length N with free ends and zero external field,

$$H = -J \sum_{i=1}^{N-1} \sigma_i \sigma_{i+1} \qquad (1)$$

The partition function is given by

$$Z_N = \sum_{\sigma_1=\pm1} \cdots \sum_{\sigma_N=\pm1} \exp\left\{\beta J \sum_{i=1}^{N-1} \sigma_i \sigma_{i+1}\right\} \qquad (2)$$

The last spin occurs only once in the sum in the exponential and we have, independently of the value of σ_{N-1},

$$\sum_{\sigma_N=\pm1} e^{\beta J \sigma_{N-1} \sigma_N} = 2 \cosh \beta J$$

giving

$$Z_N = 2 \cosh \beta J \, Z_{N-1}$$

We can repeat this process to obtain

$$Z_N = (2 \cosh \beta J)^{N-2} Z_2$$

$$Z_2 = \sum_{\sigma_1=\pm1} \sum_{\sigma_2=\pm1} e^{\beta J \sigma_1 \sigma_2} = 4 \cosh \beta J$$

so that we finally obtain

$$Z_N = 2(2 \cosh \beta J)^{N-1} \qquad (3)$$

The free energy is then

$$G = -k_B T \ln Z_N = -k_B T[\ln 2 + (N - 1) \ln (2 \cosh \beta J)]$$

In the thermodynamic limit only the term proportional to N is important and

$$G = -N k_B T \ln (2 \cosh \beta J) \tag{4}$$

We can also find the free energy in the presence of a magnetic field. To avoid end effects (which do not matter in the thermodynamic limit) we assume periodic boundary conditions, that is, assume that the Nth spin is connected to the first so that the chain forms a ring. Then

$$H = -J \sum_{i=1}^{N} \sigma_i \sigma_{i+1} - h \sum_{i=1}^{N} \sigma_i$$

where the spin labels run modulo N (i.e., $N + i = i$). The Hamiltonian can be rewritten

$$H = -\sum_{i=1}^{N} \left[J \sigma_i \sigma_{i+1} + \frac{h}{2}(\sigma_i + \sigma_{i+1}) \right] \tag{5}$$

giving for the partition function

$$Z_N = \sum_{\sigma_1 = \pm 1} \cdots \sum_{\sigma_N = \pm 1} \exp \left\{ \beta \sum_{i=1}^{N} \left[J \sigma_i \sigma_{i+1} + \frac{h}{2}(\sigma_i + \sigma_{i+1}) \right] \right\}$$

$$= \sum_{\{\sigma_i\}} \prod_{i=1}^{N} \exp \left\{ \beta \left[J \sigma_i \sigma_{i+1} + \frac{h}{2}(\sigma_i + \sigma_{i+1}) \right] \right\} \tag{6}$$

It is convenient to introduce the 2×2 *transfer matrix*

$$\mathbf{P} = \begin{bmatrix} P_{11} & P_{1-1} \\ P_{-11} & P_{-1-1} \end{bmatrix} \tag{7}$$

where

$$P_{11} = e^{\beta(J+h)} \qquad P_{-1-1} = e^{\beta(J-h)}$$

$$P_{1-1} = P_{-11} = e^{-\beta J}$$

giving

$$Z_N = \sum_{\{\sigma_i\}} P_{\sigma_1 \sigma_2} P_{\sigma_2 \sigma_3} \cdots P_{\sigma_N \sigma_1} = \text{Tr } \mathbf{P}^N \tag{8}$$

The matrix $\mathbf{P}$ can be diagonalized and the eigenvalues λ_1 and λ_2 are the roots of the secular determinant

$$|\mathbf{P} - \lambda \mathbf{I}| = 0 \tag{9}$$

Similarly, the matrix $\mathbf{P}^N$ has eigenvalues λ_1^N, λ_2^N and the trace of $\mathbf{P}^N$ is the sum of the eigenvalues

$$Z_N = \lambda_1^N + \lambda_2^N \tag{10}$$

The solution of (9) is

$$\lambda_{1,2} = e^{\beta J} \cosh \beta h \pm (e^{2\beta J} \sinh^2 \beta h + e^{-2\beta J})^{1/2} \qquad (11)$$

We note that λ_1 associated with the positive root is always larger than λ_2. The free energy is

$$G = -\frac{1}{\beta} \ln (\lambda_1^N + \lambda_2^N) = -\frac{1}{\beta}\left\{ N \ln \lambda_1 + \ln \left[1 + \left(\frac{\lambda_2}{\lambda_1}\right)^N \right] \right\}$$

$$\rightarrow -Nk_B T \ln\lambda_1 \qquad \text{as } N \rightarrow \infty$$

This gives for the free energy in the thermodynamic limit,

$$G = -Nk_B T \ln [e^{\beta J} \cosh \beta h + (e^{2\beta J} \sinh^2 \beta h + e^{-2\beta J})^{1/2}] \qquad (12)$$

For the special case $h = 0$ we obtain the previous result (4). We may compute the magnetization from

$$m = \langle \sigma_0 \rangle = -\frac{1}{N}\frac{\partial G}{\partial h} = \frac{1}{\beta \lambda_1}\frac{\partial \lambda_1}{\partial h}$$

After some straightforward manipulations we find

$$m = \frac{\sinh \beta h}{(\sinh^2 \beta h + e^{-4\beta J})^{1/2}} \qquad (13)$$

We see that for $h = 0$ there is no spontaneous magnetization at any nonzero temperature. However, in the limit of low temperature

$$\sinh^2 \beta h >> e^{-4\beta J}$$

for any $h \neq 0$ and only a very minute field is needed to produce saturation magnetization. The zero-field free energy will, in the limit $T \rightarrow 0$, approach the value $G(T \rightarrow 0+) = -NJ$ corresponding to completely aligned spins. We can thus say that we have a phase transition at $T = 0$, while for $T \neq 0$ the free energy is an analytic function of its variables. This behavior contrasts with that of mean field (Section 3.A) or Bragg–Williams approximations (Section 3.B) in which a phase transition (or coexistence) line extending from $T = 0$ to $T = T_c$ separates regions of positive and negative order parameter, with a discontinuity of the order parameter across the line. It is interesting to compare in more detail the exact and the mean field solution. For this reason we plot in Figure 3.6 the energy calculated exactly and in the Bragg–Williams approximation for different values of the external field. In Figure 3.7 we plot the susceptibility for different fields, as a function of temperature, in the two approximations, while the specific heat is shown in Figure 3.8. Results from the Bethe approximation are not shown since they agree with the exact ones in this case.

In Sections 3.D and 3.E we introduced the spin pair distribution function

$$g(j) = \langle \sigma_0 \sigma_j \rangle \qquad (14)$$

and argued that in mean field theories one neglects long-range *correlations* be-

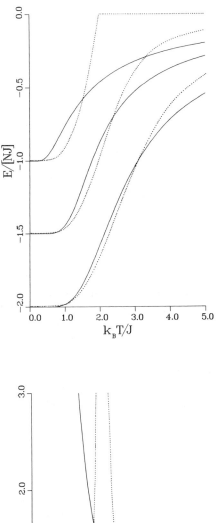

Figure 3.6 Comparison between exact and Bragg-Williams results for the internal energy of the one-dimensional Ising chain. Solid line, exact theory; dotted line, mean field theory. The three sets of curves correspond to $h = 0$, $h = 0.5J$, and $h = J$, respectively.

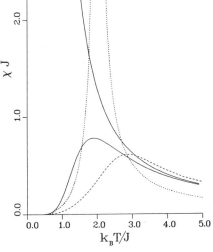

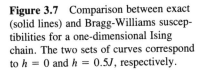

Figure 3.7 Comparison between exact (solid lines) and Bragg-Williams susceptibilities for a one-dimensional Ising chain. The two sets of curves correspond to $h = 0$ and $h = 0.5J$, respectively.

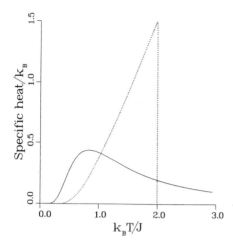

Figure 3.8 Specific heat calculated exactly (solid line) and in the Bragg-Williams approximation (dotted line) for the one-dimensional Ising chain. Only results for $h = 0$ shown. For $h \neq 0$ the difference between exact and approximate calculation is similar to the case of the susceptibilitiy shown in Figure 3.7.

tween spins. In the simplest mean field theory, one makes the approximation

$$\langle \sigma_0 \sigma_j \rangle = \langle \sigma_0 \rangle \langle \sigma_j \rangle$$

The error introduced by this approximation can be analyzed by studying the *spin correlation function*

$$\Gamma(j) = \langle \sigma_0 \sigma_j \rangle - \langle \sigma_0 \rangle \langle \sigma_j \rangle \tag{15}$$

This function can be calculated quite straightforwardly for the Ising chain. For simplicity we consider only the zero-field case ($h = 0$). Since there is no spin ordering for $T \neq 0$ we have $\langle \sigma_0 \rangle = 0$ and $\Gamma(j) = g(j)$ in this case. We assume an Ising chain with free ends and assume that the central spin σ_0 is far away from the ends. We have

$$\langle \sigma_0 \sigma_j \rangle = \frac{1}{Z_N} \sum_{\{\sigma_j\}} \sigma_0 \sigma_j \exp \left\{ \beta \sum_{i=1}^{N-1} J_i \sigma_i \sigma_{i+1} \right\} \tag{16}$$

$$Z_N = 2 \prod_{i=1}^{N-1} (2 \cosh \beta J_i) \tag{17}$$

where the exchange constant J_i at each bond will be set equal to J at the end of the calculation. Since $\sigma_i^2 = 1$,

$$\begin{aligned}
\langle \sigma_0 \sigma_j \rangle &= \frac{1}{Z_N} \sum_{\{\sigma_j\}} (\sigma_0 \sigma_1)(\sigma_1 \sigma_2) \cdots (\sigma_{j-1} \sigma_j) \exp \left\{ \beta \sum_{i=1}^{N-1} J_i \sigma_i \sigma_{i+1} \right\} \\
&= \frac{1}{Z_N \beta^j} \frac{\partial^j Z_N(J_0 \cdots J_N)}{\partial J_0 \cdots \partial J_{j-1}} \bigg|_{J_i = J} \\
&= (\tanh \beta J)^j = e^{-j/\xi}
\end{aligned} \tag{18}$$

where we define

$$\xi = -[\ln (\tanh \beta J)]^{-1}$$

as the *correlation length*. Since tanh $\beta J < 1$, we have $\xi > 0$, and the spin-pair correlation will decay exponentially with increasing j for all nonzero temperatures. The concept of correlation length will prove most useful later. At low temperatures

$$\ln (\tanh \beta J) \approx -2e^{-2\beta J} \qquad (19)$$

and

$$\xi = \tfrac{1}{2} e^{2\beta J} \qquad (20)$$

and we see that the correlation length can become quite large. The divergence of the correlation length is a universal feature of systems undergoing continuous phase transitions.

3.G LANDAU THEORY OF PHASE TRANSITIONS

In 1936, Landau constructed a general theory of phase transitions. The crucial hypothesis is that in the vicinity of the critical point we may expand the free energy in a power series in the order parameter, which we denote by m. The equilibrium value of m is then the value that minimizes the free energy. It is worth pointing out immediately that the basic assumption that the free energy is an analytic function of m at $m = 0$ is not correct. Nevertheless, Landau theory is of great utility as a qualitative tool and also plays an important role, after suitable generalization, in the modern renormalization theory of Wilson.

We begin by discussing a system in which the Gibbs free energy has the simple symmetry $G(m, T) = G(-m, T)$. We have tacitly assumed that the field h, which is conjugate to the order parameter m, is zero. With this symmetry the most general expansion of $G(m, T)$ is

$$G(m, T) = a(T) + \tfrac{1}{2} b(T)m^2 + \tfrac{1}{4} c(T)m^4 + \tfrac{1}{6} d(T)m^6 + \cdots \qquad (1)$$

where the fractional coefficients have been introduced in view of later manipulations. We have already encountered this type of expansion in the mean field treatment of the Ising model (Section 3.B), but formula (1) is more general than a specific instance of mean field theory.

The coefficients $b(T)$, $c(T)$, $d(T)$, . . . are at this point unspecified, and we will investigate the consequences of different types of behavior of these functions. The first case we wish to consider is when $c, d, e, . . . > 0$ and $b(T)$ changes sign at some temperature T_c. We write

$$b(T) = b_0 (T - T_c)$$

in the vicinity of $T = T_c$. In this case the function $G(m, T)$ takes the form shown in Figure 3.9 for various values of T.

For $T < T_c$ the point $m = 0$ corresponds to a local maximum of the free energy and the equilibrium state is one of two states of spontaneously broken

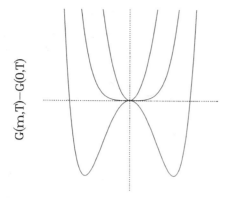

Order parameter m

Figure 3.9 Free energy when c, $d, \ldots > 0$ and $b(T) = b_0(T - T_c)$.

symmetry for which G has an absolute minimum. It is easy to work out the temperature dependence of the order parameter

$$\left(\frac{\partial G}{\partial m}\right)_T = 0 = bm + cm^3 + dm^5 + \cdots \tag{2}$$

Ignoring the term dm^5, we find that

$$m = \pm \left[\frac{b_0}{c(T_c)}\right]^{1/2} (T_c - T)^{1/2} \tag{3}$$

for $T \to T_c^-$. We may also obtain the behavior of the specific heat

$$C = T\left(\frac{\partial S}{\partial T}\right)$$

We let a prime indicate differentiation with respect to T and obtain

$$S = -\frac{\partial G}{\partial T} = -a' - \frac{b'}{2}m^2 - \frac{c'}{4}m^4 - \cdots - \frac{b}{2}(m^2)' - \frac{c}{4}(m^4)' - \cdots \tag{4}$$

As $T \to T_c^-$,

$$C \to -Ta'' - Tb'(m^2)' - \frac{Tc(m^4)''}{4} \tag{5}$$

$$(m^2)' \to \frac{b_0}{c}$$

$$b' \to b_0 \tag{6}$$

$$(m^4)'' \to 2\left(\frac{b_0}{c}\right)^2$$

$$C \to -Ta'' + \frac{\frac{1}{2}Tb_0^2}{c} \qquad T \to T_c^-$$

$$\to -Ta'' \qquad\qquad T \to T_c^+$$

(7)

We see that the order parameter and specific heat have the same form which we obtained previously in our mean field treatment of the Ising model.

We now consider a slightly different situation. Assume that c changes sign at some temperature while $d(T) > 0$ and b is a decreasing function of the temperature which is still positive in the region of interest. The free energy in this case will be as depicted in Figure 3.10. In this situation a discontinuous jump in the order parameter is expected. To see this, let $m_0 \neq 0$ be the location of a minimum of G. We must show that when $G(m_0, T_c) = G(0, T_c)$, $b(T_c) > 0$; that is, there is a local minimum at the point $m = 0$. The equilibrium condition is

$$\frac{\partial G}{\partial m}\bigg|_{m_0} = 0 = bm_0 + cm_0^3 + dm_0^5 + \cdots$$

(8)

$$G(m_0) - G(0) = 0 = \frac{b}{2}m_0^2 + \frac{c}{4}m_0^4 + \frac{d}{6}m_0^6$$

(9)

Solving for the nontrivial value of m_0^2, we obtain

$$m_0^2 = -\frac{3c(T_c)}{4d}$$

(10)

and

$$b(T_c) = \frac{3c^2}{16d} > 0$$

(11)

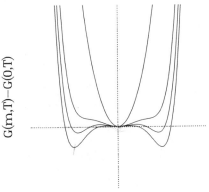

Figure 3.10 Free energy for a sequence of temperatures in the case that $c(T)$ changes sign above the temperature at which $b(T)$ changes sign.

Order parameter m

which justifies the claim that the first-order transition occurs before a continuous transition can take place. The case where b and c approach zero at the same temperature seems at this stage rather unlikely. In Section 3.I we shall see that these special points can occur when the coefficients depend on other thermodynamic parameters besides the temperature and more than one type of ordering can occur.

3.H EXAMPLE OF SYMMETRY CONSIDERATIONS: THE MAIER–SAUPE MODEL

In this section we discuss the situation when $G(m, T) \neq G(-m, T)$; that is, there are terms of odd order in m in the Landau expansion. Later, we illustrate this situation by considering the Maier–Saupe model for nematic liquid crystals, but first consider the general case. Any term in the Landau expansion which is linear in the order parameter can be eliminated by making the transformation $m \rightarrow m + \Delta$ and choosing Δ to make the linear term vanish. We therefore assume that the leading term of odd order in m is cubic and write

$$G(m, T) = a(T) + \tfrac{1}{2}b(T)m^2 - \tfrac{1}{3}c(T)m^3 + \tfrac{1}{4}d(T)m^4 + \cdots \qquad (1)$$

We assume for stability that $d(T) > 0$ and that $c(T) > 0$; $c(T) < 0$ corresponds simply to changing the sign of the order parameter. We also assume as before that $b(T)$ is a decreasing function of T which changes sign at some temperature T^*. With these assumptions the free energy will have the general form shown in Figure 3.11.

As we shall see, a first-order transition will again preempt the second-order transition. At the transition point T_c we have

$$G(m_0, T_c) = G(0, T_c) \qquad (2)$$

and

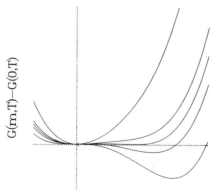

Order parameter m

Figure 3.11 Free energy for different temperatures in the case where the Landau expansion contains a cubic term.

$$\frac{\partial G}{\partial m_0} = 0 = bm_0 - cm_0^2 + dm_0^3 \tag{3}$$

Solving, we obtain

$$m_0 = \frac{2c}{3d} \tag{4}$$

and

$$b(T_c) = \frac{2c^3}{9d} > 0 \tag{5}$$

We see, therefore, that the appearance of a cubic term in the Landau expansion signals a first-order phase transition. This prediction of the theory has been found to hold for three-dimensional systems, but the result turns out to be incorrect in the case of the three-state Potts model in two dimensions [see Section 6.D(b)]. This is another indication that mean field theory is not reliable for low-dimensional systems.

An example of a model that gives rise to a Landau expansion of the form (1) is the Maier–Saupe model for the isotropic to nematic transition in liquid crystals.[4] We consider a system of anisotropic molecules with a symmetry axis. The center of the ith molecule is taken to be at $\mathbf{r}_i$ and the unit vector pointing in the direction of the symmetry axis is denoted by $\hat{n}_i$.

We further assume that the directions $\hat{n}_i$ and $-\hat{n}_i$ are equivalent. The interaction between the molecules is represented by a pair potential $W(\mathbf{r}_j - \mathbf{r}_i, \hat{n}_j, \hat{n}_i)$. We assume that the number of molecules ρ per unit volume is constant and let $f(\hat{n})$ be the probability density that a molecule is oriented along $\hat{n}$, and define $\mathbf{r}_{ji} = \mathbf{r}_j - \mathbf{r}_i$. We introduce the pair distribution function $g(\mathbf{r}_{ji}, \hat{n}_j, \hat{n}_i)$ as the conditional probability that there is a molecule at $\mathbf{r}_i$ with orientation $\hat{n}_i$ given that there is a molecule at $\mathbf{r}_j$ with orientation $\hat{n}_j$. In analogy with our discussion of the Weiss molecular field in Section 3.A we write for the term in the total energy which depends on the orientation $\hat{n}_i$:

$$\epsilon(\hat{n}_i) = \text{const.} + \rho \int d^3 r_{ji} \int d\Omega_j \, W(\mathbf{r}_{ji}, \hat{n}_i, \hat{n}_j) f(\hat{n}_j) g(\mathbf{r}_{ji}, \hat{n}_j, \hat{n}_i) \tag{6}$$

where the integration over Ω_j extends over the solid angle of $\hat{n}_j$. The fundamental approximation of the Maier–Saupe model consists of ignoring the dependence of the pair distribution function on orientation. This allows us to write

$$\epsilon(\hat{n}_i) = \text{const.} + \rho \int d\Omega_j f(\hat{n}_j) \int d^3 r_{ji} \, W(\mathbf{r}_{ji}, \hat{n}_j, \hat{n}_i) g(\mathbf{r}_{ji})$$
$$= \text{const.} + \rho \int d\Omega_j f(\hat{n}_j)(\gamma - UP_2(\hat{n}_i \cdot \hat{n}_j) + O[P_4(\hat{n}_j \cdot \hat{n}_i)]) \tag{7}$$

where γ and U are constants and $U > 0$ corresponds to the situation where it

---

[4] See Maier and Saupe [1959, 1960]. For an elementary introduction to the properties of liquid crystals, we recommend de Gennes [1974], Priestley et al. [1974] and Stephens and Straley, [1974].

is energetically advantageous for the molecules to align parallel. $P_2(x) = \frac{1}{2}(3x^2 - 1)$ is the second Legendre polynomial, and P_4 is the fourth polynomial. Note that since we have assumed that the directions $\hat{n}$ and $-\hat{n}$ are equivalent, there will be no odd terms in the Legendre series (7). The second approximation in the Maier–Saupe model is to stop at second order in the Legendre expansion. We define

$$\sigma_{i\alpha\beta} = \frac{1}{2}(3n_{i\alpha}n_{i\beta} - \delta_{\alpha\beta}) \tag{8}$$

where α, $\beta = x$, y, z, $n_{i\alpha}$ is a Cartesian component of $\hat{n}_i$ and $\delta_{\alpha\beta}$ is the Kronecker delta. Using the identity

$$P_2(\hat{n}_j \cdot \hat{n}_i) = \frac{2}{3} \sum_{\alpha,\beta=1}^{3} \sigma_{i\alpha\beta}\sigma_{j\beta\alpha}$$

and defining

$$Q_{\alpha\beta} = \langle \sigma_{\alpha\beta} \rangle = \int d\Omega_j \, \sigma_{j\alpha\beta} f(\hat{n}_j) \tag{9}$$

we can rewrite (7) to obtain

$$\epsilon(\hat{n}_i) = -\frac{2}{3}\rho U \sum_{\alpha,\beta} Q_{\alpha\beta}\sigma_{i\beta\alpha} \tag{10}$$

where we have omitted terms that do not depend on particle orientation. Taking into account the double counting when summing the pair potential over all molecules, we find for the orientational contribution to the internal energy,

$$E = -\frac{1}{3}\rho UN \sum_{\alpha,\beta} Q_{\alpha\beta}Q_{\beta\alpha} \tag{11}$$

We write for the orientational contribution to the entropy

$$S_{\text{or}} = -Nk_B \int d\Omega \, f(\hat{n}) \ln [f(\hat{n})] \tag{12}$$

It then follows from the argument used in Section 2.E that the single-particle distribution function

$$f(\hat{n}) = \frac{\exp\{-\beta\epsilon(\hat{n})\}}{\int d\Omega \exp\{-\beta\epsilon(\hat{n})\}} \tag{13}$$

will minimize the free energy $G = E - TS_{\text{or}}$, and the order parameter can then be determined from the self-consistency criterion

$$Q_{\alpha\beta} = \int d\Omega \sigma_{\alpha\beta} f(\hat{n}) \tag{14}$$

In analogy with the Weiss molecular field theory there will always be a solution $Q_{\alpha\beta} = 0$, and we identify such solutions with the isotropic high-temperature phase. At low temperatures nonzero solutions of (14) will appear. Each such solution corresponds to a preferential orientation of the *director field* $\hat{n}$ and we call this phase *nematic*. The nematic phase does not exhibit long-range spatial order, and we distinguish it from the more complicated *smectic* phases, which exhibit varying degrees of translational ordering. The nonzero solutions to (14) will not be unique, because of the overall rotational symmetry of the system. However, since $Q_{\alpha\beta}$ is real and symmetric, there will always be a principal-axis frame in which $Q_{\alpha\beta}$ is diagonal. We let θ and ϕ be the polar angles of $\hat{n}$ in such a system

$$n_x = \sin\theta\cos\phi \qquad n_y = \sin\theta\sin\phi \qquad n_z = \cos\theta$$

Defining

$$p = \tfrac{3}{2}\sin^2\theta\cos 2\phi \qquad q = \tfrac{1}{2}(3\cos^2\theta - 1)$$

$$P = \langle p \rangle \qquad\qquad Q = \langle q \rangle \tag{15}$$

we find after some algebra

$$Q_{\alpha\beta} = \begin{pmatrix} -\tfrac{1}{2}(Q - P) & & \\ & -\tfrac{1}{2}(Q + P) & \\ & & Q \end{pmatrix} \tag{16}$$

$$\epsilon(\hat{n}) = -\rho U(Qq + \tfrac{1}{3}Pp)$$

$$E = -\tfrac{1}{2}\rho U(Q^2 + \tfrac{1}{3}P^2) \tag{17}$$

If we choose the z axis to be the preferred axis, we have $P = 0$, and with $\mu = \cos\theta$,

$$\frac{G}{N} = -\frac{1}{\beta}\ln\left(4\pi\int_0^1 d\mu\,\exp\left\{\frac{1}{2}\rho U\beta[(3\mu^2 - 1)Q - Q^2]\right\}\right) \tag{18}$$

It is now a straightforward matter to obtain a Taylor expansion of (18) in powers of Q, and after some algebra we obtain the expansion

$$\frac{G}{N} = -\frac{1}{\beta}\ln(4\pi) + Q^2(1 - 0.4\beta) - \frac{8}{105}\beta^2 Q^3 + \frac{4}{175}\beta^3 Q^4 + \cdots \tag{19}$$

where we have set ρU equal to 1. This expansion is indeed of the form (1). We leave it as an exercise to the reader (Problem 3.10) to work out further details of the model.

In Figure 3.12 we plot the order parameter obtained by minimizing (19) as a function of the temperature. We also plot the order parameter resulting from the self-consistent equation (14). In the principal-axis frame with the molecules aligned preferentially along the z axis, this equation takes the form

$$Q = \frac{1}{2}\langle 3\mu^2 - 1\rangle = \frac{\tfrac{1}{2}\int_0^1 d\mu\,(3\mu^2 - 1)\exp\{3\beta\rho UQ\mu^2/2\}}{\int_0^1 d\mu\,\exp\{3\rho\beta UQ\mu^2/2\}} \tag{20}$$

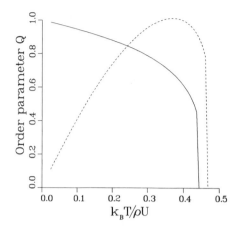

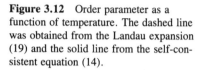

Figure 3.12 Order parameter as a function of temperature. The dashed line was obtained from the Landau expansion (19) and the solid line from the self-consistent equation (14).

Equation (20) is most easily solved by choosing a value for $x = 3\beta\rho UQ/2$, evaluating Q numerically and then using the calculated value of Q to obtain the temperature.

Figure 3.12 illustrates that while the Landau theory gives a correct qualitative picture, there are difficulties associated with using the Landau expansion for quantitative purposes in the case of a first-order transition. The problem is that the jump in the order parameter is not necessarily small and the expansion (1) to low order may not be accurate.

We note that the simple mean field theory (20) predicts that the discontinuity in the order parameter Q at the transition should be 0.43. Both smaller and larger values have been observed experimentally. When steric effects are included in the theory, as in the van der Waals theory of Section 3.L, one tends to get a more strongly first-order transition. On the other hand, when one takes into account the fact that actual molecules do not have cylindrical symmetry (see, e.g., Straley, 1974) the order parameter discontinuity tends to be smaller than in the Maier–Saupe theory.

It is of interest to extend the Maier–Saupe theory to include the effect of a magnetic (or electric) field. The magnetic susceptibility will in general be anisotropic and we let $\chi_{\parallel}$ and $\chi_{\perp}$ correspond to orientations of the field parallel and perpendicular to the molecular axis. The energy of a molecule in the field is then

$$-\chi_{\parallel}(\hat{n} \cdot \mathbf{H})^2 - \chi_{\perp}[H^2 - (\hat{n} \cdot \mathbf{H})^2] = -\tfrac{1}{3}H^2\{(\chi_{\parallel} + 2\chi_{\perp}) + (\chi_{\parallel} - \chi_{\perp}) \\ [3(\hat{n} \cdot \hat{H})^2 - 1]\} \quad (21)$$

We assume for simplicity that $\Delta\chi = (\chi_{\parallel} - \chi_{\perp}) > 0$ (i.e., that there is a tendency for the molecules to align parallel to the field). The field direction then becomes the preferred axis. We drop the first term in (21), which is independent of the molecular orientation, and write for the energy per molecule in the presence of a field,

$$\epsilon_H(\hat{n}) = -\rho U (Q + \gamma)q \quad (22)$$

where

$$\gamma = \frac{2\,\Delta\chi\,H^2}{3\rho U} \tag{23}$$

The self-consistent equation now becomes

$$Q = \frac{\int_0^1 d\mu\,\tfrac{1}{2}(3\mu^2 - 1)\,\exp\{3\rho U\beta(Q + \gamma)\mu^2/2\}}{\int_0^1 d\mu\,\exp\{3\rho U\beta(Q + \gamma)\mu^2/2\}} \tag{24}$$

The order parameter Q can easily be evaluated numerically for a given value of $x = 3\rho U\beta(Q + \gamma)/2$; once Q is determined, we can solve for the effective temperature. The results are depicted in Figure 3.13.

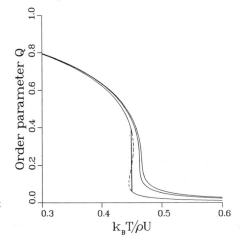

Figure 3.13 Order parameter as a function of temperature for different fields when $\Delta\kappa > 0$. Solid curves, $\gamma = 0.005$, $\gamma = 0.011 = \gamma_c$, and $\gamma = 0.013$. Dashed curve for the lowest γ corresponds to the unphysical region discussed in the text.

We see that for γ below a critical value γ_c there is a narrow temperature range for which Q is a triple-valued function of T. Not all these values correspond to a minimum in the free energy. This can be seen by plotting G/N and T as the parameter x is varied. The resulting curves are plotted in Figure 3.14. The region that is not a minimum of the free energy is the loop in the top curve of Figure 3.14. This region corresponds to the dashed curve in Figure 3.13. A

Figure 3.14 G/N as a function of T for different values of γ.

first-order transition occurs at the intersection point. The point T_c, γ_c at which the loop has degenerated into a point is an ordinary critical point where the transition is second order. We refer the interested reader to Wojtowicz and Sheng [1974] and Palffy-Muhoray and Dunmur [1983] for further details (the latter authors also discuss the interesting case $\Delta\chi < 0$, for which there is a tricritical point and a biaxial phase occurs). The full phase diagram is given by Frisken et al. [1987].

3.I LANDAU THEORY OF TRICRITICAL POINTS

In Section 3.G we pointed out that it is conceivable that the coefficients b and c in the Landau expansion may approach zero simultaneously and that this could lead to new types of critical behavior. This situation is likely to occur when one or more parameters as well as the temperature control the critical behavior of the system and there is more than one order parameter. As an example we consider a mixture of ^{3}He and ^{4}He in the liquid phase. As the temperature is lowered, pure ^{4}He undergoes a transition to the superfluid state (see Section 7.B). This transition is continuous and is known as the λ transition because of the λ-like shape of the specific heat singularity. If ^{3}He is added to the system, the transition temperature is lowered and the mixture remains homogeneous for low ^{3}He concentrations both below and above the critical temperature. At concentrations greater than $x_t = n_3/(n_3 + n_4) = 0.670$, the transition is discontinuous and accompanied by phase separation. One of the coexisting phases is a ^{4}He-rich superfluid, the other a ^{3}He-rich normal fluid. The dividing point x_t, T_t is the tricritical point alluded to above. Other examples of systems exhibiting tricritical points are the antiferromagnet $FeCl_2$, which undergoes a continuous transition in low applied magnetic fields and a first-order transition to a mixed phase (coexisting antiferromagnetic and ferromagnetic phases) at sufficiently high magnetic fields; the solid NH_4Cl, whose orientational transition changes from second to first order as a function of pressure; the ferroelectric $KDPO_4$; ternary liquid mixtures; and a number of liquid crystals. For a general review of tricritical phenomena see Lawrie and Sarbach [1984]. Below we discuss the general Landau theory and then carry out a mean field treatment of a simple model for the ^{3}He–^{4}He system.

   We denote, as is our practice, the order parameter of the system by m (in ^{3}He–^{4}He this would be the superfluid wave function ψ) and the field that couples to m by h. The field that couples to the subsidiary order parameter x is denoted by Δ. In the case of ^{3}He–^{4}He mixtures, $\Delta = \mu_3 - \mu_4$, the difference in chemical potentials and x is the fractional concentration $n_3/(n_4 + n_3)$ (see Problem 1.5). In the case of $FeCl_2$,

$$m = M_Q = \sum_\mathbf{r} S_\mathbf{r} \exp\{i\,\mathbf{Q}\cdot\mathbf{r}\}$$

is a staggered magnetization, with $S_\mathbf{r}$ the spin at site $\mathbf{r}$, h a staggered magnetic field which is not realizable in the laboratory, x a uniform magnetization, and Δ an applied uniform magnetic field. In the case of uniaxial–biaxial transitions in liquid crystals, m and x may be the order parameters P and Q of (3.H.15) and two associated fields (e.g., one electric and one magnetic) in two orthogonal directions. We assume the following form for the free energy:

$$G(m, \Delta, T) = a(T, \Delta) + \tfrac{1}{2}b(T, \Delta)m^2 + \tfrac{1}{4}c(T, \Delta)m^4 + \tfrac{1}{6}d(T, \Delta)m^6 \tag{1}$$

The line of critical points is given by $b(T, \Delta) = 0$, which defines a curve in the T–Δ plane (Figure 3.15). The tricritical point is given by $b(T, \Delta) = c(T, \Delta) = 0$, which, in general, is a unique point (Δ_t, T_t).

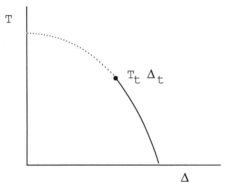

Figure 3.15 Phase behavior near a tricritical point. Solid line, first-order transitions; dotted line, critical points.

We assume that when the temperature is lowered and $\Delta > \Delta_t$, the coefficient c approaches zero before b. The usual equations for first-order transitions will then apply. For $\Delta < \Delta_t$, the transition is continuous. We first show that the line of first-order transitions joins the line of critical points in a smooth fashion. The equation for the first-order line is given by (3.G.11)

$$b(\Delta, T) - \frac{3c^2(\Delta, T)}{16d(\Delta, T)} = 0 \tag{2}$$

The equation $b(\Delta, T) = 0$ for the line of critical points yields for the slope of this line at $b = 0$,

$$\left. \frac{d\Delta}{dT} \right|_{\text{crit}} = - \frac{\partial b/\partial T \,|_\Delta}{\partial b/\partial \Delta \,|_T} \equiv - \frac{b_T}{b_\Delta} \tag{3}$$

We let the subscript indicate partial differentiation and find from (2) for the slope of the first-order line,

$$\left. \frac{d\Delta}{dT} \right|_{\substack{\text{first} \\ \text{order}}} = - \frac{b_T d + d_T b - \tfrac{3}{8}cc_T}{b_\Delta d + d_\Delta b - \tfrac{3}{8}cc_\Delta} \tag{4}$$

As $(T, \Delta) \to (T_t, \Delta_t)$, $c \to 0$, $b \to 0$, and equations (3) and (4) become the same.

It is also easy to see that the first-order transition for $\Delta > \Delta_t$ implies coexistence of two phases with different values of the density x. We suppose that the expectation value of x may be obtained from the free energy through

$$Nx = -\frac{\partial G}{\partial \Delta}\bigg|_T \tag{5}$$

where the minus sign implies a suitable sign convention for Δ. The first-order transition occurs when

$$G(m, T, \Delta) = G(0, T, \Delta) \tag{6}$$

and

$$\frac{\partial G}{\partial m}\bigg|_{T,\Delta} = 0 \tag{7}$$

We leave it as an exercise to prove that (5)-(7) yield an equation for the discontinuity in x. As the tricritical point is approached along the first-order line, the discontinuity in x takes the form

$$\delta x = -\frac{3c}{8d}b_\Delta + O(c^2) \tag{8}$$

where we have used (3.G.10). Therefore, in the T–x plane the phase diagram has the shape shown in Figure 3.16.

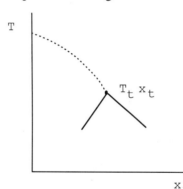

Figure 3.16 Phase diagram in the T-x plane near the tricritical point.

It is also of interest to calculate the asymptotic behavior of the order parameter and specific heat as the tricritical point is approached. Solving (7) we obtain

$$m^2 = \left(\frac{c^2}{4d^2} - \frac{b}{d}\right)^{1/2} - \frac{c}{2d} \tag{9}$$

There are two different asymptotic forms of this function, depending on how the tricritical point is approached. Since both b and c approach zero linearly, we expect that in most cases

$$\left|\frac{b}{d}\right| >> \frac{c^2}{4d^2}$$

If $b(T, \Delta) = b_0(T - T_t)$, we obtain

$$m(T) \approx \left[\frac{b_0}{d(T_t)} \right]^{1/4} (T_t - T)^{1/4} \tag{10}$$

The exponent β at a tricritical point is therefore $\frac{1}{4}$ rather than the value $\frac{1}{2}$ found near a critical point. There is, however, a narrow region in the T–Δ plane given by

$$\left| \frac{b}{d} \right| < \left(\frac{c}{2d} \right)^2$$

in which the asymptotic behavior above does not hold. In this region all terms in (9) are proportional to $T_t - T$ and

$$m \propto (T_t - T)^{1/2} \tag{11}$$

A rough sketch of the critical and tricritical regions is shown in Figure 3.17.

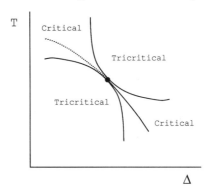

Figure 3.17   Critical and tricritical regions near a tricritical point.

The exponents for a path of approach that lies in the tricritical region are often subscripted with a t, those for a path in the critical region with a u. Thus $\beta_t = \frac{1}{4}$, $\beta_u = \frac{1}{2}$. We leave it as an exercise to the reader to show that $\gamma_u = 2$, $\gamma_t = 1$, $\alpha_t = \frac{1}{2}$, $\alpha_u = -1$. It is a remarkable fact that in three dimensions the predictions regarding the tricritical exponents are exact (to within logarithmic corrections, Wegner [1973]). In Section 3.J we present self-consistency arguments which indicate the reasons for this result and which also show that the Landau-Ginzburg theory of critical points will be correct for spatial dimensionality $d \geq 4$.

We first discuss, however, a simple model for the $^3\mathrm{He}$–$^4\mathrm{He}$ system described above. This model, known as the Blume–Emery–Griffith model (Blume et al., 1971), is a classical lattice gas model which ignores the quantum-statistical nature of the λ transition (Section 7.B) but does take into account the effect of nonordering impurities ($^3\mathrm{He}$) on the transition. The simplest version of the BEG model has the Hamiltonian

$$H = -J \sum_{\langle ij \rangle} S_i S_j + \Delta \sum_i S_i^2 - \Delta N \tag{12}$$

where $S_i = 0$ or ± 1 and the spins occupy a three-dimensional lattice with coordination number q. The connection between this "magnetic" Hamiltonian and the ^3He–^4He system is made by identifying $S_i = \pm 1$ with a ^4He atom on site i (or in cell i) and $S_i = 0$ with a ^3He atom. The parameter Δ controls the number of ^3He atoms and represents the difference $\mu_3 - \mu_4$ in chemical potentials. The concentration of ^3He atoms is given by

$$x = 1 - \langle S_i^2 \rangle \tag{13}$$

and it is clear that $x \to 0$ as $\Delta \to -\infty$ and $x \to 1$ as $\Delta \to +\infty$. The normal to superfluid transition is modeled by the transition from a high-temperature paramagnetic phase, $\langle S_i \rangle = 0$, to an ordered ferromagnetic phase, $\langle S_i \rangle = m$, and we now proceed to construct a mean field theory for this transition. The expectation value $m = \langle S_i \rangle$ is, of course, the order parameter in this theory. We use the variational method and approximate the density matrix by a direct product of a single site density matrices

$$\rho = \prod_i \rho_i \tag{14}$$

where ρ_i is a 3×3 matrix with only diagonal elements. We then have

$$G(T, \Delta) = -\tfrac{1}{2} N q J \, (Tr \, \rho_i S_i)^2 + N\Delta \, Tr \, (\rho_i S_i^2) + N k_B T \, Tr \, (\rho_i \ln \rho_i) - N\Delta \tag{15}$$

By minimizing G with respect to variations in ρ_i subject to $Tr \, \rho_i = 1$, we obtain (Problem 3.13)

$$\rho_i(S_i) = \frac{e^{\beta(qJmS_i - \Delta S_i^2)}}{1 + 2e^{-\beta\Delta} \cosh \beta q Jm} \tag{16}$$

Substituting (16) into (15), we obtain the approximate free energy

$$\frac{G(T, \Delta, m)}{N} = -\tfrac{1}{2} q Jm^2 + \Delta \langle S_i^2 \rangle$$
$$+ k_B T \left[\frac{e^{\beta(qJm-\Delta)}}{D} \ln \frac{e^{\beta(qJm-\Delta)}}{D} + \frac{e^{-\beta(qJm+\Delta)}}{D} \ln \frac{e^{-\beta(qJm+\Delta)}}{D} - \frac{1}{D} \ln D \right] - \Delta \tag{17}$$

where $D = 1 + 2e^{-\beta\Delta} \cosh \beta q Jm$ and

$$\langle S_i^2 \rangle = \frac{2e^{-\beta\Delta} \cosh \beta q Jm}{D}$$

The expression for G may be simplified by decomposing the logarithms and we obtain

$$\frac{G(T, \Delta, m)}{N} = \tfrac{1}{2} q Jm^2 - k_B T \ln \left(1 + 2e^{-\beta\Delta} \cosh \beta q Jm \right) - \Delta \tag{18}$$

This function must still be minimized with respect to the parameter m in order to obtain the equilibrium state for each (Δ, T). We proceed by obtaining the

Landau expansion for the free energy

$$\frac{G(T, \Delta, m)}{N} = a(T, \Delta) + \tfrac{1}{2}b(T, \Delta)m^2 + \tfrac{1}{4}c(T, \Delta)m^4 + \tfrac{1}{6}d(T, \Delta)m^6$$

By comparing terms (Problem 3.13), one finds

$$a(T, \Delta) = -k_B T \ln(1 + 2e^{-\beta\Delta}) - \Delta$$

$$b(T, \Delta) = qJ\left(1 - \frac{qJ}{\delta k_B T}\right) \tag{19}$$

$$c(T, \Delta) = \frac{qJ}{2\delta^2}(\beta qJ)^3\left(1 - \frac{\delta}{3}\right)$$

where $\delta = 1 + \tfrac{1}{2}e^{\beta\Delta}$. At the phase transition $m = 0$ and

$$x(T, \Delta) = \frac{1}{1 + 2e^{-\beta\Delta}} = \frac{\delta - 1}{\delta}$$

giving

$$\frac{T_c(x)}{T_c(0)} = 1 - x$$

and $x_t = \tfrac{2}{3}$.

The tricritical concentration $x_t = \tfrac{2}{3}$ is in remarkable agreement with the experimental value $x_t = 0.670$. The predicted linear dependence of the transition temperature on concentration is not observed. The actual transition temperature varies as $(1 - x)^{2/3}$ for small x and the discrepancy is a consequence of the use of classical rather than quantum statistics in our model. The tricritical temperature in the BEG model $[T_t/T_c(0) = \tfrac{1}{3}]$ is for this reason somewhat lower than the observed value of 0.4.

The nomenclature "tricritical point" is a consequence of the fact that at this point three lines of critical points meet. In our treatment of the BEG model we have not considered the effect of the field h which couples to the order parameter m and have therefore found only the line of λ transitions. The other two lines emerge in a symmetrical fashion from the tricritical point in the $\pm h$ directions. The interested reader is encouraged to consult the original paper by Blume et al. [1971] for a more general treatment of the problem in which the full structure of the critical surface is apparent.

3.J ROLE OF FLUCTUATIONS: LANDAU–GINZBURG THEORY

We have pointed out in Section 3.E that the failure of mean field theories at a critical point is due to the neglect of long-range correlations. It is possible to generalize the Landau theory to incorporate fluctuations at least in an approximate fashion. Instead of the Gibbs free energy[5] which we have used throughout this chapter, we now make a Legendre transformation to a Helmholtz free energy which in the homogeneous case is $A(M, T) = G + hM$ with $dA = -S\, dT + hdM$. We will allow the independent variable to have spatial variation

$$M = \int d^3r\, m(\mathbf{r})$$

and assume that the free energy can be written

$$A(\{m(\mathbf{r})\}, T) = \int d^3r \left\{ a(T) + \frac{b}{2} m^2(\mathbf{r}) + \frac{c}{4} m^4(\mathbf{r}) + \cdots + \frac{f}{2} [\nabla m(\mathbf{r})]^2 \right\} \quad (1)$$

The first three terms are a simple generalization of (3.G.1); the last term expresses the fact that the free energy is raised by fluctuations in the order parameter. The coefficient f can thus be assumed to be positive. We have in (1) expressed the free energy as a volume integral over the free-energy density. This is valid even for discrete systems, such as spins on a lattice, as long as $m(\mathbf{r})$ varies significantly only over sufficiently large distances that we may "coarse grain" dynamical variables. Near a critical point this approximation will certainly be valid. However, the fundamental objection to the Landau theory of Section 3.G, namely that the free energy is not necessarily an analytic function of the order parameter, applies equally to the inhomogeneous form (1).

In the homogeneous case we have

$$h = \left. \frac{\partial A}{\partial M} \right|_T$$

In the inhomogeneous case the generalization is the functional derivative

$$h(r) = \frac{\delta A}{\delta m(\mathbf{r})}$$

We construct the variation of A,

$$\delta A = \int d^3r \left\{ \delta m(\mathbf{r})[bm(\mathbf{r}) + cm^3(\mathbf{r}) + dm^5(\mathbf{r}) + \cdots] \right.$$
$$\left. + f\, \nabla\delta\, m(\mathbf{r}) \cdot \nabla m(\mathbf{r}) \right\} \quad (2)$$

The last term can be simplified by carrying out an integration by parts and demanding that $\delta m(\mathbf{r}) = 0$ at the surface of the sample. We then obtain

$$h(\mathbf{r}) = bm(\mathbf{r}) + cm^3(\mathbf{r}) + dm^5(\mathbf{r}) + \cdots - f\, \nabla^2 m(\mathbf{r}) \quad (3)$$

From this equation we may recover the results of the homogeneous Landau theory by letting $h(\mathbf{r}) = 0$ and $\nabla m(\mathbf{r}) = 0$. Near a second-order transition the (uniform) order parameter then obeys the equation

$$m_0^2 = -\frac{b}{c} \qquad T < T_c \quad (4)$$

---

[5]We could equally well use the Gibbs potential
$$G(\{h(\mathbf{r})\}, T, \{m(\mathbf{r})\}) = A - \int d^3r\, h(\mathbf{r})m(\mathbf{r})$$
and treat $m(\mathbf{r})$ as a variational parameter. The resulting expressions (3)–(9) are identical.

which is familiar from Section 3.G. Imagine now that a localized perturbation $h_0 \delta(\mathbf{r})$ is applied to the material. Equation (3) allows us to calculate the effect of this perturbation throughout the system. Let $m(\mathbf{r}) = m_0(T) + \varphi(\mathbf{r})$. Discarding terms nonlinear in φ we write $m^3(\mathbf{r}) = m_0^3 + 3m_0^2 \varphi(\mathbf{r})$. With these approximations we obtain

$$\nabla^2 \varphi(\mathbf{r}) - \frac{b}{f} \varphi(\mathbf{r}) - 3\frac{c}{f} m_0^2 \varphi(\mathbf{r}) - \frac{b}{f} m_0 - \frac{c}{f} m_0^3 = -\frac{h_0}{f} \delta(\mathbf{r}) \qquad (5)$$

Using $m_0 = 0$ for $T > T_c$ and (4) for $T < T_c$, we find

$$\nabla^2 \varphi - \frac{b}{f} \varphi = -\frac{h_0}{f} \delta(\mathbf{r}) \qquad T > T_c$$

$$\nabla^2 \varphi + 2\frac{b}{f} \varphi = -\frac{h_0}{f} \delta(\mathbf{r}) \qquad T < T_c \qquad (6)$$

These equations are easily solved in spherical coordinates

$$\varphi = \frac{h_0}{4\pi f} \frac{1}{r} e^{-r/\xi} \qquad (7)$$

with

$$\xi(T) = \left[\frac{f}{b(T)}\right]^{1/2} \qquad T > T_c \qquad (8)$$

and

$$\xi(T) = \left[-\frac{f}{2b(T)}\right]^{1/2} \qquad T < T_c \qquad (9)$$

The function $\xi(T)$ is the *correlation length* and with

$$b(T) = b'(T - T_c)$$

we see that it diverges as $T \rightarrow T_c$ from both the low- and high-temperature sides. In this theory

$$\xi(T) \propto |T - T_c|^{-1/2}$$

Experimentally and in more exact theories,

$$\xi(T) \propto |T - T_c|^{-\nu}$$

with the critical exponent ν both model and dimensionality dependent.

We may relate the function $\varphi(r)$ to a *correlation function*. Assuming that a term

$$-\int d^3 r \, m(\mathbf{r}) h(\mathbf{r})$$

is included in the Hamiltonian, we have

$$\langle m(\mathbf{r}) \rangle = \frac{\text{Tr } m(\mathbf{r}) \exp\{-\beta[H_0 - \int d^3 r' h(\mathbf{r}')m(\mathbf{r}')]\}}{\text{Tr } \exp\{-\beta[H_0 - \int d^3 r' h(\mathbf{r}')m(\mathbf{r}')]\}} \qquad (10)$$

where H_0 refers to the part of H which is independent of $h(\mathbf{r})$. We see that

$$\frac{\delta \langle m(\mathbf{r}) \rangle}{\delta h(0)} = \varphi(\mathbf{r})/h_0 = \beta (\langle m(\mathbf{r})m(0) \rangle - \langle m(\mathbf{r}) \rangle \langle m(0) \rangle) = \beta \Gamma(\mathbf{r}) \quad (11)$$

The function $\varphi(\mathbf{r})$ is thus proportional to the two-particle or order parameter–order parameter correlation function. The susceptibility (compressibility in the case of a fluid) is given by

$$\chi = \int d^3 r \, \varphi(\mathbf{r})$$

and it is easily shown that the usual mean field result

$$\chi \propto |T - T_c|^{-1}$$

is recovered.

The results above allow us to establish a self-consistency criterion for mean field (or Landau) theories known as the Ginzburg criterion. We first generalize the analysis to systems of spatial dimensionality d. Equations (6)–(7) are quite general—one simply replaces the operator ∇^2 by the d-dimensional operator and the δ function by the appropriate d-dimensional δ function. The solutions to (6)–(7) are generally not of the simple form (8). However, one can show that in arbitrary dimension, d, for $r << \xi$, $\varphi \propto r^{-d+2}$, while for $r >> \xi$, $\varphi \propto e^{-r/\xi}$. For the purpose of order of magnitude estimates we can thus write

$$\varphi(\mathbf{r}) \approx \frac{e^{-r/\xi}}{r^{d-2}} \quad (12)$$

In mean field theories we always crudely approximate the correlation functions (11) at large distances. Therefore, one might expect that such approximations would be valid if the ratio

$$\frac{\int_{\Omega(\xi)} d^d r \, [\langle m(\mathbf{r})m(0) \rangle - \langle m(\mathbf{r}) \rangle \langle m(0) \rangle]}{\int_{\Omega(\xi)} d^d r \, m_0^2} << 1 \quad (13)$$

where the integral is carried out over a d-dimensional hypersphere of radius ξ. This criterion can be used to estimate the range of temperatures over which mean field theory adequately describes the system (Kadanoff et al., 1967). We shall instead use it to estimate the dimensionality d at which Landau theory describes exactly the critical behavior of the system. To do this we substitute the asymptotic form, as calculated from the Landau model, for the various functions appearing in (13). Substitution of

$$m_0^2 \approx |T - T_c|^{2\beta}$$

and

$$\langle m(\mathbf{r})m(0) \rangle - \langle m(\mathbf{r}) \rangle \langle m(0) \rangle \approx \frac{\exp\{-r/\xi\}}{r^{d-2}}$$

and carrying out the integration in spherical coordinates, we obtain the condition

$$\frac{Bd \int_0^\xi dr\, r^{d-1}\, e^{-r/\xi} r^{-(d-2)}}{B\xi^d\, |T - T_c|^{2\beta}} << 1 \tag{14}$$

where Br^d is the volume of a d-dimensional sphere of radius r. Letting $r = \xi x$ in the numerator produces

$$(d \int_0^1 dx\, xe^{-x})|T - T_c|^{(d\nu - 2\beta - 2\nu)} << 1$$

The first factor is simply a constant of order unity and the inequality will be satisfied as $T \to T_c$, if and only if $d\nu - 2\beta - 2\nu > 0$, or

$$d > 2 + \frac{2\beta}{\nu} \tag{15}$$

At critical points the Landau theory value of β is $\frac{1}{2}$, $\nu = \frac{1}{2}$, and we obtain $d_c \geq 4$. At tricritical points we have $\beta_t = \frac{1}{4}$, $\nu_t = \frac{1}{2}$, and hence $d_t \geq 3$. The borderline values $d_c = 4$ and $d_t = 3$ are called upper critical dimensionalities and play an important role in the development of the renormalization group approach to critical phenomena. At these marginal dimensionalities there are small (almost unmeasurable) corrections to the Landau critical exponents. The Landau theory of tricritical points thus provides an excellent representation of the correct cooperative effect.

Another application of the Ginzburg criterion is that estimates of the correlation length can be used to determine the range of temperatures near T_c where critical fluctuations play an important role. We return to this question in Section 7.C(d) where we argue that in the case of superconductivity the temperature range is too small to be significant. A further example of Landau–Ginzburg theory is given in Section 4.E where we study properties of liquid–vapor interfaces.

3.K MULTICOMPONENT ORDER PARAMETERS: THE *n*-VECTOR MODEL

In many cases of physical interest the ground state of the system has a degeneracy which is greater than the twofold degeneracy of the zero-field Ising model. An example of such a system is the Heisenberg model with the Hamiltonian

$$H = -\sum_{i<j} J_{ij}\mathbf{S}_i \cdot \mathbf{S}_j \tag{1}$$

in the absence of an applied field. The dynamical variables are the three-dimensional spin operators

$$\mathbf{S}_i = (S_{xi}, S_{yi}, S_{zi})$$

obeying the usual angular momentum commutation relations. The Hamiltonian (1) favors the parallel alignment of neighboring spins if $J_{ij} > 0$, and it is easily shown that the ground state has all the spins aligned in the same direction, which we label z, $S_{zi} = S$. The Hamiltonian (1) is rotationally invariant in spin space, and the z direction may be taken to be any direction. The application of a magnetic field breaks this symmetry, but the nature of the correlated fluctuations that determine the critical behavior of the system depend essentially on the existence of rotational symmetry. In this section we only wish to demonstrate the appropriate generalizations of Landau theory to take such symmetries into account. The equilibrium state of the Heisenberg model may be expressed in terms of the three thermal expectation values

$$m_x = \frac{1}{N}\langle \sum_i S_{xi}\rangle \qquad m_y = \frac{1}{N}\langle \sum_i S_{yi}\rangle \qquad m_z = \frac{1}{N}\langle \sum_i S_{zi}\rangle \tag{2}$$

The rotational symmetry of (1) can then be incorporated into the Landau theory by constructing an expansion that is invariant under arbitrary rotations of the vector $\mathbf{m}$, that is,

$$G = a + \tfrac{1}{2}b(T)(m_x^2 + m_y^2 + m_z^2) + \tfrac{1}{4}c(T)(m_x^2 + m_y^2 + m_z^2)^2 + \cdots \tag{3}$$

The general n-vector model in which the three-component order parameter of the Heisenberg model is replaced by an n-component order parameter will thus have its Landau expansion in terms of the quantity

$$m^2 = \sum_{\alpha=1}^{n} m_\alpha^2 \tag{4}$$

Similarly, in the case of a nematic liquid crystal where the order parameter is the symmetric and traceless tensor $Q_{\alpha\beta}$ defined by (3.H.9), the Landau expansion must be expressible in terms of the two invariants which can be formed from such a tensor, namely

$$\sum_{\alpha,\beta} Q_{\alpha\beta}Q_{\beta\alpha} \qquad \sum_{\alpha,\beta,\gamma} Q_{\alpha\beta}Q_{\beta\gamma}Q_{\gamma\alpha} \tag{5}$$

It is clear from the foregoing that specific forms of symmetry breaking may easily be incorporated in the Landau form. For example, a magnet in a cubic lattice is in general subject to a crystal field of cubic symmetry. For a Heisenberg model on a cubic lattice the appropriate form of the Landau free energy is

$$G(\{m\}, T) = a + \frac{b}{2}(m_x^2 + m_y^2 + m_z^2)$$

$$+ \frac{c}{4}(m_x^2 m_y^2 + m_x^2 m_z^2 + m_y^2 m_z^2) + \frac{d}{4}(m_x^4 + m_y^4 + m_z^4) + \ldots \tag{6}$$

where in the absence of crystal fields $c(T) = 2d(T)$. The equilibrium behavior of the system is obtained for an n-component order parameter from the equations

$$\frac{\partial G}{\partial m_\alpha} = 0 \qquad \alpha = 1, 2, \ldots, n$$

Other examples of systems with a multicomponent order parameter include superfluid ^{4}He (two components), superconductors (two components), and the q-state Potts model ($q - 1$ components) which describes the critical behavior of a number of two- and three-dimensional materials [see, e.g., Section 6.D(b)].

3.L MEAN FIELD THEORY OF FLUIDS: VAN DER WAALS APPROACH

With the exception of the Maier–Saupe model, our examples of mean field theories to this point have been lattice models. We now turn to the case of a fluid consisting of particles interacting via a pair potential which contains a hard core, preventing the particles from overlapping, and a weak attractive tail. We wish to obtain an approximate equation of state for such a system in the spirit of mean field theory. Various modifications of the ideal gas law have been put forward to take into account the effect of interparticle interaction. One approach, with considerable physical appeal, was put forward by van der Waals about a hundred years ago. The van der Waals equation of state can be derived through many different routes; perhaps the simplest approach is through the following observations:

1. The internal energy of the ideal gas is purely kinetic in origin and independent of the volume. The entropy (2.A.14) can be written

$$S = Nk_\text{B} \ln V + \text{terms independent of volume}$$

The Helmholtz free energy is thus

$$A = -Nk_\text{B} T \ln \frac{V}{N} + \text{terms independent of volume}$$

This form of the free energy can be used to derive the equation of state for the pressure

$$P = -\frac{\partial A}{\partial V} = \frac{Nk_B T}{V} \tag{1}$$

2. In a first approximation, the attraction between the particles reduces the internal energy per particle by an amount proportional to the average number of surrounding particles (i.e., to the density). This allows us to approximate the volume-dependent part of the internal energy

$$E = -a\left(\frac{N}{V}\right)N \tag{2}$$

where a is a constant that depends on molecular properties.

3. Short-distance repulsion prevents particles from approaching each other too closely. This has no direct effect on the internal energy, but reduces the free volume available to each particle. Let b be the excluded volume per particle. The total free volume is thus $V_f = V - Nb$. It is in the spirit of the derivation of the expression (2.A.14) for the entropy to interpret the volume dependence as being due to the free volume, while the energy in (2.A.14) is the kinetic energy. With this interpretation we obtain for the free energy

$$A = -\frac{aN^2}{V} - Nk_BT \ln \frac{V - Nb}{N} + \text{terms independent of volume} \tag{3}$$

The van der Waals equation of state follows by differentiation as in (1). After rearranging terms, we have

$$\left[P + a\left(\frac{N}{V}\right)^2\right](V - Nb) = Nk_BT \tag{4}$$

This equation crudely describes the condensation of a gas into a liquid. For an extremely dilute gas $N/V \to 0$, $Nb \ll V$ and (4) reduces to the ideal gas equation of state. We are here concerned with lower temperatures and higher densities.

In Figure 3.18 we plot the behavior predicted by the van der Waals equation of state in the $P-V$ plane. The critical isotherm is characterized by an infinite compressibility at the critical temperature, that is,

$$\left.\frac{\partial P}{\partial V}\right|_{N,T} = 0$$

at $T = T_c$, $V = V_c$. The isotherms predicted by (4) will, below $T = T_c$, have both a maximum and a minimum, that is, for certain values of the pressure below the critical point there will be three real roots when solving for the volume. As $T \to T_c$ from below, the maximum and minimum of the isotherm merge and we obtain an inflection point. The critical point is therefore given by

$$\left(\frac{\partial P}{\partial V}\right)_T = \left(\frac{\partial^2 P}{\partial V^2}\right)_T = 0 \tag{5}$$

These equations yield

$$V_c = 3Nb \qquad P_c = \frac{a}{27b^2} \qquad k_BT_c = \frac{8a}{27b} \tag{6}$$

Using these values of the critical parameters, it is possible to rewrite the van der Waals equation in a parameter-independent way. Defining the reduced

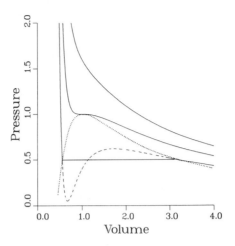

Figure 3.18 Isotherms and coexistence region according to van der Waals theory. Solid line, isotherm; dashed line, unphysical part of isotherm in coexistence region; dotted line, coexistence region. Pressure and volume in units of P_c and V_c.

dimensionless variables

$$v = \frac{V}{V_c} \qquad p = \frac{P}{P_c} \qquad t = \frac{T}{T_c} \qquad (7)$$

and substituting the reduced quantities into the van der Waals equations gives the *law of corresponding states*,

$$\left(p + \frac{3}{v^2}\right)\left(v - \frac{1}{3}\right) = \frac{8t}{3} \qquad (8)$$

We must now deal with the coexistence region. For $t < 1$, the system undergoes a first-order phase transition from the gas to the liquid phase. The unphysical behavior of the isotherms given by (8) in this region (mechanical stability does not allow $\partial p/\partial v > 0$) is a characteristic of mean field theory. There is a simple method known as the equal-area or Maxwell construction for removing the unphysical regions. The coexisting regions must be at the same pressure and on the same isotherm for reasons of mechanical and thermal equilibrium. Consider the chemical potential (or Gibbs free energy per particle)

$$\mu = \frac{G}{N} = \frac{A + PV}{N}$$

$$d\mu = -\frac{S}{N}\,dT + \frac{V}{N}\,dP$$

Two coexisting phases must have the same chemical potential and we must therefore have along an isotherm

$$\int_1^2 d\mu = \mu(2) - \mu(1) = \frac{1}{N}\int V\,dP = 0 \qquad (9)$$

where the coexisting phases have been labeled 2 and 1. In Figure 3.19 we have

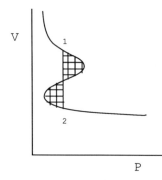

Figure 3.19 Maxwell construction.

exchanged the axes of the plot of Figure 3.18 and it is easy to see that (9) implies that the areas of the two shaded regions must be the same—hence the name "equal-area construction."

We note that for $T < T_c$ the equation $\left(\dfrac{\partial P}{\partial v}\right)_T = 0$ defines a curve known as the *spinodal*. Van der Waals suggested that the states between the coexistence curve and the spinodal are metastable single phase states. In the case of phase separation discussed in 3.C, the spinodal is given by $\left.\dfrac{\partial^2 A}{\partial c^2}\right|_T = 0$. Spinodals also occur in the Maier-Saupe model of 3.H. They are a general feature of first order transitions in mean field theory.

Since (7) does not contain any free parameters, the law of corresponding states implies that when expressed in terms of the reduced variables, all fluids should exhibit similar behavior (i.e., the coexistence region in reduced units should look the same for all fluids). Experimental evidence (Figure 3.20) indicates that the law of corresponding states is a valid concept, but that the van der Waals equation of state does not provide a good quantitative approximation to it.

The van der Waals theory outlined above can be generalized in a number

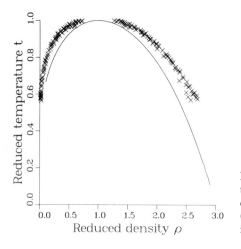

Figure 3.20 Density versus temperature in reduced units. Curve from van der Waals law of corresponding states. (Data points experimental results quoted in Guggenheim [1967].)

of different ways. Guggenheim [1965], Thiele [1963], and Longuet-Higgins and Widom [1965] all have proposed different phenomenological equations of state which are parametrized in terms of the van der Waals constants a and b, and which give better agreement with observed corresponding states than the original van der Waals theory.

The van der Waals approach can also be generalized to more complicated systems. Consider first an isotropic mixture containing N_i molecules of species i. We can obtain a van der Waals theory of mixing by writing, in the spirit of the Bragg–Williams approximation in Section 3.B, for the configurational part of the internal energy per particle

$$E = \frac{1}{2} \sum_{ij} N_i \rho_j e_{ij} \tag{10}$$

where $\rho_j = N_j/V$ is the number density of species j and the e_{ij}'s are constants. If we write for the entropy of mixing

$$S_m = -k_B \sum_i N_i \ln \frac{N_i}{V_f^i} \tag{11}$$

and for the free volume of particles of species i,

$$V_f^i = V(1 - \tfrac{1}{2} \sum_{ij} \rho_j v_{ij}) \tag{12}$$

we may approximate the Helmholtz free energy by

$$A = E + k_B T \sum_i N_i \ln \frac{N_i}{V_f^i}$$

It is now straightforward to compute the pressure from $P = -\partial A/\partial V$ to obtain the equation of state. Depending on the values of the parameters, this theory produces a rich variety of possible phase diagrams. Phase coexistence curves can be obtained by imposing the condition of mechanical and thermodynamic stability (i.e., that the pressure and chemical potential $\mu_i = \partial A/\partial N_i$ of each species be the same in each phase). For a discussion of this type of calculation, we refer the interested reader to Hicks and Young [1977].

It is also possible to construct van der Waals theories for anisotropic systems such as nematics in a similar spirit (Flapper and Vertogen, 1981a, b; Palffy-Muhoray and Bergersen, 1987). The van der Waals theory for interfaces is discussed in Section 4.E.

PROBLEMS

3.1. *Solid–Solid Solutions.* A crystalline solid is composed of constituents A and B. The energies associated with nearest-neighbor pairs of different types are, respectively, e_{AA}, e_{BB}, and e_{AB}. Assume that

$$\epsilon = \tfrac{1}{2} e_{AA} + \tfrac{1}{2} e_{BB} - e_{AB} < 0$$

and that each site has q nearest neighbors.

(a) Calculate the Helmholtz free energy in the Bragg–Williams approximation for a homogeneous system in which the concentration of type A atoms is c_A and that of B atoms is $c_B = 1 - c_A$.

(b) Show that the system will phase separate when $c_A = c_B = \tfrac{1}{2}$ below the temperature

$$k_B T_c = \tfrac{1}{2} q |\epsilon|$$

(c) For $c_A \neq c_B$, show that phase separation will occur at a lower temperature than T_c and that the phase transition is discontinuous.

(d) Find the coexistence curve numerically and plot the result in the $k_B T/q|\epsilon|$ versus c_A plane.

3.2. *Antiferromagnetic Ising Model in the Bragg–Williams Approximation.* Consider a system described by the Hamiltonian (3.C.15) with $J > 0$.

(a) Complete the minimization of the Helmholtz free energy to obtain an expression of the form (3.C.12) assuming a density matrix of the form (3.C.21).

(b) Solve (3.C.10) to obtain a plot of m as a function of $k_B T/qJ$ for (i) $c_A = 0.5$ and (ii) $c_A = 0.45$.

3.3. *One-Dimensional Ising Model in Bethe Approximation.* Calculate the free energy for the one-dimensional Ising model in a magnetic field using (3.D.15) and appropriate generalizations of (3.D.16) and (3.D.17). Compare with results of Section 3.F to check if the result is still exact. Discuss possible reasons for this behavior.

3.4. *Critical Exponents in Mean Field Theory.*

(a) Fill in the missing steps to obtain equations (3.E.3)–(3.E.5).

(b) Show that the specific heat C_h at $h = 0$ is discontinuous in the Bethe approximation at $T = T_c$ for $q > 2$.

(c) Show that in the Bethe approximation to the Ising model $m(h = 0) \propto |T - T_c|^{1/2}$ near T_c.

(d) Show that the exponent for the critical isotherm in the Bethe approximation for the Ising model satisfies $\delta = 3$.

3.5. *Cluster Approximation for Two-Dimensional Ising Model.* The Bethe approximation can be modified to consider clusters of a more general type. Consider as an example the two-dimensional Ising model on the square and triangular lattices. Divide the lattice into blocks of four and three spins as shown in Figure 3.21. Treat the interactions within a block exactly, while using the molecular field approximation for the interactions between spins in different blocks. Calculate the

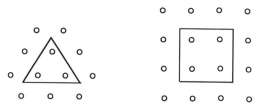

Figure 3.21 Clusters for the two-dimensional Ising model on the square and triangular lattices.

critical temperature for (a) the $\triangle$ lattice and (b) the $\square$ lattice and compare with the exact values

$$\frac{J}{k_B T_c} = 0.441 \ldots (\square) \qquad \frac{J}{k_B T_c} = 0.275 \ldots (\triangle)$$

and with results from the simplest molecular field theory.

3.6. *Application of One-Dimensional Ising Model to a Polymer Problem.* Use the one-dimensional Ising model to describe the following observation: The length l of molecules in a dilute solution of long-chain-like polymer molecules is found to change with the temperature T as shown in Figure 3.22.

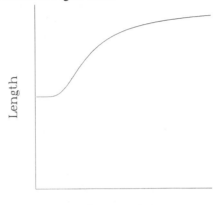

Figure 3.22 Temperature dependence of the average length l of long chain-like polymer molecules.

3.7. *One-Dimensional Ising Model with Spin 1.* Calculate the internal energy of the one-dimensional Ising model defined by the Hamiltonian

$$H = -J \sum_i \sigma_i \sigma_{i+1} \qquad \sigma_i = 0, \pm 1, \quad J > 0$$

The solution of this problem requires differentiation of the root of a cubic equation. You may wish to do this numerically.

3.8. *Generalized Random Walk Problem.* Use the transfer matrix formalism to solve the following generalized random walk problem: By observing drunken sailors, one may notice that the completely random walk is only a crude first approximation in describing the motion. Inertia plays an important role in determining which way to take the next step. For this reason the next step will take place with greater probability in the direction of the previous one. A simple model for the motion can be constructed by assuming that there is a correlation between nearest-neighbor steps, namely that the next step will be in the same direction as the previous one with probability p and in the opposite direction with probability $(1 - p)$.

Calculate the mean-square displacement after N steps. The motion can be assumed to be one-dimensional.

3.9. *Clusters of Spins for the Ising Chain.*

(a) Consider the one-dimensional Ising model in a magnetic field h subject to periodic boundary conditions. Suppose that spin j is in a specific state, either up or down. Using the transfer matrix of Section 3.F, calculate the probability that spins $j + 1, j + 2, \ldots, j + n$ will be in the same state as spin j and that spin $j + n + 1$ will be in the opposite state.

(b) Remove the restriction on spin $j + n + 1$ and again calculate the probability.

3.10. *Maier–Saupe Model of a Liquid Crystal.*

  (a) Using the Maier–Saupe expression (3.H.18) for the free energy, derive the Landau expansion (3.H.19).

  (b) Find the transition temperature and the discontinuity in the order parameter at the transition predicted by the expansion found in part (a).

  (c) Express the self-consistent equation for the order parameter in terms of the complex error function and evaluate the transition temperature and the discontinuity of the order temperature and entropy at the transition. (One of the weaknesses of the Maier–Saupe approximation is that the predicted latent heat is much too large.)

3.11. *Latent Heat of a First-Order Transition.* Consider the Landau free energy

$$G(m, T) = a(T) + \frac{b}{2}m^2 + \frac{c}{3}m^4 + \frac{d}{4}m^6$$

and assume that $b > 0$, $c < 0$, so that a first-order transition takes place. Derive an expression for the latent heat of transition.

3.12. *Asymptotic Behavior near a Tricritical Point.*

  (a) Derive the result (3.I.8) for the discontinuity of the order parameter x across the first-order line near the tricritical point.

  (b) Show that $\gamma_t = 1$, $\alpha_t = \frac{1}{2}$ in the tricritical region.

  (c) Show that in the critical region the exponents predicted by Landau theory are $\gamma_u = 2$, $\alpha_u = -1$.

3.13. *Blume–Emery–Griffith Model.*

  (a) By minimizing the trial free energy (3.I.15) with respect to the density matrix ρ_i derive (3.I.16).

  (b) Derive the formulas (3.I.19) for the Landau coefficients.

3.14. *Heisenberg Model in a Crystal Field.* In Section 3.K the Landau free energy in the presence of a cubic crystal field was given by equation (3.K.6). Assuming that the coefficients of higher than fourth order in m are all positive, determine the nature of the ordered phase. You may assume that the system will order in a (100), (111), or (110) preferred spin orientation and minimize the free energy with respect to a simple amplitude. In which situations will the transition be discontinuous?

4

Dense Gases and Liquids

In this chapter we discuss selected topics in the theory of nonideal gases and liquids, a subject with a lengthy history and one in which a considerable degree of understanding of the basic phenomena has been attained. One of the earliest theories of dense gases is the well-known van der Waals equation which we discussed in Chapter 3 as an example of mean field theory. We shall not return to the van der Waals theory of bulk liquids in this chapter but rather concentrate on more general theories of gas and liquid phases. In the case of atomic and molecular gases and fluids we may, except in the case of very light constituents such as hydrogen or helium, safely neglect quantum effects and concentrate on the evaluation of the classical partition function

$$Z_c = \frac{1}{N!h^{3N}} \int d^{3N}p\, d^{3N}r \; e^{-\beta H} \tag{1}$$

where

$$H = \sum_i \frac{p_i^2}{2m} + \sum_{i<j} U(|\mathbf{r}_i - \mathbf{r}_j|) \tag{2}$$

in the case of a simple atomic gas or liquid. At this point it is worth pointing out that even for a rare-gas system, such as liquid argon, the Hamiltonian (2) is not complete. Three-body interactions, $V(\mathbf{r}_1, \mathbf{r}_2, \mathbf{r}_3)$, play an important role in the thermodynamic properties of the system. In molecular fluids the potential energy will generally be a function of the relative orientation of the molecules as well as of their separation. The determination of an appropriate intermolecular potential is a difficult, and only partially solved problem in quantum chemistry. The reader is referred to the review article by Barker and Henderson [1976] for a discussion of the role and parametrization of intermolecular potentials and to the classic monograph of Hirshfelder et al. [1954] or the newer book by Maitland et al. [1981] for a more detailed treatment of this topic. In this chapter we invariably assume that our system can be described by the Hamiltonian (2) with only central two-body forces between the constituents. The calculation of Z_c can then be reduced, after an integration over the momentum variables, to

$$Z_c = \lambda^{-3N} \frac{1}{N!} \int d^{3N}r \exp\left\{-\beta \sum_{i<j} U(r_{ij})\right\} \tag{3}$$

where $\lambda = [h^2/(2m\pi k_B T)]^{1/2}$ is the thermal wavelength introduced in (2.D.9). The remaining integral will be denoted by $Q_N(V, T)$ and is called the *configuration integral*:

$$Q_N(V, T) = \frac{1}{N!} \int d^{3N}r \exp\left\{-\beta \sum_{i<j} U(r_{ij})\right\} \tag{4}$$

The evaluation of this expression is the central problem in the theory of dense gases and liquids. We shall describe, in the following sections, a number of different approaches that have been devised for this purpose. In Section 4.A we discuss the virial expansion for $Q_N(V, T)$. In Section 4.B we focus on the reduced distribution functions and summarize some of the more successful approximation schemes for the solution of the Ornstein–Zernike equation. Section 4.C contains a brief description of simulation techniques which are, with the development of supercomputers, becoming a more and more powerful tool. In Section 4.D we discuss perturbation theories, and in Section 4.E we turn to the topic of inhomogeneous fluids. In this section we finally return to van der Waals theory when we construct a Landau–Ginzburg theory of the liquid–vapor interface.

A number of excellent general references for the material of this chapter exist. Among them are Barker and Henderson [1976] and the book by Hansen and McDonald [1986].

4.A THE VIRIAL EXPANSION

The ideal gas equation of state provides a reasonable approximation to the properties of interacting atoms or molecules only in the dilute limit. A systematic approach to the effects of increasing density, or lower temperature, is the virial expansion in which one expands the pressure in a power series in the density

$$\frac{P}{k_B T} = \frac{N}{V}\left(1 + B_2(T)\frac{N}{V} + B_3(T)\left(\frac{N}{V}\right)^2 + \cdots\right) \tag{1}$$

The coefficients B_j are known as virial coefficients. The most elegant method of deriving the virial coefficients utilizes the grand canonical ensemble, rather than the canonical, and is due to J. E. Mayer. The pressure is given by

$$\frac{P}{k_B T} = \frac{1}{V} \ln Z_G = \frac{1}{V} \ln \left[\sum_{N=0}^{\infty} e^{\beta \mu N} \left(\frac{2m\pi k_B T}{h^2} \right)^{3N/2} Q_N(V, T) \right] \quad (2)$$

with

$$Q_N(V, T) = \frac{1}{N!} \int d^{3N}r \, \exp \left\{ -\beta \sum_{i < j \leq N} U(r_{ij}) \right\}$$

The potential $U(r_{ij})$ depends on the particular system in question, but typically, for a neutral system, will be sharply repulsive at short distances due to overlap of electronic wave functions and will be weakly attractive at larger separations. A potential frequently used to describe rare gases and fluids is the Lennard-Jones or 6–12 potential, which is of the form

$$U(r) = 4\epsilon \left[\left(\frac{\sigma}{r} \right)^{12} - \left(\frac{\sigma}{r} \right)^{6} \right] \quad (3)$$

This potential has a minimum value of $-\epsilon$ at $r = 2^{1/6}\sigma$. For argon, appropriate values of ϵ and σ are $\epsilon/k_B = 120$ K, $\sigma = 3.4$ Å.

The function $e^{-\beta U(r)}$ which appears in the configuration integral has the undesirable property of approaching unity rather than zero as r goes to infinity. To construct an expansion in powers of the density, we need a function of the potential which is significant only if groups of atoms are close to each other. Such a function is the Mayer function

$$f_{ij} = \exp \left\{ -\beta U(r_{ij}) \right\} - 1 \quad (4)$$

which is sketched in Figure 4.1. In terms of this function the configuration integral becomes

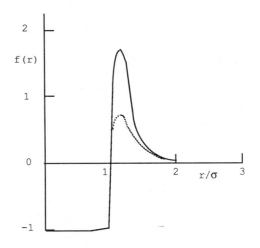

Figure 4.1 $f_{ij}(r)$ for the Lennard-Jones potential for $\beta\epsilon = 1$. Dotted line, $\beta\epsilon = 0.5$.

$$Q_N(V, T) = \frac{1}{N!} \int d^{3N}r \prod_{j<m} (1 + f_{jm}) \tag{5}$$

The expansion of the product (5) results in a series

$$Q_N(V, T) = \frac{1}{N!} \int d^{3N}r \left(1 + \sum_{j<m} f_{jm} + \sum_{j<m, r<s} f_{jm}f_{rs} + \cdots \right) \tag{6}$$

The evaluation of the various terms in (6) is greatly facilitated by a graphical notation. We identify a graph with each term. A particle is denoted by a heavy dot $\cdot$, the function f_{jm} by a line connecting particles j and m. Thus each term in (6) consists of a graph with N dots and a variable number of lines joining pairs of dots. Some simple graphs with the corresponding term in the integrand of (6) are

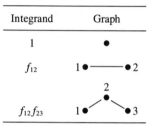

The integrations over different disconnected pieces of a graph can clearly be carried out independently. We therefore focus on the different connected subgraphs and define a *cluster integral* for each topologically distinct connected subgraph. In view of later simplifications, we define the cluster integral $b_j(T)$ for graph j to be

$$b_j(T) = \frac{1}{n_j!V} \sum \int d^3r_1 \, d^3r_2 \ldots d^3r_{nj} \left(\prod_{i,j} f_{ij}\right) \tag{7}$$

In formula (7), n_j is the number of vertices (particles) in graph j and the product over Mayer functions is the appropriate combination determined by the connectivity of the graph. The summation in (7) is a sum over distinct permutations of the labels on the vertices which occur in the expansion (6). This requires some comment and we first illustrate the notion with a few simple examples. The three particle graphs $1 \cdot \underline{\quad 2 \quad} \cdot 3$ and $3 \cdot \underline{\quad 2 \quad} \cdot 1$ are not distinct from each other but *are* distinct from $2 \cdot \underline{\quad 1 \quad} \cdot 3$ and $1 \cdot \underline{\quad 3 \quad} \cdot 2$. The four-particle graph $\square$ has the three distinct assignments of labels (1234), (1324), (1243) (counterclockwise from the lower left-hand corner), and no others. Any other assignment of labels, such as (1432), corresponds to the same product of Mayer functions as one of the three previous assignments and this product appears only once in the configuration integral (6). We note that since the particle coordinates are integrated out in (7), the cluster integral depends only on the topology of the graph and we may henceforth drop the labeling of the vertices. Moreover, since the function f is short ranged, one of the vertices of a connected graph, or equivalently the center of mass coordinate, can be freely integrated over the volume. Thus $b_j(T)$ is independent of the volume. The cluster integrals of the first few graphs are listed in Table 4.1.

TABLE 4.1

j	Graph	$b_j(T)$
1	•	1
2	•—•	$\dfrac{1}{2!\,V}\int d^3r_1\,d^3r_2\,f(r_{12}) = \dfrac{1}{2}\int d^3x\,f(x)$
3		$\dfrac{1}{3!\,V}\int d^3r_1\,d^3r_2\,d^3r_3\,(f_{12}f_{23} + f_{12}f_{13} + f_{13}f_{23})$
		$= \dfrac{1}{2}\int d^3x\,d^3y\,f(x)\,f(y) = 2b_2^2(T)$
4		$\dfrac{1}{3!\,V}\int d^3r_1\,d^3r_2\,d^3r_3\,f_{12}f_{23}f_{13}$
		$= \dfrac{1}{3!}\int d^3x\,d^3y\,f(x)f(y)f(|\mathbf{x} - \mathbf{y}|)$
5		$\dfrac{1}{4!\,V}\int d^3r_1\,d^3r_2\,d^3r_3\,d^3r_4\,(f_{12}f_{23}f_{14}f_{34}$
		$+ f_{13}f_{14}f_{23}f_{24} + f_{12}f_{24}f_{34}f_{13})$

The most general term in (6), expressed in terms of cluster integrals, is

$$\frac{1}{N!}\left\{ \frac{N!}{\prod_j m_j!(n_j!)^{m_j}} \prod_s [Vn_s!\,b_s(T)]^{m_s} \right\} \qquad (8)$$

In equation (8), m_j is the number of times that the unlabeled graph j appears in the given term, n_j is the number of vertices of graph j, and $Vn_s!\,b_s(T)$ is the contribution of graph s to Q_N when a specific set of particles is assigned to the vertices of the graph. The combinatorical factor

$$\frac{N!}{\prod_j m_j!\,(n_j!)^{m_j}}$$

is the number of ways that the N particles can be assigned to the set of disconnected graphs and $N = \Sigma_j\, m_j n_j$.

We show explicitly that this formula produces the correct contribution in a specific simple case. One term in the configuration integral is (graph 4)

$$\frac{1}{N!}\sum_{i,j,k}\int d^{3N}r\,f_{ij}f_{jk}f_{ki}$$

$$= \frac{N(N-1)(N-2)}{3!}\frac{V^{N-3}}{N!}\int d^3r_1\,d^3r_2\,d^3r_3\,f(r_{12})\,f(r_{13})\,f(r_{23})$$

by direct counting. In expression (8), we have $m_4 = 1$, $m_1 = N - 3$, $n_1 = 1$, and $n_4 = 3$ and we obtain

$$\frac{N(N-1)(N-2)V^{N-3}}{N!}[Vb_4(T)]$$

which is the same as the previous expression. The reader is encouraged to check a few other simple cases.

We now return to expression (2) for the pressure:

$$\frac{P}{k_B T} = \frac{1}{V}\ln\sum_{N=0}^{\infty} z^N \lambda^{-3N} Q_N(V, T)$$

$$= \frac{1}{V}\ln\sum_{N=0}^{\infty} z^N \lambda^{-3N} \sum_{\{m_j\}}\prod_{j=1}^{\infty}\frac{[Vb_j(T)]^{m_j}}{m_j!} \tag{9}$$

where $z = e^{\beta\mu}$ and $\sum_{\{m\}}$ indicates a sum over all possible combinations of graphs subject to the restriction $\sum m_j n_j = N$. The sum (9) may be decomposed into a product of sums over m_j. Using $N = \sum m_j n_j$, we obtain

$$\frac{P}{k_B T} = \frac{1}{V}\ln\left\{\prod_j\sum_{m_j=0}^{\infty}\frac{[(z\lambda^{-3})^{n_j}Vb_j(T)]^{m_j}}{m_j!}\right\} \tag{10}$$

$$= \frac{1}{V}\ln\prod_j\exp\left[(z\lambda^{-3})^{n_j}Vb_j(T)\right] = \sum_j (z\lambda^{-3})^{n_j}b_j(T) \tag{11}$$

To complete the virial expansion we must still express the chemical potential in terms of the density $n = N/V$ and the temperature. This is accomplished by constructing an expansion of the density

$$N = z\left(\frac{\partial}{\partial z}\ln Z_G\right)_{T,V} = Vz\frac{\partial}{\partial z}\left(\sum_{j=1}^{\infty}(z\lambda^{-3})^{n_j}b_j(T)\right)_{T,V}$$

or

$$n = \frac{N}{V} = \sum_{j=1}^{\infty} n_j(z\lambda^{-3})^{n_j}b_j(T) \tag{12}$$

Substituting $z = a_1 n + a_2 n^2 + a_3 n^3 + \cdots$ and solving for the a_j's, one finally obtains, on substituting into (11), the virial expansion

$$\frac{P}{k_B T} = n + B_2(T)n^2 + B_3(T)n^3 + \cdots \tag{13}$$

The completion of this task is left as an exercise. We simply quote the result:

$$B_2(T) = -\frac{1}{2}\int d^3r\, f(r)$$

$$B_3(T) = -\frac{1}{3V}\int d^3r_1\, d^3r_2\, d^3r_3\, f_{12}f_{13}f_{23} \tag{14}$$

The reader will note that only the graph $\triangle$ contributes to the third virial coefficient B_3. The contributions from the graph $\wedge$ have canceled in the process of eliminating z. This result is a particular manifestation of a general theorem: The virial expansion can be expressed to all orders in terms of cluster integrals of *stars*. A star graph is a graph that cannot be separated into disjoint pieces by cutting through a single vertex. The difference between stars and other graphs is that the cluster integral of non stars can be expressed as the product of cluster integrals of the separable pieces. Such is not the case with stars. An example of this decomposition has already appeared in Table 4.1 in the case of graph 3 and a general proof is straightforward. The general expression of the virial coefficients in terms of star graph cluster integrals is derived in the book by Mayer and Mayer [1940] and can also be found in that of Uhlenbeck and Ford [1963].

It is clear that the evaluation of even the third virial coefficient presents computational difficulties for realistic potentials. For the hard-sphere potential the virial coefficients up to B_7 have been computed, either analytically or numerically. For the $6-12$ potential the virial coefficients up to B_5 have been obtained. In Figure 4.2 we show the equations of state for a system of hard spheres obtained from the virial expansion using more and more terms. The dots represent the results of computer simulations. We see that the agreement is rather good except at higher densities. The results are taken from Barker and Henderson [1971, 1976].

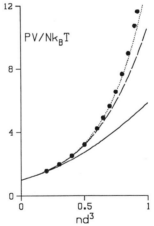

PV/Nk_BT

Figure 4.2 Virial equation of state for hard spheres of diameter d. Solid curve, two virial coefficients. Dashed curve, four coefficients; dotted curve, six coefficients. (Filled circles are Monte Carlo data of Barker and Henderson [1971].)

nd^3

From the virial series for the $6-12$ potential one can obtain a series of estimates for the critical temperature by requiring that $(\partial P/\partial V)$ and $(\partial^2 P/\partial V^2)$ both be zero at $T = T_c$. The results are given below for the dimensionless critical temperature $(k_B T_c/\epsilon)$ from Temperley et al. [1968].

	$k_B T_c/\epsilon$
B_3	1.445
B_4	1.300
B_5	1.291

Argon, which is thought to be a good example of a Lennard-Jones system, has an experimental value of 1.26 for this parameter. We see that the virial expansion does seem to be converging although rather slowly.

4.B DISTRIBUTION FUNCTIONS

(a) The Pair Correlation Function

One of the most useful approaches to the theory of liquids has been the study of reduced distribution functions and, in particular, the calculation by a number of sophisticated approximation schemes of the pair correlation function. Consider the function

$$P(\mathbf{r}_1, \mathbf{r}_2, \ldots, \mathbf{r}_N) = \frac{1}{N! \, Q_N(V, T)} \exp\{-\beta W(\mathbf{r}_1, \mathbf{r}_2, \ldots, \mathbf{r}_N)\} \tag{1}$$

with $W(\mathbf{r}_1, \mathbf{r}_2, \ldots, \mathbf{r}_N) = \sum_{i<j} U(\mathbf{r}_i - \mathbf{r}_j)$. This function is the probability density that the N particles are at positions $\mathbf{r}_1, \ldots, \mathbf{r}_N$. The function P provides far more information than is necessary for the calculation of thermodynamic functions. To proceed systematically, we define a sequence of reduced distribution functions:

$$n_1(\mathbf{x}) = \sum_{i=1}^{N} \langle \delta(\mathbf{x} - \mathbf{r}_i) \rangle \tag{2}$$

$$n_2(\mathbf{x}_1, \mathbf{x}_2) = \sum_{i \neq j} \langle \delta(\mathbf{x}_1 - \mathbf{r}_i) \, \delta(\mathbf{x}_2 - \mathbf{r}_j) \rangle \tag{3}$$

and, in general,

$$n_s(\mathbf{x}_1, \mathbf{x}_2, \ldots, \mathbf{x}_s) = \sum_{i \neq j \neq \ldots m} \langle \delta(\mathbf{x}_1 - \mathbf{r}_i) \, \delta(\mathbf{x}_2 - \mathbf{r}_j) \ldots \delta(\mathbf{x}_s - \mathbf{r}_m) \rangle \tag{4}$$

In a homogeneous system the reduced distribution function n_1 is simply the density:

$$n_1(\mathbf{r}_1) = N \frac{\int d^3r_2 \ldots d^3r_N \exp\{-\beta \sum_{i<j} U(\mathbf{r}_i - \mathbf{r}_j)\}}{\int d^3r_1 \ldots d^3r_N \exp\{-\beta \sum_{i<j} U(\mathbf{r}_i - \mathbf{r}_j)\}} = \frac{N}{V} \tag{5}$$

which is easily seen by letting $\mathbf{r}_j = \mathbf{r}_1 + \mathbf{x}_j$ for $j \neq 1$, integrating over $\mathbf{x}_j$ and noting that the integrand in the denominator becomes independent of $\mathbf{r}_1$. The two-particle distribution

$$n_2(\mathbf{x}_1, \mathbf{x}_2) = N(N - 1)$$

$$\frac{\int d^3r_3 d^3r_4 \ldots d^3r_N \exp\{-\beta W(\mathbf{x}_1, \mathbf{x}_2, \mathbf{r}_3, \ldots \mathbf{r}_N)\}}{\int d^3r_1 \, d^3r_2 \ldots d^3r_N \exp\{-\beta W(\mathbf{r}_1, \mathbf{r}_2, \ldots \mathbf{r}_N)\}} \tag{6}$$

is the probability that two particles occupy the positions $\mathbf{x}_1$ and $\mathbf{x}_2$ and, as $|\mathbf{x}_1 - \mathbf{x}_2| \to \infty$, approaches the limiting value $N(N-1)/V^2$. It is easy to see that the expectation value of the interaction energy may be expressed in terms of $n_2(\mathbf{x}_1, \mathbf{x}_2)$:

$$\langle U \rangle = \sum_{i<j} \langle U(\mathbf{r}_i - \mathbf{r}_j) \rangle$$

$$= \tfrac{1}{2} \sum_{i \neq j} \int d^3x_1 \, d^3x_2 \, \langle \delta(\mathbf{x}_1 - \mathbf{r}_i) \, \delta(\mathbf{x}_2 - \mathbf{r}_j) \, U(\mathbf{x}_1 - \mathbf{x}_2) \rangle$$

$$= \tfrac{1}{2} \sum_{i \neq j} \int d^3x_1 \, d^3x_2 \, U(\mathbf{x}_1 - \mathbf{x}_2) \, \langle \delta(\mathbf{x}_1 - \mathbf{r}_i) \delta(\mathbf{x}_2 - \mathbf{r}_j) \rangle \qquad (7)$$

$$= \tfrac{1}{2} \int d^3x_1 \, d^3x_2 \, U(\mathbf{x}_1 - \mathbf{x}_2) n_2(\mathbf{x}_1, \mathbf{x}_2)$$

In a homogeneous system $n_2(\mathbf{x}_1, \mathbf{x}_2) = n_2(|\mathbf{x}_1 - \mathbf{x}_2|)$. It is conventional to define another function $g(|\mathbf{x}_1 - \mathbf{x}_2|)$ called the pair distribution function through

$$n_2(|\mathbf{x}_1 - \mathbf{x}_2|) \equiv \left(\frac{N}{V}\right)^2 g(|\mathbf{x}_1 - \mathbf{x}_2|) \qquad (8)$$

In the thermodynamic limit $N \to \infty$, $N/V = $ constant, we have $N(N-1) \approx N^2$ and as the separation becomes large $g(|\mathbf{x}_1 - \mathbf{x}_2|) \to 1$. Therefore, from (7) and (8), we have

$$\langle U \rangle = \frac{N^2}{2V} \int d^3r \, U(r) g(r) \qquad (9)$$

The Fourier transform of the pair distribution function is intimately related to the *structure factor* $S(q)$, which we define as follows:

$$S(q) - 1 = \frac{N}{V} \int d^3r \, \{g(r) - 1\} e^{i\mathbf{q} \cdot \mathbf{r}} \qquad (10)$$

By substituting the definitions (3) and (8), we find

$$S(q) = \frac{1}{N} \left\langle \sum_{i,j} \exp\{i\mathbf{q} \cdot (\mathbf{r}_i - \mathbf{r}_j)\} \right\rangle - N\delta_{\mathbf{q},0} \qquad (11)$$

where $\delta_{\mathbf{q},0}$ is the three-dimensional Kronecker delta:

$$\delta_{\mathbf{q},0} = \frac{1}{V}(2\pi)^3 \delta(\mathbf{q})$$

Equation (11) can be used as an alternative definition of the structure factor. The last term is sometimes not included, but removes an uninteresting singularity at $\mathbf{q} = 0$. The structure factor plays an important role in the interpretation of elastic scattering experiments employing e.g., neutrons or light. To see this, consider the situation in which an incoming beam can be described in terms of plane waves

$$\Psi_k(\mathbf{r}) = \frac{1}{\sqrt{V}} e^{i\mathbf{k}\cdot\mathbf{r}}$$

One then detects an elastically scattered outgoing wave with wave vector $\mathbf{k}' = \mathbf{k} + \mathbf{q}$,

$$\Psi_{k'}(\mathbf{r}) = \frac{1}{\sqrt{V}} e^{i\mathbf{k}'\cdot\mathbf{r}}$$

We assume that the interaction between the probe and the particles of the system can be expressed as a sum of contributions from the individual particles:

$$\sum_i u(\mathbf{r} - \mathbf{r}_i)$$

Let the Fourier transform of this potential be given by

$$u(\mathbf{q}) = \int d^3r\, u(\mathbf{r})\, e^{-i\mathbf{q}\cdot\mathbf{r}}$$

The golden rule transition rate for elastic scattering to the state $\mathbf{k}' = \mathbf{k} + \mathbf{q}$ is given by

$$W_{\mathbf{k}\to\mathbf{k}'} = \frac{2\pi}{\hbar} \left| \langle \mathbf{k} + \mathbf{q} | \sum_i u(\mathbf{r} - \mathbf{r}_i) | \mathbf{k} \rangle \right|^2 \delta(\epsilon(\mathbf{k}) - \epsilon(\mathbf{k}')) \qquad (12)$$

The thermal average of the squared matrix element in (12), for $\mathbf{q} \neq 0$, is given by

$$\frac{N}{V^2} |u(\mathbf{q})|^2 S(\mathbf{q})$$

where we have used (11). We thus have, for the scattered intensity,

$$I(\mathbf{k}' - \mathbf{k} = \mathbf{q}) \propto f(\mathbf{q})S(\mathbf{q})I_0 \qquad (13)$$

where I_0 is the intensity of the incoming beam and the *form factor* $f(\mathbf{q}) = |u(\mathbf{q})|^2$. In the special case of neutron scattering, the potential $u(\mathbf{r} - \mathbf{r}_i)$ will be essentially a δ-function potential and the form factor will thus be a slowly varying function of $\mathbf{q}$. Hence $S(\mathbf{q})$ will be proportional to the intensity of the scattered beam. The functions g and S have the general appearance shown in Figure 4.3 for a dense fluid.

Knowledge of the pair distribution function thus allows us to predict the internal energy (9) and the results of elastic scattering experiments. We next show that the pair distribution function is also intimately related to the compressibility $K_T = -1/V(\partial V/\partial P)_T$.

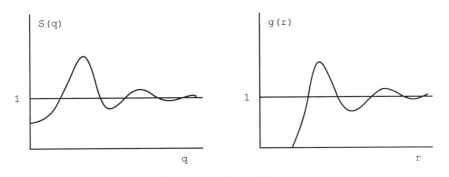

Figure 4.3   Sketch of $g(r)$ and $S(q)$ for a typical liquid.

For a homogeneous system with a fixed number, N, of particles, we have

$$\int d^3x_1\, n_2\left(|\mathbf{x}_1 - \mathbf{x}_2|\right) = \frac{N(N-1)}{V} \tag{14}$$

If the number of particles is allowed to fluctuate, as in the grand canonical ensemble, the right-hand side will be replaced by $\langle N(N-1)\rangle/V$ and the definition (8) changes to

$$g\left(|\mathbf{x}_1 - \mathbf{x}_2|\right) = \frac{V^2}{\langle N\rangle^2}\, n_2\left(|\mathbf{x}_1 - \mathbf{x}_2|\right)$$

We thus have

$$\int d^3r\,(g(r) - 1) = \frac{V}{\langle N\rangle^2}\langle N(N-1)\rangle - V = V\frac{(\Delta N)^2}{\langle N\rangle^2} - \frac{V}{\langle N\rangle} \tag{15}$$

where

$$(\Delta N)^2 = \langle (N - \langle N\rangle)^2\rangle$$

If we now recall the relationship (2.C.22) between the particle fluctuation ΔN and the compressibility, we find the *fluctuation* or *compressibility* equation of state

$$n\int d^3r\,[g(r) - 1] = -1 + nk_B T K_T \tag{16}$$

The quantity

$$h(r) = g(r) - 1 \tag{17}$$

is commonly referred to as the *pair correlation function.*

The use of the grand canonical definition of $g(r)$ deserves some comment. A mechanical measurement of the compressibility would be carried out at fixed N. However, as we have shown above, the pair correlation function is related to the intensity of elastically scattered radiation and is normally obtained from such a scattering experiement. In a scattering experiment the beam samples a fraction of the total volume and in this subvolume the number of particles, while macroscopic, fluctuates. Thus the use of the grand canonical ensemble is appropriate.

The pair correlation function is a measure of the extent to which the spatial distribution of particles is correlated. For an ideal gas, $g(r) = 1$ or $h(r) = 0$ and the fluctuation equation (16) gives the compressibility directly:

$$K_T = \frac{V}{Nk_B T} \tag{18}$$

The same result can be obtained from the ideal gas equation of state

$$PV = Nk_B T \tag{19}$$

and the definition of the isothermal compressibility:

$$K_T = -\frac{1}{V}\left(\frac{\partial V}{\partial P}\right)_{N,T} \tag{20}$$

Consider next an ideal solid in which the particles sit at fixed positions R_i. In this case

$$n \int d^3r\, g(r) = \frac{1}{N} \sum_{i \neq j} \int d^3r\, \delta(\mathbf{r} - \mathbf{R}_i + \mathbf{R}_j) = N - 1 \tag{21}$$

Thus

$$n \int d^3r\, [g(r) - 1] = -1$$

and the fluctuation equation of state gives $K_T = 0$. Physically, this means that if the atoms are not allowed to vibrate about their equilibrium positions, the compressibility is zero.

In self-condensed systems, such as solids and liquids far from the critical point, the compressibility will be much smaller than that of an ideal gas

$$0 < nk_B T K_T << 1$$

Thus for nearly incompressible systems

$$n \int d^3r\, [g(r) - 1] \approx -1$$

Another situation in which the fluctuation equation of state offers valuable insight is in the discussion of the critical region (i.e., the region of the phase diagram near the liquid–vapor critical point). At the critical point $(\partial P/\partial V) = 0$, so that $K_T = \infty$. This means that

$$\int d^3r[g(r) - 1] = \int d^3r\, h(r) \longrightarrow \infty \tag{22}$$

as the system approaches the critical point. The divergence is due to a long tail in $h(r)$. Conversely, the structure factor $S(q)$ becomes very large for small q. This feature is responsible for the phenomenon of critical opalescence observed in light-scattering studies near critical points.

The reader will note that the left-hand side of (16) is equal to

$$\lim_{q \to 0} [S(q) - 1]$$

In the canonical ensemble this limit is equal to -1, but as we have already pointed out above, a scattering experiment invariably samples a fluctuating number of particles and we have indicated in Figure 4.3 that $S(q)$ approaches a nonzero limit as q becomes small.

We next show how to obtain from the pair distribution function an equation of state that generalizes the ideal gas law (19). As before, we assume a system of particles interacting pairwise with a velocity-independent potential energy. First we note that if p_i, q_i are the generalized coordinates of the particles

$$\left\langle \sum_i p_i q_i \right\rangle$$

must be independent of time for a system in equilibrium. Thus

$$0 = \frac{d}{dt}\left\langle \sum_i p_i q_i \right\rangle = \left\langle \sum_i \dot{p}_i q_i \right\rangle + \left\langle \sum_i p_i \dot{q}_i \right\rangle \tag{23}$$

Now

$$\langle p_i \dot{q}_i \rangle = \frac{1}{Z_c} \int d\Omega \, p_i \frac{\partial H}{\partial p_i} e^{-\beta H} = -\frac{k_B T}{Z_c} \int d\Omega \, p_i \frac{\partial}{\partial p_i} e^{-\beta H}$$

where Z_c is the partition function and

$$d\Omega = \frac{d^{6N}x}{h^{3N}N!}$$

Integrating by parts and noting that as $p_i \to \infty$, $e^{-\beta H} \to 0$, we find that

$$\langle p_i \dot{q}_i \rangle = \frac{k_B T}{Z_c} \int d\Omega \, e^{-\beta H} = k_B T \tag{24}$$

Thus

$$\left\langle \sum_i p_i \dot{q}_i \right\rangle = 3N k_B T \tag{25}$$

Using (25) we now rewrite (23):

$$3N k_B T = -\left\langle \sum_i \dot{p}_i q_i \right\rangle = -\left\langle \sum_i F_i q_i \right\rangle \tag{26}$$

where F_i is the generalized force acting on the ith generalized coordinate. Clearly,

$$\sum_{i=1}^{3N} q_i F_i = \sum_{j=1}^{N} \mathbf{r}_j \cdot \mathbf{F}_j \tag{27}$$

where $\mathbf{F}_j$ is the force on the jth particle, located at $\mathbf{r}_j$. We split this force into contributions from internal (other particles) forces and external (wall) forces:

$$\mathbf{F}_j = \mathbf{F}_j^{\text{ext}} + \mathbf{F}_j^{\text{int}} \tag{28}$$

The external force produces the pressure exerted on the system by the container. Let A be the surface area of the container of volume V which confines the system. Then

$$-\left\langle \sum_{j=1}^{N} \mathbf{r}_j \cdot \mathbf{F}_j^{\text{ext}} \right\rangle = P \int_A \mathbf{r} \cdot d\mathbf{A} \tag{29}$$

where $\mathbf{r}$ is the position vector of the element of surface area $d\mathbf{A}$. The negative sign in (29) is due to the fact that the pressure force acts inward, whereas the positive direction of the vector $d\mathbf{A}$ is outward. Using Gauss's theorem, we have

$$-\left\langle \sum_{j=1}^{N} \mathbf{r}_j \cdot \mathbf{F}_j^{\text{ext}} \right\rangle = P \int_V (\nabla \cdot \mathbf{r}) \, d^3 r = 3PV \tag{30}$$

The internal force on the ith particle due to the jth particle is given by

$$-\frac{\partial U (\mathbf{r}_i - \mathbf{r}_j)}{\partial \mathbf{r}_i} \tag{31}$$

Summing over i and j and using the definition (3), we obtain

$$-\left\langle \sum_j \mathbf{r}_j \cdot \mathbf{F}_j^{\text{int}} \right\rangle = \tfrac{1}{2} \int d\mathbf{x}_1 \int d\mathbf{x}_2 \, n_2 (\mathbf{x}_1, \mathbf{x}_2)(\mathbf{x}_2 - \mathbf{x}_1) \cdot \nabla_2 U (\mathbf{x}_2 - \mathbf{x}_1) \tag{32}$$

where the factor of $\tfrac{1}{2}$ compensates for double counting. Collecting terms and taking the thermodynamic limit, we finally obtain

$$PV = Nk_{\text{B}} T \left[1 - \frac{n}{6k_{\text{B}} T} \int d^3 r \, (\mathbf{r} \cdot \nabla U (\mathbf{r})) \, g (\mathbf{r}) \right] \tag{33}$$

Equation (33) is known both as the *virial* equation of state and as the *pressure* equation of state. Both (33) and (16) are exact for a system of classical particles with pairwise forces between them. The compressibility obtained by differentiating (33) should therefore agree with the result obtained from (16). This will, of course, hold when an exact pair distribution function is used. However, it turns out to be difficult to achieve agreement between the two expressions when an approximate form of $g(r)$ is used. Comparison of the two equations thus offers a useful check on the validity of approximate calculations.

Having demonstrated the central role of the pair correlation function, we now discuss methods of calculating this function.

(b) The BBGKY Hierarchy

We proceed to derive a set of equations for the reduced distribution functions introduced at the beginning of this section. This hierarchy is an equilibrium version of the BBGKY (Born, Bogoliubov, Green, Kirkwood, Yvon) hierarchy for the evolution of time-dependent distribution functions (see e.g. Balescu, 1975). Consider the function $\nabla n_1(\mathbf{x})$:

$$\nabla n_1(\mathbf{x}) = \frac{N}{QN!} \nabla \int d^3r_2 \, d^3r_3 \ldots d^3r_N \exp\left\{-\beta\left(\sum_{i\neq 1} U(\mathbf{x} - \mathbf{r}_i)\right.\right.$$

$$\left.\left. + \sum_{i,j\neq 1} U(\mathbf{r}_i - \mathbf{r}_j)\right)\right\}$$

$$= -\frac{\beta N(N-1)}{QN!} \int d^3r_2 \nabla_x U(\mathbf{x} - \mathbf{r}_2) \int d^3r_3 \ldots d^3r_N$$

$$\exp\{-\beta W(\mathbf{x}_1, \mathbf{r}_2, \ldots, \mathbf{r}_N)\}$$

$$= -\beta \int d^3r_2 \, [\nabla_x U(\mathbf{x} - \mathbf{r}_2)] n_2(\mathbf{x}, \mathbf{r}_2) \tag{34}$$

where $W(\mathbf{r}_1, \mathbf{r}_2, \ldots, \mathbf{r}_N) = \sum_{i<j} U(\mathbf{r}_{ij})$.

For a homogeneous system both sides of equation (34) are zero and this derivation serves only to indicate how a coupled set of integrodifferential equations relating the reduced distribution functions may be obtained. Proceeding in a similar fashion, we find

$$\nabla_1 n_2(\mathbf{x}_1 - \mathbf{x}_2)$$

$$= \frac{N(N-1)}{QN!} \nabla_1 \exp\{-\beta U(\mathbf{x}_1 - \mathbf{x}_2)\}$$

$$\int d^3r_3 \ldots d^3r_N \exp\{-\beta W(\mathbf{x}_1, \mathbf{x}_2, \ldots, \mathbf{r}_N)\}$$

$$= -\beta[\nabla_1 U(\mathbf{x}_1 - \mathbf{x}_2)] n_2(\mathbf{x}_1, \mathbf{x}_2) -$$

$$\beta \int d^3r_3 \, [\nabla_1 U(\mathbf{x}_1 - \mathbf{r}_3)] n_3(\mathbf{x}_1, \mathbf{x}_2, \mathbf{r}_3) \tag{35}$$

Converting to the pair and triplet distribution functions

$$n_2(\mathbf{x}_1, \mathbf{x}_2) = \left(\frac{N}{V}\right)^2 g(\mathbf{x}_1, \mathbf{x}_2) \qquad n_3(\mathbf{x}_1, \mathbf{x}_2, \mathbf{x}_3) = \left(\frac{N}{V}\right)^3 g_3(\mathbf{x}_1, \mathbf{x}_2, \mathbf{x}_3)$$

we have

$$-k_B T \nabla_1 g(\mathbf{x}_1, \mathbf{x}_2) = [\nabla_1 U(\mathbf{x}_1 - \mathbf{x}_2)] g(\mathbf{x}_1, \mathbf{x}_2)$$

$$+ n \int d^3x_3 \, [\nabla_1 U(\mathbf{x}_1 - \mathbf{x}_3)] g_3(\mathbf{x}_1, \mathbf{x}_2, \mathbf{x}_3) \qquad (36)$$

Equation (36) is the first of an infinite series of equations known as the BBGKY hierarchy. These equations link low-order distribution functions to functions of successively higher order and may be solved to jth order by approximating g_{j+1} in some fashion. The best known of these approximations is the Kirkwood superposition approximation. In this theory one writes $g_3(\mathbf{x}_1, \mathbf{x}_2, \mathbf{x}_3) = g(\mathbf{x}_1, \mathbf{x}_2) g(\mathbf{x}_1, \mathbf{x}_3) g(\mathbf{x}_2, \mathbf{x}_3)$. This converts equation (36) into a closed nonlinear equation for the pair distribution function which is known as the Born–Green–Yvon (BGY) equation. This equation has been solved for a system of hard spheres and the results are in good agreement with numerical simulations at low density. It is clear from the nature of the decoupling that the superposition approximation can be valid only at low density. This is born out by a calculation of the virial coefficients (Section 4.A) resulting from the BGY equation. The first two terms in the density expansion of $g(r)$ are correct, the higher coefficients are approximate. We shall not discuss the BBGKY hierarchy further. The reader is referred to Barker and Henderson [1976] for a discussion of the merits of this type of approach.

(c) Ornstein–Zernike Equation

An equation that has been much used to develop approximate theories of dense gases and fluids is the Ornstein–Zernike equation. One defines a function, the *direct correlation function*, $C(\mathbf{r}_1, \mathbf{r}_2)$, by demanding that it be a solution of the integral equation

$$h(\mathbf{r}_1, \mathbf{r}_2) = C(\mathbf{r}_1, \mathbf{r}_2) + n \int d^3r_3 \, h(\mathbf{r}_1, \mathbf{r}_3) C(\mathbf{r}_3, \mathbf{r}_2) \qquad (37)$$

where $h(\mathbf{r}_1, \mathbf{r}_2) = g(\mathbf{r}_1, \mathbf{r}_2) - 1$. The origin of the term "direct correlation" function is clear from this equation—in the limit of low density, $C(\mathbf{r}_1, \mathbf{r}_2)$ is precisely the correlation function for particles at positions r_1 and r_2 and is simply the Mayer function $f(|\mathbf{r}_1 - \mathbf{r}_2|)$. The second term on the right-hand side of (37) contains the effect of three or more particles on the function h. Equation (37) obviously cannot be solved since it contains two unknown functions. It can, however, be closed by expressing C in terms of h on the basis of some physically appealing approximation. One of the most useful of these approximations is the Percus–Yevick (PY) approximation, which we now briefly discuss. The pair distribution function is given by

$$g(\mathbf{r}_1, \mathbf{r}_2) = V^2 \frac{\int d^3r_3 \, d^3r_4 \ldots d^3r_4 \exp\{-\beta W(\mathbf{r}_1, \mathbf{r}_2, \ldots, \mathbf{r}_N)\}}{\int d^{3N}r \exp\{-\beta W(\mathbf{r}_1, \mathbf{r}_2, \ldots, \mathbf{r}_N)\}} \qquad (38)$$

This equation can be used to derive a virial expansion for g. The first term in the expansion is simply $\exp\{-\beta U(r_{12})\}$ and one may define

$$y(r) \equiv \exp\{\beta U(\mathbf{r}_1 - \mathbf{r}_2)\}g(\mathbf{r}_1, \mathbf{r}_2) = 1 + \sum_{j \geq 1} y_j(\mathbf{r}_1 - \mathbf{r}_2)n^j \qquad (39)$$

where the y_j's can be expresed in terms of cluster integrals in much the same fashion as we expressed the pressure in Section 4.A. We leave the derivation of $y_1(\mathbf{r}_{12})$ as an exercise and simply quote the result

$$y_1(\mathbf{r}_1 - \mathbf{r}_2) = \int d^3r_3 f(\mathbf{r}_1 - \mathbf{r}_3)f(\mathbf{r}_3 - \mathbf{r}_2) \qquad (40)$$

where $f(r) = \exp\{-\beta U(r)\} - 1$. Noting that

$$h(r) = f(r) + \sum_{j \geq 1} n^j y_j(r)[1 + f(r)]$$

and substituting in (37), we obtain

$$C(r) = \sum_{j=0}^{\infty} n^j C_j(r)$$

with

$$\begin{aligned} C_0(r) &= f(r) \\ C_1(r) &= f(r)y_1(r) \end{aligned} \qquad (41)$$

We now make the approximation, correct to first order in the density, that

$$C(r) = f(r)y(r) = (1 - e^{-\beta U(r)})g(r) \qquad (42)$$

which when substituted in equation (37) provides a nonlinear integral equation known as the Percus–Yevick equation (see, e.g., Balescu [1975] for further details). This equation can be solved analytically in three dimensions for hard spheres (Wertheim, 1963; Thiele, 1963) and by numerical methods for arbitrary interaction potentials. The importance of this equation lies in the fact that it provides an excellent representation of the pair correlation function for hard spheres which also form a simple and remarkably successful approximation for real liquids (Verlet, 1968). In Figure 4.4 we compare the pair distribution functions for hard spheres obtained from the PY equation and from molecular dynamics calculations. The agreement between the two is remarkably good. For more realistic interatomic potentials such as the Lennard-Jones potential, the Percus–Yevick equation is not quite as successful, particularly when it comes to predictions of thermodynamic properties. Nevertheless, the attractive part of the real interatomic potential is, in many approximations, seen to provide only a small perturbation and it is this insight that led to the successful modern perturbation theories discussed in Section 4.D.

There are a number of other approximate closures of the Ornstein–Zernike equation that have been developed and a thorough review of these

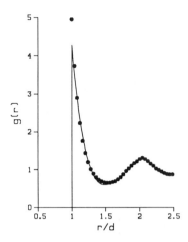

Figure 4.4 Comparison of $g(r)$ from the PY equation with computer simulations for $\pi d^3 n/6 = 0.463$, where n is the density and d the hard-sphere diameter. (Data points from Alder and Hecht [1969].)

theories may be found in Barker and Henderson [1976]. We mention only the mean spherical approximation and the hypernetted chain approximation. The mean spherical approximation consists of the ansatz

$$g(r) = 0 \qquad r < d$$
$$C(r) = -\beta U(r) \qquad r > d \tag{43}$$

where d is the hard-sphere diameter, $C(r)$ the direct correlation function, and $U(r)$ the interaction potential at separations larger than d. This approximation, when substituted into the Ornstein–Zernike equation, provides an exactly solvable equation, both for charged hard spheres (Waisman and Lebowitz, 1970, 1972) and dipolar hard spheres (Wertheim, 1971).

Another frequently used closure of the Ornstein–Zernike equation is the hypernetted chain approximation (HNC). In this scheme one writes

$$C(r) = h(r) - \beta U(r) - \ln [1 + f(r)] \tag{44}$$

When analyzed in terms of a virial expansion, the HNC seems, at first glance, to be a more satisfactory approximation than the PY equation. However, it turns out that for hard spheres and for other short-range potentials, the PY equation gives better results. For systems, such as electrolytes or the classical Coulomb gas, in which the interparticle potential is of long range, the HNC approximation is preferable.

4.C COMPUTER SIMULATIONS

We have seen in previous sections that analytic methods of calculating the configuration integral or the reduced distribution functions have inherent limitations. In the case of the virial expansion the calculation of high-order terms becomes extremely difficult and, in the case of the approximate integral equations discussed in Section 4.B, one is faced with making approximations with undeterminable consequences at some stage of the calculation. The superposition approximation for the three-particle correlation function in the BBGKY scheme and the PY approximation in the Ornstein–Zernike equation are two such approximations. Numerical simulations have provided a valuable check on approximate theories and have also provided the intuition so vital to the construction of better models of the liquid state.

The approach in numerical simulations is to model the behavior of a macroscopic system of 10^{24} particles by a small system of 10^2 to 10^5 particles. At first glance this might seem extremely naive. However, one must remember that a given particle is affected only by other particles within a distance of roughly the correlation length of the system. Except very close to the critical point of the system this correlation length is of the order of a few nearest neighbor separations and one can expect to get sensible results, away from the critical point, even with ensembles of 100 to 200 particles. There are two important numerical techniques which have been widely used in numerical studies of fluids: the Monte Carlo method and the molecular dynamics approach. In each of these schemes sophisticated tricks have been invented to minimize the effects of surfaces and to enhance convergence. We shall not discuss these details and only outline the basic methods below. Excellent reviews of both methods exist in the literature (see, e.g., Binder, 1986 and Hansen and McDonald, 1986) and the interested reader is invited to consult them.

(a) The Monte Carlo Method

The Monte Carlo method is a technique by which one samples the configuration integral. If one can generate states of the finite system with a probability given by the canonical probability density for this system, one may estimate averages of dynamical variables by the prescription

$$\overline{\phi}_M = \frac{1}{M} \sum_{m=1}^{M} \phi(m) \tag{1}$$

where $\phi(m)$ is the value of the dynamical variable ϕ in the mth state generated by the algorithm. Letting M become larger and larger produces better and better estimates of

$$\overline{\phi} = \lim_{M \to \infty} \overline{\phi}_M$$

The technique that allows us to generate states of the system with the correct probability is the method of Markov processes. We discuss this method for a system with a finite number, N, of states—the generalization to a continuous system such as a liquid is immediate. We define a single-step transition probability $p_{i,j}, j = 1, 2, \ldots, N$ which is the probability, per unit "time," that the system in state j will make a transition to state i. Clearly, $\sum_{i=1}^{N} p_{i,j} = 1$. The probability $p_{i,j}$ is fixed (i.e., independent of time and of the previous history of the system). These are the essential ingredients of Markov processes. The system is started in an arbitrary initial state, say state i_0. The probability that the system will be in state m after n Markov steps is given by

$$p_{m,i_0}^{(n)} = \sum_{i_1,i_2,\ldots,i_{n-1}} p_{m,i_{n-1}}\, p_{i_{n-1},i_{n-2}} \cdots p_{i_1,i_0} \tag{2}$$

It can be shown (see, e.g., Feller, 1957) that under certain conditions the n-step transition probability $p_{m,i}^{(n)}$ approaches a limiting distribution which is independent of the initial state i, that is,

$$\lim_{n \to \infty} p_{m,i}^{(n)} = \Pi_m \tag{3}$$

The most important of the conditions alluded to above is that no subset of the N states must be able to act as a trap. It must be possible to reach each state of the system with a finite probability from any initial state in a finite number of steps.

In the case of a statistical ensemble, we know the limiting distribution:

$$\Pi_m = \frac{e^{-BE(m)}}{\sum_i e^{-\beta E(i)}} \tag{4}$$

and therefore need only find transition probabilities $p_{i,k}$ which will ensure that the long time distribution of states obeys (4). Notice that since

$$p_{m,1}^{(n+1)} = \sum_j p_{m,j}\, p_{j,1}^{(n)} \tag{5}$$

we have, letting $n \to \infty$,

$$\Pi_m = \sum_j p_{m,j}\, \Pi_j \tag{6}$$

We are free to choose a form for the transition probabilities as long as the choice leads to the correct limiting distribution (4), and if we choose $p_{j,m}\,\Pi_m = p_{m,j}\,\Pi_j$ for all m and j, we have

$$\Pi_m = \sum_j p_{j,m}\,\Pi_m = \sum_j p_{m,j}\,\Pi_j \tag{7}$$

and we have satisfied equation (6). A simple choice for the transition probabilities is thus

$$p_{ji} = \frac{1}{N} \qquad\qquad \text{if } \Pi_j > \Pi_i \text{ (i.e., } E_i > E_j)$$

$$p_{ji} = \frac{\Pi_j}{N\,\Pi_i} = \frac{1}{N}\, e^{-\beta(E_j - E_i)} \qquad \text{if } \Pi_j < \Pi_i \text{ (i.e., } E_j > E_i) \tag{8}$$

$$p_{ii} = 1 - \sum_{j \neq i} p_{ji}$$

In practice, one starts the system in a given configuration, say i. A new configuration, j, is selected from the N available configurations—the factor

$1/N$ in (8)—and the new configuration is accepted with probability 1 if $E_j < E_i$ and accepted with probability $\exp\{-\beta(E_j - E_i)\}$ if $E_j > E_i$. This process is repeated until the available computer time is exhausted or the desired expectation values have converged. After an initial transient—due to the choice of i as the first state—the subsequent states will occur with probability Π_m and the averaging process (1) may then be carried out over the remaining states for all thermodynamic functions of interest. The convergence of the mean values $\bar{\phi}_M$ to their limiting values is unfortunately rather slow. At best,

$$|\bar{\phi}_M - \bar{\phi}| \sim \frac{1}{\sqrt{M}} \tag{9}$$

Thus, to halve the expected error, one must generate *four* times as many configurations.

The implementation of the Monte Carlo procedure in the case of liquids is quite straightforward. A system of N particles is confined to a box of volume V and each particle is initially put somewhere in the box, perhaps at a randomly chosen position. Each particle is then, in turn, moved by a random amount in a random direction and the change in total energy, ΔE, is calculated. The new position for the particle which has been moved is accepted or rejected according to rule (8). Usually, periodic boundary conditions are used and the potential energy of a particle at position r_i is given by the sum of the pair interaction energies with all the particles in the box, as well as with those in periodic replications of the box.

The procedure described above is not the only possible implementation of the Monte Carlo method. For example, it is possible to carry out a Monte Carlo sampling of an ensemble at constant pressure rather than at constant density. This alternative, and many others, are described in Binder [1986].

With modern computers one can treat very large systems by the Monte Carlo method. One of the advantages of the Monte Carlo method, in the case of liquids, lies in the fact that any potential energy function may be used. Many body and long-range interactions present only technical difficulties rather than problems of principle. This method is now a well-established tool in statistical physics for dealing with systems for which exact analytic calculations are at present impossible.

(b) Molecular Dynamics

The molecular dynamics method is based on the microcanonical ensemble rather than on the canonical ensemble. The ergodicity assumption, discussed briefly in Chapter 2, is crucial for this approach. As in the Monte Carlo method, one considers a finite system of N particles, usually subject to periodic boundary conditions. Initially, the positions and velocities of all the particles are specified. Newton's equations of motion for all the particles are then integrated forward in time, and after the disappearance of an initial transient, the thermodynamic quantities of interest are obtained as time averages over the subsequent configurations. In this process the total energy of the N particles is conserved but kinetic energy can be converted into potential energy, and vice versa. The transient alluded to above corresponds to a choice of initial configuration which is perhaps typical of only a small part of the constant energy surface. The temperature of the system of particles is given by

$$k_B T = \frac{1}{3N} \sum_i m_i \langle v_i^2 \rangle \qquad (10)$$

where

$$\langle v_i^2 \rangle = \frac{1}{L} \sum_{n=0}^{L-1} v_i^2(n) \qquad (11)$$

In (11) L is the length of the molecular dynamics run after the passage of the initial transient and $v_i(n)$ is the velocity of particle i at step n of the averaging procedure. Quantities such as the potential energy, specific heat, and pressure may also be calculated in terms of such time averages. The specific heat per particle is given by

$$C_V = \frac{k_B T^2}{\langle (\Delta T)^2 \rangle} \qquad (12)$$

where $\langle (\Delta T)^2 \rangle$ is the mean-square fluctuation in the temperature as defined through the kinetic energy. The pressure can be obtained, for example, by use of the equation of state (4.B.33)

$$P = \frac{N}{V} k_B T + \frac{1}{6V} \sum_{i \neq j} \langle (\mathbf{r}_i - \mathbf{r}_j) \cdot \mathbf{F}_{ij} \rangle$$

where $\mathbf{F}_{ij}$ is the instantaneous force exerted by particle i on particle j.

One of the great advantages of molecular dynamics is that time-dependent correlation functions (including the dynamic structure factor, which is measured in inelastic neutron scattering experiments) are a by-product of the calculation. Many important calculations for both hard spheres and continuous potentials have been carried out. With modern computers and efficient programming the equations of motion for as many as 10^5 particles (Abraham et al., 1984) can be solved over sufficiently long times to obtain both the static and dynamic properties of the system. We shall not discuss here the various algorithms which have been devised for solving Newton's equations—once again the reader is referred to Barker and Henderson [1976] and Hansen and McDonald [1986].

4.D PERTURBATION THEORY

In this section we outline the ideas on which the modern perturbation theories of liquids are based and display some of the results obtained by such methods. The basic physical idea is that in systems of atoms or molecules interacting through a potential like the 6–12 potential, the short-range repulsive piece of the interaction is responsible for most of the structure seen in the pair correlation function. One should therefore be able to use the hard-sphere system as a reference system or unperturbed system and to treat the corrections due to the attractive part of the potential perturbatively. A 6–12 potential has a soft core rather than a hard core, but we shall see that this does not present any essential difficulties. We decompose the pair potential into two pieces:

$$U(r_{ij}, \lambda) = U_0(r_{ij}) + \lambda U_1(r_{ij}) \tag{1}$$

where

$$\begin{aligned} U_0(r_{ij}) &= 0 \qquad \text{for } r_{ij} > \sigma \\ U_1(r_{ij}) &= 0 \qquad \text{for } r_{ij} < \sigma \end{aligned} \tag{2}$$

In equation (1) the case $\lambda = 1$ corresponds to the original potential. The logarithm of the configuration integral is given by

$$\ln Q_N(V, T, \lambda) = \ln \frac{1}{N!} \int d^{3N}r \, \exp \left\{ -\beta \sum_{i<j} [U_0(r_{ij}) + \lambda U_1(r_{ij})] \right\} \tag{3}$$

We now expand this function in powers of λ:

$$\ln Q_N(V, T, \lambda) = \ln Q_N(V, T, 0) + \lambda \frac{\partial}{\partial \lambda} \ln Q_N(V, T, \lambda)\big|_{\lambda=0} + O(\lambda^2) \tag{4}$$

with

$$\begin{aligned} \frac{\partial}{\partial \lambda} \ln Q_N(V, T, \lambda)\bigg|_{\lambda=0} &= -\beta \frac{N(N-1)}{2} \int d^{3N}r \, U_1(r_{12}) \frac{\exp\{-\beta \sum_{i<j} U_0(r_{ij})\}}{N! \, Q_N(V, T, 0)} \\ &= -\frac{\beta}{2} \int d^3r_1 \, d^3r_2 \, U_1(r_{12}) n_2^{(0)}(\mathbf{r}_1, \mathbf{r}_2) \\ &= -\frac{\beta}{2} \left(\frac{N}{V}\right)^2 \int d^3r_1 \, d^3r_2 \, U_1(r_{12}) g^{(0)}(\mathbf{r}_1, \mathbf{r}_2) \end{aligned}$$

where $g^{(0)}$ is the pair distribution function of the system for $\lambda = 0$ (i.e., for the reference system). Therefore, we obtain, for the Helmholtz free energy,

$$A(V, T) = A_0(V, T) - \frac{\beta N}{2} \int d^3r \, U_1(r) g^{(0)}(r) + \cdots$$

The higher-order terms can be similarly expressed as integrals over three- and higher-particle correlation functions of the *reference* system. Ideally, one would like to take the hard-sphere system as the reference system, partly because its properties are well known and partly because it would provide a common starting point for a number of different systems whose potentials are of the same general form but which may not have exactly the same reference potential $U_0(r)$. Barker and Henderson [1967] have provided a method of achieving this goal. We shall not repeat their argument. The result is that one can indeed replace $g^{(0)}(r)$ by a hard-sphere correlation function provided that the hard-sphere diameter, d, is taken to be temperature dependent. Specifically,

$$d = \int_0^{\sigma} dr[1 - \exp\{-\beta U_0(r)\}] \tag{5}$$

With this choice of effective hard-sphere diameter very good agreement with computer simulations and experiment is obtained. This is illustrated in Figure 4.5, where the pair distribution function $g(r)$ obtained from zeroth- and first-order perturbation theory is compared with the results of computer simulations. The agreement in first order perturbation theory is extremely good, indicating that the ideas behind this approach are indeed correct.

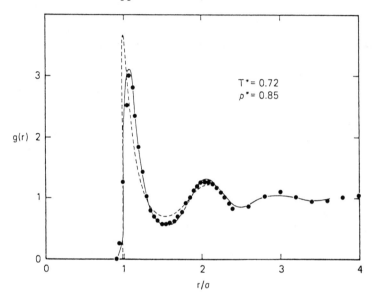

Figure 4.5 Pair distribution function of a Lennard-Jones fluid for $n = 0.85\sigma^{-3}$ and $T = 0.72\epsilon/k_{\text{B}}$ (near the triple point). Points are the results of simulations of Verlet [1968]. The dashed and solid curves correspond to zeroth- and first-order Barker–Henderson perturbation theory. (From Barker and Henderson [1976].)

In this short discussion we have only mentioned one of the perturbation theories which are useful for real liquids. Several other versions have been developed and the reader is referred to Barker and Henderson [1976] and the original articles cited therein.

4.E INHOMOGENEOUS LIQUIDS

In this section we discuss some of the properties of inhomogeneous liquids. Section 4.E(a) is devoted to the van der Waals theory of the liquid vapor interface and of the surface tension and in Section 4.E(b) we construct a simple theory of the normal modes—the capillary waves—of a free liquid surface.

(a) Liquid–Vapor Interface

In most elementary physics courses the concept of surface tension of liquids is introduced. In the present subsection we wish to relate this quantity to the statistical mechanics of the two distinct coexisting phases separated by the surface, namely the bulk fluid and the vapor. In Chapter 1 we introduced the surface tension thermodynamically by including a term $\sigma\, d\mathcal{A}$ in the expression for the work done by a system in a general change of state. Before beginning our mean field treatment of the interface we expand on the thermodynamic treatment and define more carefully the appropriate interfacial parameters. In this discussion we follow closely the treatment of Rowlinson and Widom [1982].

Consider a single-component system in a volume V at temperature T. Suppose that two phases, liquid and vapor, coexist and let the volume occupied by these phases be V_L and V_G with $V_L + V_G = V$. Let the molecular density well inside the individual phases be n_L and n_G and $n(\mathbf{r})$ be the density at point $\mathbf{r}$. We will assume that the interface between the two phases is planar and perpendicular to the z direction. The assignment of volumes V_L and V_G to the two phases is in a sense arbitrary. The density will vary between the liquid and gas densities over some distance, d, and the boundary between liquid and gas is somewhat ambiguous. We shall see that there is a convenient choice of the "dividing surface" which makes the subsequent calculations easier.

We now define the surface energy, the number of particles in the surface region, and the surface Helmholtz free energy through the equations

$$V_L n_L + V_G n_G + N_S = N$$
$$V_L e_L + V_G e_G + E_S = E \qquad (1)$$
$$V_L a_L + V_G a_G + A_S = A$$

where the quantities N, E, A are the total particle number, energy, and Helmholtz free energy of the system and the lowercase symbols refer to the corresponding bulk densities. If the boundary between liquid and vapor were mathematically sharp, and coincided with our choice of dividing surface, the number of particles N_S in the surface would be zero.

We now consider the temperature of the system to be fixed. A slight generalization of the derivation of the Gibbs–Duhem equation (1.D.3) yields

$$A = -PV + \sigma\mathcal{A} + \mu N \qquad (2)$$

where $\mathcal{A}$ is the area of the interface. Similarly,

$$A_L = -PV_L + \mu N_L$$
$$A_G = -PV_G + \mu N_G \qquad (3)$$

and from (1),

$$A_S = \mathcal{A}a_S = \sigma\mathcal{A} + \mu(N - N_L - N_G)$$
$$= \sigma\mathcal{A} + \mu N_S \tag{4}$$

From equation (4) we see that if we choose the location of the Gibbs dividing surface (at $z = 0$) to satisfy the equation

$$N_S = \mathcal{A}\int_{-\infty}^{0} dz\,[n(z) - n_L] + \mathcal{A}\int_{0}^{\infty} dz\,[n(z) - n_G] = 0 \tag{5}$$

the surface tension will be related to the excess Helmholtz free energy per unit area through

$$\sigma = a_S \tag{6}$$

We now construct an approximate theory of the interfacial region following the ideas of van der Waals [1893] and Cahn and Hilliard [1958]. We are considering a system for which the total volume (gas + liquid) is held fixed at constant temperature. Thus the appropriate free energy to minimize with respect to any variational parameters is the Helmholtz free energy. We suppose that there exists a *local* Helmholtz free energy per unit volume, $\Psi(\mathbf{r})$, which, in our geometry, can only be a function of z, $\Psi(z)$. The surface tension (6) is related to this function through

$$\sigma = \int_{-\infty}^{0} dz\,[\Psi(z) - \Psi_L] + \int_{0}^{\infty} dz\,[\Psi(z) - \Psi_G] \tag{7}$$

where $\Psi_L = A_L/V_L$, $\Psi_G = A_G/V_G$. This free-energy density is, at fixed temperature, a functional of the local density, that is, $\Psi(z) = \Psi[n(z)]$. We will assume a phenomenological free-energy density which is composed of two parts, a term $\Gamma(dn/dz)^2/2$, where Γ is a positive constant, and a term that results from analytic continuation of a mean field free energy into the unphysical density region. In Section 3.L this unphysical region was avoided by means of the Maxwell construction. We assume that the Helmholtz free-energy density resembles the curve shown in Figure 4.6 for $T < T_c$. The straight-line section connecting the bulk densities is the result of the Maxwell construction (or

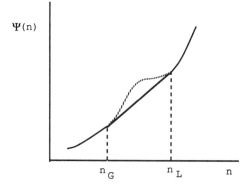

Figure 4.6 Phenomenological free-energy density.

equivalently the double tangent construction of Section 3.C). The dotted curve represents a mean field free-energy density resulting, for example, from a van der Waals equation of state (Section 3.L). Therefore, we have

$$\sigma = \int_{-\infty}^{\infty} dz \left\{ \frac{\Gamma}{2} \left(\frac{dn}{dz} \right)^2 + [\Psi_n([n(z)]) - \Psi_e(z)] \right\} \tag{8}$$

where Ψ_n is the nonequilibrium free-energy density (dotted curve in figure 4.6) and Ψ_e is the equilibrium bulk liquid or gas free-energy density depending on whether z is greater or less than zero. The density profile wil be determined by minizing (8) with respect to $n(z)$. The first term in (8) is a typical Landau–Ginzburg term which reduces the fluctuations in the free energy.

To make further progress, we must find an approximate expression for the term $\Delta\Psi = [\Psi_n([n(z)]) - \Psi_e(z)]$ in the transition region. We define $n_0 = (n_L + n_G)/2$, $\lambda = (n_L - n_G)/2$ and assume that $\Delta\Psi$ has an expansion of the form

$$\Delta\Psi = \sum_j \alpha_j [n(z) - n_0]^j = \sum_j \alpha_j \zeta^j(z) \lambda^j \tag{9}$$

where the reduced variable

$$\zeta(z) = \frac{2[n(z) - n_0]}{n_L - n_G}$$

takes on the value $\zeta = -1$ when $n(z) = n_G$ and $\zeta = 1$ when $n(z) = n_L$. We must require that $\Delta\Psi([n_G]) = \Delta\Psi([n_L]) = 0$. Furthermore, since Ψ is the free-energy density, $d\Psi(n, T)/dn = \mu$, where μ is the chemical potential, which must be the same in the two bulk phases. Also, since by the tangent construction in Figure 4.6, Ψ must match smoothly to the bulk solution, we have

$$\frac{d(\Delta\Psi)}{d\zeta} \bigg|_{-1} = \frac{d(\Delta\Psi)}{d\zeta} \bigg|_{+1} = 0 \tag{10}$$

If we truncate (9) at the fourth-order term, we find

$$\Delta\Psi = \alpha_4 \lambda^4 [1 - \zeta^2(z)]^2 \tag{11}$$

where the coefficient α_4 is undetermined but positive. Minimizing (8) with respect to $n(z)$ or, equivalently, $\zeta(z)$, we find

$$\Gamma \frac{d^2\zeta}{dz^2} + 4\alpha_4 \lambda^2 \zeta(1 - \zeta^2) = 0 \tag{12}$$

A first integral of this differential equation is easily obtained by substituting $\phi = d\zeta/dz$ and noting that $d\phi/dz = \phi \, d\phi/d\zeta$. Thus we find

$$\frac{d\phi^2}{d\zeta} = -\frac{8\alpha_4 \lambda^2}{\Gamma} \zeta(1 - \zeta^2) \tag{13}$$

and integrating from -1 to ζ yields

$$\phi^2(\zeta) - \phi^2(-1) = -\frac{\alpha_4 \lambda^2}{\Gamma}(4\zeta^2 - 2\zeta^4 - 2)$$

Since we expect

$$\phi(-1) = \frac{d\zeta}{dz}\bigg|_{z=\infty} = 0$$

we have

$$\frac{d\zeta}{dz} = -\sqrt{\frac{2\alpha_4 \lambda^2}{\Gamma}}(1 - \zeta^2)$$

where the minus sign on the right has been chosen in order to have liquid at $z = -\infty$. Integrating, we find that

$$\zeta(z) = -\tanh\left(\sqrt{\frac{2\alpha_4 \lambda^2}{\Gamma}}z\right) \tag{14}$$

and the density profile in the liquid–vapor interface is

$$n(z) = \frac{n_L + n_G}{2} - \frac{n_L - n_G}{2}\tanh\left(\sqrt{\frac{2\alpha_4 \lambda^2}{\Gamma}}z\right) \tag{15}$$

The quantity $(\Gamma/2\alpha\lambda^2)^{1/2}$ also appears in the theory of superconductivity, where it is commonly referred to as the Ginzburg–Landau coherence length. In Chapter 7 we use an argument, very similar in spirit to the derivation above, to discuss superconductivity. In the theory of the liquid–vapor interface above, it plays the role of an interfacial width and we see from (15) that if the coefficient Γ in (8) were taken to be zero, the profile would be infinitely sharp. We may now substitute (15) into (8) to find the surface tension:

$$\sigma = 2\alpha_4 \lambda^4 \int_{-\infty}^{\infty} dz \, \text{sech}^4 \sqrt{\frac{2\alpha_4 \lambda^2}{\Gamma}}z = \sqrt{2\alpha_4 \Gamma}\lambda^3 \int_{-\infty}^{\infty} dy \, \text{sech}^4 y \tag{16}$$

As the system approaches the liquid–vapor critical point

$$\lambda = \frac{n_L - n_G}{2} \sim (T_c - T)^{1/2}$$

and we see in (16) the classical scaling form of the surface tension

$$\sigma \sim (T_c - T)^\mu$$

with $\mu = \frac{3}{2}$.

A generalization of the theory above for the liquid–vapor interface which incorporates the scaling theory of critical phenomena (Chapter 5) has been provided by Fisk and Widom [1968] but is beyond the scope of this book. Our treatment has been very much phenomenological, containing two parameters, α_4 and Γ, which at this level we cannot determine microscopically as well as some possibly incorrect assumptions about the form of the free energy. Other more sophisticated mean field theories based on the inhomogeneous Ornstein–Zernike equation are parameter free in the sense that the interparticle potential, temperature, and bulk density completely determine the density profile (Plischke et al., 1985). These theories, as well as density functional theories, are capable of producing surface tensions to within roughly 10% of the experimental values except very close to the criticial point where mean field theories fail in general.

The liquid–vapor interface has also been extensively studied by computer simulations (Chapela et al., 1977, and references therein). For some time there was considerable controversy concerning the possible existence of small-amplitude oscillations superimposed on a smooth density variation such as that given by (15). For molecular liquids, such as condensed rare gases, it is now generally accepted that there are no oscillations in the density profile and that the general form (15), with the phenomenological constant $(\Gamma/2\alpha_4\lambda^2)^{1/2}$ replaced by a function $d(T)$ of the temperature, provides a very good fit to the numerical results.

Experimental measurements of the density profile can be carried out in a number of ways. Beaglehole [1979] measured the thickness of the interface in liquid argon at 90 and 120 K by ellipsometry. At 90 K he found a thickness of roughly 8 Å (defined as the distance over which the density goes from 10% to 90% of its final value). Other techniques, such as the use of synchrotron radiation incident at glancing angles (Als-Nielsen, 1985), hold out the promise that much more of the detailed structure of the interfacial region will become known in the near future.

An essential assumption in our derivation of the density $n(z)$ was that the interface was planar. If we think of the interface as a stretched membrane, we see that there will be thermally excited normal modes of vibration which will distort the interface and broaden it further. A simple treatment of these "capillary waves" forms the subject matter of the next subsection.

(b) Capillary Waves

Let us imagine that the position of the interface between the liquid and vapor phases can be specified by a function, $z(x, y)$, which may refer either to the location of an infinitely sharp dividing surface or to the center of a more diffuse profile such as (15). If the energy of a flat surface at $z = 0$ is taken to be zero, the added energy due to distortion and displacement is given by

$$\Delta E = \int dx\, dy \left[\frac{n_L - n_G}{2} gz^2(x, y) + \sigma \left\{ \sqrt{1 + \left(\frac{\partial z}{\partial x}\right)^2 + \left(\frac{\partial z}{\partial y}\right)^2} - 1 \right\} \right]$$

where σ is the surface tension and g the acceleration due to gravity. The derivation of the first term, the change in gravitational potential energy due to a displacement of the dividing surface, is left as an exercise. The second term represents the change in area of the interface due to distortion. Assuming that the distortion is small (i.e., $|\partial z/\partial x|$, $|\partial z/\partial y| \ll 1$), we may expand the second term to obtain

$$\Delta E = \int dx\, dy \left[\frac{n_L - n_G}{2} gz^2(x, y) + \frac{\sigma}{2}\left\{\left(\frac{\partial z}{\partial x}\right)^2 + \left(\frac{\partial z}{\partial y}\right)^2\right\}\right] \quad (17)$$

If we consider a liquid in a cubical container of length L on each side and apply periodic boundary conditions in the x and y directions,

$$z(x + L, y) = z(x, y + L) = z(x, y)$$

we may express the energy (17) in terms of the energies of a set of normal modes. We let

$$z(x, y) = \frac{1}{L}\sum_q \hat{z}_q e^{i\mathbf{q}\cdot\mathbf{r}} \quad (18)$$

with $\mathbf{q} = 2\pi/L\,(n_x, n_y)$ and $n_x, n_y = 0, \pm 1, \pm 2, \ldots$. Substituting, we obtain

$$\Delta E = \frac{\sigma}{2}\sum_q \left[\frac{(n_L - n_G)g}{\sigma} + q^2\right]\hat{z}_q\hat{z}_{-q} \quad (19)$$

The quantity $a^2 = 2\sigma/\{(n_L - n_G)g\}$ has the dimension of (length)^2 and its square root is given the name "capillary length." Modes with a wavelength much larger than the capillary length are called gravity waves, while modes with wavelength less than $2\pi a$ are known as capillary waves.

Decomposing $\hat{z}_q$ into its real and imaginary parts and using the equipartition theorem of classicial statistics we find, for the thermal expectation value $\langle \hat{z}_q\hat{z}_{-q}\rangle$,

$$\langle \hat{z}_q\hat{z}_{-q}\rangle = \frac{2k_B T/\sigma}{2a^{-2} + q^2} \quad (20)$$

The broadening of the interface due to thermal excitation of these modes is conveniently expressed in terms of the quantity

$$\xi^2 = \frac{1}{L^2}\int dx\, dy\, \langle [z(x, y) - \bar{z}]^2\rangle \quad (21)$$

where

$$\bar{z} = \frac{1}{L^2}\int dx\, dy\, z(x, y)$$

Substituting (18) into (21), we see that

$$\xi^2 = \frac{1}{L^2}\sum_{|q|>0}\langle \hat{z}_q\hat{z}_{-q}\rangle = \frac{2k_B T}{\sigma \mathscr{A}}\sum_{|q|>0}\frac{1}{q^2 + 2a^{-2}} \quad (22)$$

Converting the sum over q to an integral ($\Sigma_q = \mathscr{A}/(2\pi)^2 \int d^2q$) and carrying out the integration in cylindrical coordinates, we obtain

$$\xi^2 = \frac{k_B T}{\pi\sigma} \int_{2\pi/L}^{q_{max}} dq \frac{q}{q^2 + 2a^{-2}} = \frac{k_B T}{2\pi\sigma} \ln \frac{q_{max}^2 a^2/2 + 1}{2\pi^2 a^2/L^2 + 1} \tag{23}$$

The cutoff at q_{max} has been introduced because wave vectors greater than $2\pi/a_0$, where a_0 is a molecular diameter, are essentially meaningless, as the interface is not a true continuum. Taking $a_0 = 3.4$ Å (appropriate for argon) and the surface tension to be the experimental value for argon at its triple point $\sigma = 15.1$ d/cm, we find a capillary length of 1.5 mm and from equation (23) $\xi \approx 6.4$ Å for $L \gg a$. The dependence of ξ on L is quite weak due to the slow variation of the logarithm.

The divergence of the interfacial width in zero gravity due to the long-wavelength modes is found in other physical situations as well. Simple models of crystal growth, such as the solid-on-solid model (Weeks and Gilmer, 1979), display roughening of the surface due to excitation of the analogous modes. Similarly, domain walls in the three-dimensional Ising model are rough in the same way.

PROBLEMS

4.1. *Tonks Gas.* Consider a one-dimensional gas of particles of length a confined to a strip of length L. The particles interact through the potential

$$U(x_i - x_j) = \infty \text{ for } |x_i - x_j| < a$$
$$= 0 \text{ for } |x_i - x_j| > a$$

    (a) Calculate the partition function and equation of state exactly.
    (b) Evaluate the virial coefficients B_2 and B_3 and show that your results are consistent with part (a).

4.2. *Decomposition of Reducible Graphs.* Consider an arbitrary graph g with $n = n_1 + n_2 - 1$ vertices which can be cut at a point into two disjoint pieces g_1 and g_2 with n_1 and n_2 vertices, respectively. Find an expression for the cluster integral $b(g)$ in terms of the cluster integrals $b(g_1)$ and $b(g_2)$.

4.3. *Virial Expansion for the Pair Distribution Function.* Generalize the method of Section 4.A to the pair distribution function. In particular, find an expression for the function $y_1(r_{12})$ of equation (4.B.39).

4.4. *Inversion Temperature of Argon.* In the Joule–Thompson process a gas is forced from a high-pressure chamber through a porous plug into a lower-pressure chamber. The process takes place at constant enthalpy and the change in temperature of the gas, for a small pressure difference between the two chambers, is given by $dT = \mu_J \, dP$, where μ_J is the Joule–Thompson coefficient:

$$\mu_J = \left(\frac{\partial T}{\partial P}\right)_{H,N} = \frac{1}{C_P}\left[T\left(\frac{\partial V}{\partial T}\right)_{P,N} - V\right]$$

The locus $\mu_J = 0$ is called the Joule–Thompson inversion curve.
    (a) Use the first two terms of the virial equation of state

$$P = \frac{Nk_{\mathrm{B}}T}{V}\left(1 + B_2(T)\frac{N}{V}\right)$$

to obtain an equation for the inversion curve in the P–T plane.

(b) Calculate, numerically, the coefficient $B_2(T)$ for argon assuming a 6–12 potential with $\epsilon/k_{\mathrm{B}} = 120$ K and $\sigma = 3.4$ Å. Experimentally, the inversion curve has a maximum at 780 K. Compare your results with this value.

4.5 *Monte Carlo Simulation of the One-Dimensional Ising Model.* Carry out a Monte Carlo simulation for a chain of 100 Ising spins with periodic boundary conditions at temperature $k_{\mathrm{B}}T/J = 1$. This should be possible on a personal computer.

(a) Calculate the specific heat in zero magnetic field by simulation from: (1) the fluctuation in the internal energy $\langle E^2 \rangle - \langle E \rangle^2$, (2) a numerical derivative of the internal energy $\langle E \rangle$. Compare with the exact result for an infinite chain (Section 3.F)

(b) Carry out the analogous calculation for the zero field susceptibility.

5

Critical Phenomena, Part I

In this chapter and in Chapter 6, we discuss continuous phase transitions in some detail. We have already developed, in Chapter 3, several approximate methods for dealing with strongly interacting, or highly correlated systems, namely the mean field approach and its extensions as well as the Landau theory of phase transitions. In that chapter we also demonstrated the inherent limitation of mean field theories, namely the fact that the correlation function becomes very long ranged near a critical point. For this reason, theories based on an exact treatment of small clusters cannot be expected to produce the correct critical behavior of a system.

Our approach in these two chapters will commence by following the historical development of the field. In the present chapter, we begin in Section 5.A with a derivation of the Onsager solution of the two-dimensional Ising model. We then discuss, in Section 5.B, the series expansion methods developed primarily during the 1950s and 1960s to determine the critical properties of model systems. Section 5.C contains a discussion of the scaling theory of Widom [1965], Domb and Hunter [1965], and Kadanoff et al. [1967]. In Section 5.D we concern ourselves with the universality hypothesis and examine the theoretical and experimental evidence in its favor. Finally, in Section 5.E, we present a short and qualitative discussion of the Kosterlitz–Thouless mechanism for phase transitions in two-dimensional systems with continuous symmetry. At that point we shall be in a position to undertake a study of the renormalization group approach to critical phenomena developed by Wilson and others in the 1970s. This theory, which has provided a firm theoretical basis for both scaling and universality, forms the subject matter of Chapter 6.

5.A ISING MODEL IN TWO DIMENSIONS

In a landmark paper, Onsager [1944] exactly calculated the free energy of the two-dimensional ferromagnetic Ising model in zero magnetic field on the rectangular lattice. This calculation provided the first exact solution of a model that displays a phase transition. Onsager's original derivation is mathematically complex. Since his original paper, a number of more transparent solutions of the problem have appeared. Below, we present a brief account of one of these, namely that of Schultz et al. [1964]. Our motivation for including this calculation is twofold. Most of this book is concerned with approximation techniques and we feel that it is worthwhile to exhibit at least one example of a nontrivial exact calculation in statistical physics as a counterpoint to the technically straightforward examples of mean field theory and renormalization group calculations. Second, we frequently quote the exact results of Onsager for the specific heat and order parameter and feel that some readers may not feel comfortable with these results without the evidence of a derivation. Those readers not interested in the technical details may skip ahead to Section 5.A(d).

(a) The Transfer Matrix

We have already solved the one-dimensional Ising model in Section 3.F by use of the transfer matrix approach and will also apply this method in two dimensions. We first reformulate the one-dimensional problem in a slightly different way. Consider, again, the Hamiltonian

$$H = -J \sum_{i=1}^{N} \sigma_i \sigma_{i+1} - h \sum_{i=1}^{N} \sigma_i \tag{1}$$

The partition function is

$$Z = \sum_{\{\sigma\}} (e^{\beta h \sigma_1} e^{K \sigma_1 \sigma_2})(e^{\beta h \sigma_2} e^{K \sigma_2 \sigma_3}) \cdots (e^{\beta h \sigma_N} e^{K \sigma_N \sigma_1}) \tag{2}$$

where we have grouped the factors somewhat differently from (3.F.6) and where $K = \beta J$.

We now introduce two orthonormal basis states $|+1\rangle$ and $|-1\rangle$ and Pauli operators, which in this basis have the representation

$$\sigma_z = \begin{bmatrix} 1 & 0 \\ 0 & -1 \end{bmatrix} \qquad \sigma^+ = \begin{bmatrix} 0 & 1 \\ 0 & 0 \end{bmatrix} \qquad \sigma^- = \begin{bmatrix} 0 & 0 \\ 1 & 0 \end{bmatrix} \tag{3}$$

with $\sigma_x = \sigma^+ + \sigma^-$ and $\sigma_y = -i(\sigma^+ - \sigma^-)$. It is now easy to see that the Boltzmann weight $\exp\{\beta h \sigma_i\}$ can be expressed as a diagonal matrix, $\mathbf{V}_1$, in this basis:

$$\langle +1 | \mathbf{V}_1 | + 1\rangle = e^{\beta h}, \langle -1 | \mathbf{V}_1 | - 1\rangle = e^{-\beta h}$$

or

$$\mathbf{V}_1 = \exp \{\beta h \sigma_Z\} \tag{4}$$

Similarly, we define the operator $\mathbf{V}_2$ corresponding to the nearest-neighbor coupling by its matrix elements in this basis:

$$\langle +1 | \mathbf{V}_2 | + 1 \rangle = \langle -1 | \mathbf{V}_2 | - 1 \rangle = e^K$$

$$\langle +1 | \mathbf{V}_2 | - 1 \rangle = \langle -1 | \mathbf{V}_2 | + 1 \rangle = e^{-K}$$

Therefore,

$$\mathbf{V}_2 = e^K \mathbf{1} + e^{-K} \sigma_X = A(K) \exp \{K^* \sigma_X\} \tag{5}$$

where in the second step we have used the fact that $(\sigma_X)^{2n} = \mathbf{1}$. The constants $A(K)$ and K^* are determined from the equations

$$A \cosh K^* = e^K$$
$$A \sinh K^* = e^{-K} \tag{6}$$

or $\tanh K^* = \exp \{-2K\}$, $A = (2 \sinh 2K)^{1/2}$. Using these results, we write the partition function as follows:

$$Z = \sum_{\{\mu = +1, -1\}} \langle \mu_1 | \mathbf{V}_1 | \mu_2 \rangle \langle \mu_2 | \mathbf{V}_2 | \mu_3 \rangle \langle \mu_3 | \mathbf{V}_1 | \mu_4 \rangle \cdots \langle \mu_{2N} | \mathbf{V}_2 | \mu_1 \rangle$$

$$= \mathrm{Tr} \, (\mathbf{V}_1 \mathbf{V}_2)^N = \mathrm{Tr} \, (\mathbf{V}_2^{1/2} \mathbf{V}_1 \mathbf{V}_2^{1/2})^N = \lambda_1^N + \lambda_2^N \tag{7}$$

where λ_1 and λ_2 are the two eigenvalues of the Hermitian operator

$$\mathbf{V} = (\mathbf{V}_2^{1/2} \mathbf{V}_1 \mathbf{V}_2^{1/2}) = \sqrt{2 \sinh 2K} \; e^{K^* \sigma_X/2} e^{\beta h \sigma_Z} e^{K^* \sigma_X/2} \tag{8}$$

In arriving at this symmetric form of the transfer matrix $\mathbf{V}$ we have used the invariance of the trace of a product of matrices under a cyclic permutation of the factors. Clearly, in the case $h = 0$, the two eigenvalues are given by $\lambda_1 = A \exp \{K^*\}$, $\lambda_2 = A \exp \{-K^*\}$ and we recover our previous result (3.F.12).

We note, in passing, that in this procedure a one-dimensional problem in classical statistics has been transformed into a zero-dimensional (only one "site") quantum-mechanical ground-state problem (largest eigenvalue). This result is quite general. There exists a correspondence between the ground state of quantum Hamiltonians in $d - 1$ dimensions and classical partition functions in d dimensions which can sometimes be exploited, for example in numerical simulations of quantum-statistical models (Suzuki, 1976, 1985).

We now generalize this procedure to the two-dimensional Ising model and consider an $M \times M$ square lattice with periodic boundary conditions (see Figure 5.1) and the Hamiltonian

$$H = -J \sum_{r,c} \sigma_{r,c} \sigma_{r+1,c} - J \sum_{r,c} \sigma_{r,c} \sigma_{r,c+1} \tag{9}$$

where the label r refers to rows, c to columns, and $\sigma_{r+M,c} = \sigma_{r,c+M} = \sigma_{r,c}$. The

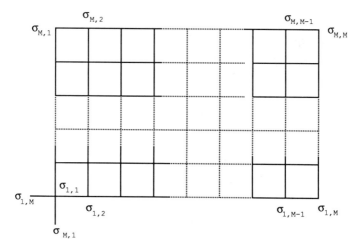

Figure 5.1 $M \times M$ square lattice with periodic boundary conditions.

first term in (9) contains only interactions in column c and is, in this sense, analogous to the magnetic field term in (1). The second term in (9) is the coupling between neighboring columns and will lead to a nondiagonal factor in the complete transfer matrix.

In analogy with the one-dimensional case, we now introduce the 2^M basis states

$$|\mu\rangle \equiv |\mu_1, \mu_2, \ldots, \mu_M\rangle \equiv |\mu_1\rangle|\mu_2\rangle \cdots |\mu_M\rangle \tag{10}$$

with $\mu_j = \pm 1$ and M sets of Pauli operators $(\sigma_{jX}, \sigma_{jY}, \sigma_{jZ})$ which act on the jth state in the product (10), that is,

$$\sigma_{jZ}|\mu_1, \mu_2, \ldots, \mu_j, \ldots, \mu_M\rangle = \mu_j|\mu_1, \mu_2, \ldots, \mu_j, \ldots, \mu_M\rangle$$

$$\sigma_j^+|\mu_1, \mu_2, \ldots, \mu_j, \ldots, \mu_M\rangle = \delta_{\mu_j,-1}|\mu_1, \mu_2, \ldots, \mu_j + 2, \ldots, \mu_M\rangle$$

$$\sigma_j^-|\mu_1, \mu_2, \ldots, \mu_j, \ldots, \mu_M\rangle = \delta_{\mu_j,1}|\mu_1, \mu_2, \ldots, \mu_j - 2, \ldots, \mu_M\rangle$$

$$\tag{11}$$

Moreover, we impose the commutation relations $[\sigma_{j\alpha}, \sigma_{m\beta}] = 0$ for $j \neq m$. For $j = m$ the usual Pauli matrix commutation relations apply.

If we think of the index μ_i as the orientation of the ith spin in a given column, we see immediately that the Boltzmann factors $\exp\{K \Sigma_r \sigma_{r,c}\sigma_{r+1,c}\}$ are given by the matrix elements of the operator $V_1 = \exp\{K \Sigma_j \sigma_{jZ}\sigma_{j+1,Z}\}$. Similarly, the matrix element

$$\langle\{\mu\}|V_2|\{\mu'\}\rangle$$

$$= \langle\mu_M, \mu_{M-1}, \ldots, \mu_1| \prod_{j=1}^{M} (e^K\mathbf{1} + e^{-K}\sigma_{jX}) | \mu_1', \mu_2', \ldots, \mu_M'\rangle \tag{12}$$

$$= \exp\{(M - 2n)K\}$$

where n of the indices $\{\mu'\}$ differ from the corresponding entries in $\{\mu\}$. Thus the partition function of the two-dimensional Ising model, in zero magnetic field, is, as can easily be verified, given by

$$
\begin{aligned}
Z &= \sum_{[\{\mu_1\},\{\mu_2\},\ldots\{\mu_{2M}\}]} \langle\mu_1 \,|\, \mathbf{V}_1 \,|\, \mu_2\rangle\langle\mu_2 \,|\, \mathbf{V}_2 \,|\, \mu_3\rangle\langle\mu_3 \,|\, \mathbf{V}_1 \,|\, \mu_4\rangle \cdots \langle\mu_{2M} \,|\, \mathbf{V}_2 \,|\, \mu_1\rangle \\
&= \mathrm{Tr}(\mathbf{V}_1\mathbf{V}_2)^M = \mathrm{Tr}(\mathbf{V}_2^{1/2}\mathbf{V}_1\mathbf{V}_2^{1/2})^M
\end{aligned}
\tag{13}
$$

In (13) the sum over each $\{\mu_j\}$ is, of course, over the entire set of 2^M basis states. Using (5) and (6), we may write

$$
\mathbf{V}_2 = (2 \ \sinh 2K)^{M/2} \exp\left\{K^* \sum_{j=1}^{M} \sigma_{jX}\right\}
\tag{14}
$$

and we have reduced the calculation of the partition function to the determination of the largest eigenvalue of the Hermitian operator

$$
\begin{aligned}
\mathbf{V} = \mathbf{V}_2^{1/2}\mathbf{V}_1\mathbf{V}_2^{1/2} &= (2 \ \sinh 2K)^{M/2} \exp\left\{\frac{K^*}{2} \sum_{j=1}^{M} \sigma_{jX}\right\} \\
&\times \exp\left\{K \sum_{j=1}^{M} \sigma_{jZ}\sigma_{j+1,Z}\right\} \exp\left\{\frac{K^*}{2} \sum_{j=1}^{M} \sigma_{jX}\right\}
\end{aligned}
\tag{15}
$$

which is still a nontrivial task since the factors in (15) do not commute with each other and since the matrix $\mathbf{V}$ becomes infinite dimensional in the thermodynamic limit.

(b) Transformation to an Interacting Fermion Problem

It is convenient for what follows to perform a rotation of the spin operators and to let $\sigma_{jZ} \to -\sigma_{jX}$, $\sigma_{jX} \to \sigma_{jZ}$ for all j. These rotations, of course, leave the eigenvalues invariant. Using $\sigma_{jZ} = 2\sigma_j^+\sigma_j^- - 1$ and $\sigma_{jX} = \sigma_j^+ + \sigma_j^-$, we arrive at the forms

$$
\begin{aligned}
\mathbf{V}_1 &= \exp\left\{K \sum_{j=1}^{M} (\sigma_j^+ + \sigma_j^-)(\sigma_{j+1}^+ + \sigma_{j+1}^-)\right\} \\
\mathbf{V}_2 &= (2 \sinh 2K)^{M/2} \exp\left\{2K^* \sum_{j=1}^{M} (\sigma_j^+\sigma_j^- - \tfrac{1}{2})\right\}
\end{aligned}
\tag{16}
$$

Schultz et al. [1964] showed that these operators can be simplified by a series of transformations, the first of which is the Jordan–Wigner transformation which converts the Pauli operators to fermion operators (see the Appendix for a discussion of second quantization). This step is useful because of subsequent canonical transformations which are not possible for angular momentum operators. One writes

$$\sigma_j^+ = \exp\left\{\pi i \sum_{m=1}^{j-1} c_m^+ c_m\right\} c_j^+$$

$$\sigma_j^- = c_j \exp\left\{-\pi i \sum_{m=1}^{j-1} c_m^+ c_m\right\} = \exp\left\{\pi i \sum_{m=1}^{j-1} c_m^+ c_m\right\} c_j \tag{17}$$

where the operators c, c^+ obey the commutation relations

$$[c_j, c_m^+]_+ = c_j c_m^+ + c_m^+ c_j = \delta_{jm}$$

$$[c_j, c_m]_+ = [c_j^+, c_m^+]_+ = 0$$

The operator $c_m^+ c_m$ is the fermion number operator for site m with integer eigenvalues 0 or 1. Since $e^{i\pi m} = e^{-i\pi m}$ the last step of (17) follows. To see that the spin commutation relations are preserved under this transformation consider, for $n > j$,

$$[\sigma_j^-, \sigma_n^+] = \exp\left\{\pi i \sum_{m=j+1}^{n-1} c_m^+ c_m\right\}(c_j e^{\pi i c_j^+ c_j} c_n^+ - c_n^+ e^{\pi i c_j^+ c_j} c_j)$$

Noting that $\exp\{\pi i c_j^+ c_j\} c_j = c_j$ and $c_j \exp\{\pi i c_j^+ c_j\} = -c_j$, we have $[\sigma_j^-, \sigma_n^+] = 0$ for $n \neq j$. We also immediately see that the on-site anticommutator

$$[\sigma_j^-, \sigma_j^+]_+ = [c_j, c_j^+]_+ = 1$$

The verification of further commutation relations and the derivation of the inverse of the transformation (17) is left as an exercise. Using (17), we can express the operators $\mathbf{V}_1$ and $\mathbf{V}_2$ in terms of the fermion operators. $\mathbf{V}_2$ presents no difficulties and is immediately given by

$$\mathbf{V}_2 = (2 \sinh 2K)^{M/2} \exp\left\{2K^* \sum_{j=1}^{M} (c_j^+ c_j - \tfrac{1}{2})\right\} \tag{18}$$

In the case of $\mathbf{V}_1$ there is a slight difficulty due to the periodic boundary conditions. We first note that for $j \neq M$ the term

$$(\sigma_j^+ + \sigma_j^-)(\sigma_{j+1}^+ + \sigma_{j+1}^-) = c_j^+ c_{j+1}^+ + c_j^+ c_{j+1} + c_{j+1}^+ c_j + c_{j+1} c_j$$

For the specific case $j = M$,

$$(\sigma_M^+ + \sigma_M^-)(\sigma_1^+ + \sigma_1^-) = \exp\left\{\pi i \sum_{j=1}^{M-1} c_j^+ c_j\right\} c_M^+(c_1^+ + c_1)$$

$$+ \exp\left\{\pi i \sum_{j=1}^{M-1} c_j^+ c_j\right\} c_M(c_1^+ + c_1)$$

$$= \exp\left\{\pi i \sum_{j=1}^{M} c_j^+ c_j\right\} [e^{\pi i c_M^+ c_M}(c_M^+ + c_M)(c_1^+ + c_1)]$$

$$= (-1)^n (c_M - c_M^+)(c_1^+ + c_1)$$

where $n = \sum_j c_j^+ c_j$ is the total fermion number operator. The operator n commutes with $\mathbf{V}_2$ but not with $\mathbf{V}_1$. On the other hand, $(-1)^n$ commutes with both $\mathbf{V}_1$ and $\mathbf{V}_2$ as the various terms in $\mathbf{V}_1$ change the total fermion number by 0 or ± 2. Thus if we consider separately the subspaces of even and odd total number of fermions, we may write $\mathbf{V}_1$ in a simple universal way, that is,

$$\mathbf{V}_1 = \exp\left\{ K \sum_{j=1}^{M} (c_j^+ - c_j)(c_{j+1}^+ + c_{j+1}) \right\} \tag{19}$$

where

$$
\begin{aligned}
c_{M+1} &\equiv -c_1, \ c_{M+1}^+ \equiv -c_1^+ &&\text{for } n \text{ even} \\
c_{M+1} &\equiv c_1, \ c_{M+1}^+ \equiv c_1^+ &&\text{for } n \text{ odd}
\end{aligned}
\tag{20}
$$

With this choice of boundary condition on the fermion creation and annihilation operators, we have recovered translational invariance and now carry out the *canonical* transformation

$$
\begin{aligned}
a_q &= \frac{1}{\sqrt{M}} \sum_{j=1}^{M} c_j e^{-iqj} \\
a_q^+ &= \frac{1}{\sqrt{M}} \sum_{j=1}^{M} c_j^+ e^{iqj}
\end{aligned}
\tag{21}
$$

with inverse

$$
\begin{aligned}
c_j &= \frac{1}{\sqrt{M}} \sum_q a_q e^{iqj} \\
c_j^+ &= \frac{1}{\sqrt{M}} \sum_q a_q^+ e^{-iqj}
\end{aligned}
\tag{22}
$$

To reproduce the boundary conditions (20), we take $q = j\pi/M$ with

$$
\begin{aligned}
j &= \pm 1, \pm 3, \ldots, \pm(M-1) &&\text{for } n \text{ even} \\
j &= 0, \pm 2, \pm 4, \ldots, \pm(M-2), \mathrm{M} &&\text{for } n \text{ odd}
\end{aligned}
$$

and where we have also assumed, without loss of generality, that M is even. It is easy to see that the operators a_q, a_q^+ obey fermion commutation relations, that is, $[a_q, a_{q'}^+]_+ = \delta_{q,q'}$ and $[a_q, a_{q'}]_+ = [a_q^+, a_{q'}^+]_+ = 0$ for all q and q'. Substituting into (18) and (19), we find for n even,

$$\mathbf{V}_2 = (2 \sinh 2K)^{M/2} \exp\left\{ 2K^* \sum_{q>0} (a_q^+ a_q + a_{-q}^+ a_{-q} - 1) \right\} \tag{23}$$

$$= (2 \sinh 2K)^{M/2} \prod_{q>0} \mathbf{V}_{2q}$$

and

$$\mathbf{V}_1 = \exp\left\{2K\sum_{q>0}[\cos q\,(a_q^+ a_q + a_{-q}^+ a_{-q}) - i\sin q\,(a_q^+ a_{-q}^+ + a_q a_{-q})]\right\}$$

$$\tag{24}$$

$$= \prod_{q>0}\mathbf{V}_{1q}$$

where, in (23) and (24), we have combined the terms corresponding to q and $-q$ and recognized, in writing the resulting operators as products, that bilinear operators with different wave vectors commute. This is a great simplification since the eigenvalues of the transfer matrix can now be written as a product of eigenvalues of, as we shall see, at most 4×4 matrices. For the case of odd n we also need the operators $\mathbf{V}_{1q}$ and $\mathbf{V}_{2q}$ for $q = \pi$ and $q = 0$. These are given by

$$\mathbf{V}_{10} = \exp\{2Ka_0^+ a_0\} \qquad \mathbf{V}_{20} = \exp\{2K^*(a_0^+ a_0 - \tfrac{1}{2})\}$$

$$\mathbf{V}_{1\pi} = \exp\{-2Ka_\pi^+ a_\pi\} \qquad \mathbf{V}_{2\pi} = \exp\{2K^*(a_\pi^+ a_\pi - \tfrac{1}{2})\}$$

$$\tag{25}$$

which are already in diagonal form and, of course, commute with each other.

(c) Calculation of Eigenvalues

We next proceed to calculate the eigenvalues of the operator

$$\mathbf{V}_q = \mathbf{V}_{2q}^{1/2}\mathbf{V}_{1q}\mathbf{V}_{2q}^{1/2}$$

for $q \neq 0$ and $q \neq \pi$. Since we are dealing with fermions, we have only four possible states: $|0\rangle$, $a_q^+|0\rangle$, $a_{-q}^+|0\rangle$, and $a_q^+ a_{-q}^+|0\rangle$, where $|0\rangle$ is the zero particle state defined by $a_q|0\rangle = a_{-q}|0\rangle = 0$. These states are already eigenstates of $\mathbf{V}_2$, and since the operator $\mathbf{V}_1$ has nonzero off-diagonal matrix elements only between states that differ by two in fermion number, the problem reduces to finding the eigenvalues of $\mathbf{V}_q$ in the basis $|0\rangle$ and $|2\rangle = a_q^+ a_{-q}^+|0\rangle$. We note that

$$\mathbf{V}_{1q}a_{\pm q}^+|0\rangle = \exp\{2K\cos q\}\,a_{\pm q}^+|0\rangle \tag{26}$$

and

$$\mathbf{V}_{2q}^{1/2}|0\rangle = \exp\{-K^*\}|0\rangle \qquad \mathbf{V}_{2q}^{1/2}|2\rangle = \exp\{K^*\}|2\rangle \tag{27}$$

To obtain the matrix elements of $\mathbf{V}_{1q}$ in the basis $|0\rangle$, $|2\rangle$, we let

$$\mathbf{V}_{1q}|0\rangle = \alpha(K)|0\rangle + \beta(K)|2\rangle$$

Differentiating this expression with respect to K, we obtain

$$\frac{d\alpha}{dK}|0\rangle + \frac{d\beta}{dK}|2\rangle = 2[\cos q\,\{a_q^+ a_q + a_{-q}^+ a_{-q}\}$$

$$- i\sin q\,\{a_q^+ a_{-q}^+ + a_q a_{-q}\}](\alpha|0\rangle + \beta|2\rangle)$$

$$= 2i\sin q\,\beta|0\rangle + [4\cos q\,\beta - 2i\sin q\,\alpha]|2\rangle \tag{28}$$

or

$$\frac{d\alpha}{dK} = 2i \sin q \, \beta(K)$$

$$\frac{d\beta}{dK} = 4 \cos q \, \beta(K) - 2i \sin q \, \alpha(K) \tag{29}$$

which we solve subject to the boundary conditions $\alpha(0) = 1$, $\beta(0) = 0$. The result is

$$\langle 0 \,|\, \mathbf{V}_{1q} \,|\, 0 \rangle = \alpha(K) = e^{2K \cos q}(\cosh 2K - \sinh 2K \cos q)$$

$$\langle 2 \,|\, \mathbf{V}_{1q} \,|\, 0 \rangle = \beta(K) = -ie^{2K \cos q} \sin q \sinh 2K \tag{30}$$

By the same method we can find the matrix elements $\langle 2 \,|\, \mathbf{V}_{1q} \,|\, 2 \rangle$ and $\langle 0 \,|\, \mathbf{V}_{1q} \,|\, 2 \rangle = \langle 2 \,|\, \mathbf{V}_{1q} \,|\, 0 \rangle^*$ and obtain the matrix

$$\mathbf{V}_{1q} = e^{2K \cos q} \begin{bmatrix} \cosh 2K - \cos q \sinh 2K & i \sin q \sinh 2K \\ -i \sin q \sinh 2K & \cosh 2K + \cos q \sinh 2K \end{bmatrix} \tag{31}$$

and

$$\mathbf{V}_q = \begin{bmatrix} \exp\{-K^*\} & 0 \\ 0 & \exp\{K^*\} \end{bmatrix} [\mathbf{V}_{1q}] \begin{bmatrix} \exp\{-K^*\} & 0 \\ 0 & \exp\{K^*\} \end{bmatrix} \tag{32}$$

The eigenvalues of this matrix are easily determined. Since we wish eventually to take the logarithm of the largest eigenvalue of the complete transfer matrix in order to calculate the free energy, we write the eigenvalues in the form

$$\lambda_q^{\pm} = \exp\{2K \cos q \pm \epsilon(q)\} \tag{33}$$

and after a bit of algebra, we obtain the equation

$$\cosh \epsilon(q) = \cosh 2K \cosh 2K^* + \cos q \sinh 2K \sinh 2K^* \tag{34}$$

for $\epsilon(q)$. By convention we choose $\epsilon(q) \geq 0$. We see that the minimum of the right-hand side of (34) occurs as $q \to \pi$ and that, for all q,

$$\epsilon(q) > \epsilon_{\min} = \lim_{q \to \pi} \epsilon(q) = 2\,|K - K^*| \tag{35}$$

and also note that

$$\lim_{q \to 0} \epsilon(q) = 2(K + K^*) \tag{36}$$

We are now in a position to combine all this information. Consider first the subspace in which all states contain an even number of fermions. In this case the allowed wave vectors do not include $q = 0$ or $q = \pi$, and comparing (33) and (26), we see that the largest eigenvalue of $\mathbf{V}_q$, for each q is λ_q^+. Thus the largest eigenvalue in this subspace, Λ_e, is given by

$$\Lambda_e = (2 \sinh 2K)^{M/2} \prod_{q > 0} \lambda_q^+ = (2 \sinh 2K)^{M/2} \exp \left\{ \sum_{q > 0} [2 \cos q + \epsilon(q)] \right\} \tag{37}$$

$$= (2 \sinh 2K)^{M/2} \exp \left\{ \frac{1}{2} \sum_q \epsilon(q) \right\}$$

where, in the last step, we have used $\Sigma_q \cos q = 0$ and have also extended the summation over the entire range $-\pi < q < \pi$.

The other subspace must be examined more carefully. For $q \neq 0$ and $q \neq \pi$ the maximum possible eigenvalue is λ_q^+. The corresponding eigenstates each are states with $(-1)^n = 1$. To make the overall state have $(-1)^n = -1$, we occupy the $q = 0$ state and leave the $q = \pi$ state empty and obtain a contribution of $\exp\{2K\}$ to the eigenvalue Λ_o. Thus the largest eigenvalue in the odd subspace is

$$\Lambda_o = (2 \sinh 2K)^{M/2} \exp \left\{ 2K + \tfrac{1}{2} \sum_{q \neq 0, \pi} \epsilon(q) \right\} \qquad (38)$$

Since the wave vectors in the two subspaces are not identical, a direct comparison between the two largest eigenvalues is somewhat complicated. However, we note that

$$\lim_{q \to 0} \tfrac{1}{2} \epsilon(q) + \lim_{q \to \pi} \tfrac{1}{2} \epsilon(q) = |K - K^*| + (K + K^*)$$
$$= 2K \qquad \text{for } K > K^*$$
$$= 2K^* \qquad \text{for } K^* > K$$

Thus if $K > K^*$, it is quite plausible, and can be shown rigorously in the thermodynamic limit $M \to \infty$, that Λ_o and Λ_e are degenerate. A little reflection will convince the reader that unless such a degeneracy exists, the order parameter $m_0(T)$ will be strictly zero. Therefore, the critical temperature of the two-dimensional Ising model is given by the equation $K = K^*$, or using the identity [from (6)]

$$\sinh 2K \sinh 2K^* = 1 \qquad (39)$$

by the more usual expression

$$\sinh \frac{2J}{k_B T_c} = 1 \qquad (40)$$

or $k_B T_c/J = 2.269185. \ldots$

The degeneracy of the two largest eigenvalues of the transfer matrix contributes only an additive term of $\ln 2$ to the dimensionless free energy and is therefore negligible. Therefore, at any temperature the free energy is given by

$$\frac{\beta G(0, T)}{M^2} = \beta g(0, T) = -\frac{1}{2} \ln (2 \sinh 2K) - \frac{1}{2M} \sum_q \epsilon(q)$$
$$= -\frac{1}{2} \ln (2 \sinh 2K) - \frac{1}{4\pi} \int_{-\pi}^{\pi} dq\, \epsilon(q) \qquad (41)$$

where we have converted the sum over wave vectors to an integral.

(d) Thermodynamic Functions

With a bit more algebra, we can simplify the expression (41) for the zero-field free energy. Using (39) and cosh $2K^* = \coth 2K$, which follows from (6), we have

$$\cosh \{\epsilon(q)\} = \cosh 2K \coth 2K + \cos q \tag{42}$$

Consider, now, the function

$$f(x) = \frac{1}{2\pi} \int_0^{2\pi} d\phi \ln (2 \cosh x + 2 \cos \phi) \tag{43}$$

Differentiating with respect to x and evaluating the resulting integral by contour integration, we find

$$\frac{df(x)}{dx} = \text{sign}(x) \quad \text{or} \quad f(x) = |x| \tag{44}$$

Taking $x = \epsilon(q)$, we obtain the integral representation:

$$\epsilon(q) = \frac{1}{\pi} \int_0^{\pi} d\phi \ln (2 \cosh 2K \coth 2K + 2 \cos q + 2 \cos \phi) \tag{45}$$

We define

$$I = \frac{1}{2\pi} \int_0^{\pi} dq \, \epsilon(q) = \frac{1}{2\pi^2} \int_0^{\pi} dq \int_0^{\pi} d\phi \ln (2 \cosh 2K \coth 2K$$
$$+ 2 \cos q + 2 \cos \phi) \tag{46}$$

Using the trigonometric identity

$$\cos q + \cos \phi = 2 \cos \frac{q + \phi}{2} \cos \frac{q - \phi}{2}$$

and changing the variables of integration to

$$\omega_1 = \frac{q - \phi}{2} \qquad \omega_2 = \frac{q + \phi}{2}$$

we have

$$I = \frac{1}{\pi^2} \int_0^{\pi} d\omega_2 \int_0^{\pi/2} d\omega_1 \ln (2 \cosh 2K \coth 2K + 4 \cos \omega_1 \cos \omega_2) \tag{47}$$

The integration over ω_2 is almost in the form (43) and we can put it into this form by writing

$$I = \frac{1}{\pi^2} \int_0^{\pi} d\omega_2 \int_0^{\pi/2} d\omega_1 \ln (2 \cos \omega_1)$$
$$+ \frac{1}{\pi^2} \int_0^{\pi/2} d\omega_1 \int_0^{\pi} d\omega_2 \ln \left(\frac{\cosh 2K \coth 2K}{\cos \omega_1} + 2 \cos \omega_2 \right) \tag{48}$$

$$= \frac{1}{\pi} \int_0^{\pi/2} d\omega_1 \ln (2 \cos \omega_1) + \frac{1}{\pi} \int_0^{\pi/2} d\omega_1 \cosh^{-1} \frac{\coth 2K \cosh 2K}{2 \cos \omega_1}$$

But $\cosh^{-1} x = \ln [x + (x^2 - 1)^{1/2}]$ and hence

$$I = \frac{1}{2} \ln (2 \cosh 2K \coth 2K) + \frac{1}{\pi} \int_0^{\pi} d\theta \ln \frac{1 + \sqrt{1 - q^2(K) \sin^2 \theta}}{2} \tag{49}$$

where

$$q(K) = \frac{2 \sinh 2K}{\cosh^2 2K} \tag{50}$$

Substituting in (41), we finally arrive at the form

$$\beta g(0, T) = -\ln (2 \cosh 2K) - \frac{1}{\pi} \int_0^{\pi/2} d\theta \ln \frac{1 + \sqrt{1 - q^2 \sin^2 \theta}}{2} \tag{51}$$

for the free energy per spin.

The function $q(K)$, defined in (50), has the form shown in Figure 5.2. It takes on a maximum value, $q = 1$, at $\sinh 2K = 1$, and it is clear that the integral on the right-hand side of equation (51) can only be nonanalytic at that point since the term inside the square root cannot vanish for $q < 1$. The internal energy per spin of the system is given by

$$u(T) = \frac{d}{d\beta} [\beta g(T)] = -J \coth 2K \left[1 + \frac{2}{\pi} (2 \tanh^2 2K - 1) K_1(q) \right] \tag{52}$$

where

$$K_1(q) = \int_0^{\pi/2} \frac{d\phi}{\sqrt{1 - q^2 \sin^2 \phi}}$$

is the complete elliptic integral of the first kind. As $q \to 1$, the term $(2 \tanh^2 2K - 1) \to 0$, and the internal energy is continuous at the transition. The specific heat $C(T)$ can be obtained by differentiating once more with respect to temperature. Some analysis (Problem 5.2) shows that

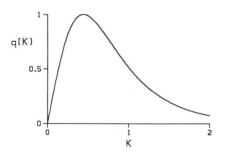

Figure 5.2 Function $q(K)$ defined in (50).

$$\frac{1}{k_B} C(T) = \frac{4}{\pi} (K \coth 2K)^2 \left\{ K_1(q) - E_1(q) \right.$$

$$\left. - (1 - \tanh^2 2K) \left[\frac{\pi}{2} + (2 \tanh^2 2K - 1) K_1(q) \right] \right\}$$

(53)

where

$$E_1(q) = \int_0^{\pi/2} d\phi \sqrt{1 - q^2 \sin^2 \phi}$$

is the complete elliptic integral of the second kind. Near T_c the specific heat (53) is given, approximately, by

$$\frac{1}{k_B} C(T) \approx -\frac{2}{\pi} \left(\frac{2J}{k_B T_c} \right)^2 \ln \left| 1 - \frac{T}{T_c} \right| + \text{const.}$$

(54)

The internal energy and specific heat are shown in Figure 5.3.

The difference between the exact specific heat and that obtained in Chapter 3 from mean field and Landau theories is striking. Instead of a discontinuity in $C(T)$, we find a logarithmic divergence. In modern theories of critical phenomena, the form assumed for the specific heat is

$$C(T) \sim \left| 1 - \frac{T}{T_c} \right|^{-\alpha}$$

(55)

Onsager's result is a special case of this power law behavior. The limiting form of the function

$$\lim_{\alpha \to 0} \frac{1}{\alpha} (X^{-\alpha} - 1) = -\ln X$$

The form (54) is thus seen to be a special case of the power law singularity with $\alpha = 0$.

The calculation of the spontaneous magnetization is a nontrivial extension of the present derivation and may be found in Schultz et al. [1964]. The result is

$$m_0(T) = -\lim_{h \to 0} \frac{\partial}{\partial h} g(h, T)$$

$$= \left[1 - \frac{(1 - \tanh^2 K)^4}{16 \tanh^4 K} \right]^{1/8} \qquad T < T_c$$

(56)

$$= 0 \qquad\qquad\qquad\qquad T > T_c$$

As $T \to T_c$ from below, the limiting form of the spontaneous magnetization is given by

$$m_0(T) \approx (T_c - T)^{1/8} \equiv (T_c - T)^{\beta}$$

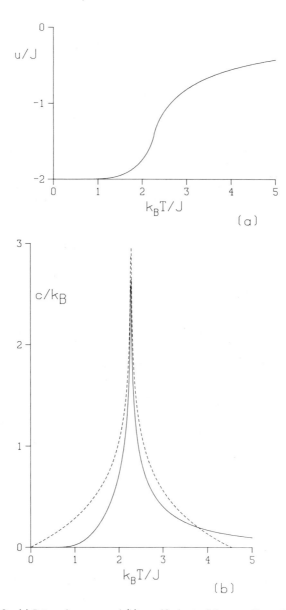

Figure 5.3 (a) Internal energy and (b) specific heat of the two-dimensional Ising model on the square lattice. The dotted curve corresponds to the approximation (54).

As in mean field theories, the order parameter has a power law singularity at the critical point but the exponent $\beta = \frac{1}{8}$, not $\frac{1}{2}$ as obtained from mean field and Landau theories. The derivation of (56) was first published by Yang [1952], but Onsager had previously announced the result at a conference. The asymptotic form as $T \to T_c$ of the zero-field susceptibility is also known (Mc-Coy and Wu, 1973):

$$\chi(0, T) = \lim_{h \to 0} \frac{\partial m(h, T)}{\partial h} \sim |T - T_c|^{-7/4} \approx |T - T_c|^{-\gamma} \qquad (57)$$

The exponents $\gamma = \frac{7}{4}$ in (57) again is to be compared with the classical value $\gamma = 1$. It is clear from the exact results described above that the form of the free energy near a critical point is quite different from that postulated in the Landau theory.

(e) Concluding Remarks

The reader who has worked through the details of the preceding subsections will appreciate the difficulty of calculating even the zero-field free-energy exactly. One can easily write down the transfer matrix of the two-dimensional Ising model in a finite magnetic field and arrive at a generalization of (15). However, the subsequent transformation to fermion operators yields a transfer matrix which is not bilinear in fermion operators and which cannot be diagonalized, at least by presently known techniques.

Similarly, one can construct the transfer matrix of the three-dimensional Ising model. In this case the matrix **V** is of dimension $2^L \times 2^L$, where $L = M^2$ if the lattice is an $M \times M \times M$ simple cubic lattice. The reader can verify that the difficulty here is not this increase in the dimensionality of the transfer matrix but rather that the Jordan–Wigner transformation (17) does not produce a bilinear form in fermion operators.

Since Onsager's solution appeared, a small number of other two-dimensional problems have been solved exactly. The reader is referred to the book by Baxter [1982] for an account of this work. Since exact results near the critical point were so elusive, workers in the field therefore devised various approximate techniques to probe the critical behavior of strongly interacting systems. We next discuss the method of series expansions which initially provided the greatest amount of information on critical behavior.

5.B SERIES EXPANSIONS

The method of series expansions was first introduced by Opechowski [1937] and has proved to be, with the help of modern computers, a powerful tool for the study of critical phenomena. To motivate the approach, let us consider first a simple function $f(z)$ and its power series expansion about $z = 0$:

$$f(z) = \left(1 - \frac{z}{z_c}\right)^{-\gamma} = \sum_{n=0}^{\infty} \binom{\gamma}{n}\left(\frac{z}{z_c}\right)^n \tag{1}$$

where

$$\binom{\gamma}{n} = \frac{\gamma(\gamma + 1)(\gamma + 2) \cdots (\gamma + n - 1)}{n!}$$

The power series converges for $|z| < |z_c|$. Now suppose that we have available a certain number of terms in the power series of an unknown function. Can we infer the function from this limited information? The answer to this question is, of course, no. However, if we have reason to believe that the unknown function has a specific structure, such as the power law singularity of the function $f(z)$ in (1), we may be able to determine the parameters of interest (z_c and γ) from the available information. Let us write, in general,

$$f(z) = \sum_{n=0}^{\infty} a_n z^n \tag{2}$$

For the case (1), the ratio of successive coefficients, a_n, takes the form

$$\frac{a_n}{a_{n-1}} = \frac{\gamma + n - 1}{n}\frac{1}{z_c} = \frac{1}{z_c}\left(1 + \frac{\gamma - 1}{n}\right) \tag{3}$$

A plot of this ratio as function of $1/n$ for a number of integers thus yields a discrete set of points which asymptotically fall on a straight line with intercept $1/z_c$ at $1/n = 0$ and with slope $(\gamma - 1)/z_c$ (Figure 5.4). The reason for concentrating on the form (1) is that we believe, because of Onsager's exact solution, that thermodynamic functions have precisely this type of structure (with $z = 1/T$, $z_c = 1/T_c$) near the critical point. Of course, real thermodynamic functions will be more complicated than the simple form (1), but it may still be possible for us to extract the information of interest, namely the critical temperature and exponent, from a finite number of terms in the power series expansion. In general, the coefficients, a_n, in the power series of a function are determined, at least for large n, primarily by the nearest singularity in the complex plane to the point about which the expansion is carried out. A func-

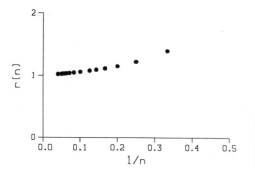

Figure 5.4 Plot of the ratios $r(n) = a_n/a_{n-1}$ for n up to 20 for the simple function $f(z) = (1 - z)^{-1.5} \exp\{z\}$. Some points have been omitted for clarity.

tion of temperature, such as the specific heat, will, in general, have singularities at various points in the complex $1/T$ plane (only real $1/T$ is of interest), but if the closest of these singularities is the physical one at $1/T_c$, then we may expect a power series about $T = \infty$ (i.e., $1/T = 0$) to have a simple structure similar to that of the function (1), at least after the first few terms. This is the basic idea behind series expansions and we will return to the analysis of series after briefly discussing the generation of such expansions.

(a) High-Temperature Expansions

The derivation of a high-temperature expansion is, in principle, straightforward. Suppose that we wish to calculate the expectation value of an operator O with respect to the canonical distribution for a system with Hamiltonian H.

$$\langle O \rangle = \frac{\text{Tr } Oe^{-\beta H}}{\text{Tr } e^{-\beta H}} = \frac{\text{Tr } O \sum\limits_{j=0}^{\infty} (-\beta H)^j / j!}{\text{Tr } \sum\limits_{j=0}^{\infty} (-\beta H)^j / j!} \tag{4}$$

For a system with a finite number, Z_0, of discrete states, such as a spin system, it is convenient to divide the numerator and denominator of (4) by $Z_0 = \text{Tr } 1$ and to define

$$\overline{O} = \frac{\text{Tr } O}{\text{Tr } 1}$$

Note that Z_0 is the partition function at $T = \infty$. $\overline{O}$ is thus the ensemble average of O at infinite temperature. At finite temperatures, we have

$$\langle O \rangle = \frac{\overline{O} - \beta \overline{OH} + \beta^2 \overline{OH^2}/2! + \cdots}{1 - \beta \overline{H} + \beta^2 \overline{H^2}/2! + \cdots}$$

$$= \overline{O} - \beta (\overline{OH} - \overline{O}\,\overline{H}) + \frac{\beta^2}{2!} (\overline{OH^2} - 2\overline{OH}\,\overline{H} + 2\overline{O}\,\overline{H}^2 - \overline{O}\,\overline{H^2}) + \cdots \tag{5}$$

Therefore, if

$$\langle O \rangle = \sum_n \frac{a_n}{T^n} \tag{6}$$

we obtain

$$a_0 = \overline{O}$$

$$a_1 = -\frac{1}{k_{\text{B}}} (\overline{OH} - \overline{O}\,\overline{H})$$

$$a_2 = \frac{1}{2!\, k_{\text{B}}^2} (\overline{OH^2} - 2\overline{OH}\,\overline{H} + 2\overline{O}\,\overline{H}^2 - \overline{O}\,\overline{H^2})$$

$$\cdots$$

(7)

As in the case of the Mayer expansion of Chapter 4, graphical methods are very useful in the construction of such series expansions. A number of different graphical methods have been developed and the most convenient, or powerful, technique depends on the specific details of the problem. We derive below a few terms in the high-temperature series of the Ising model on the square lattice and simple cubic lattices and leave a similar derivation for the Heisenberg model as an exercise (Problem 5.3).

Consider, once again, the model

$$H = -J \sum_{\langle ij \rangle} \sigma_i \sigma_j - h \sum_i \sigma_i \tag{8}$$

with $\sigma_i = \pm 1$ and where the sum over i, j is over nearest-neighbor pairs on a lattice which, at present, we do not specify. The zero-field susceptibility per spin in the disordered phase, is given by (3.E.5)

$$k_B T \chi(T) = \frac{1}{N} \frac{\sum_{i,j} \text{Tr } \sigma_i \sigma_j \exp\{-\beta H_0\}}{\text{Tr exp}\{-\beta H_0\}}$$

$$= 1 + \frac{1}{N} \frac{\sum_{i \neq j} \text{Tr } \sigma_i \sigma_j \exp\{-\beta H_0\}}{\text{Tr exp}\{-\beta H_0\}} \tag{9}$$

with $H_0 = -J \sum_{\langle ij \rangle} \sigma_i \sigma_j$. It is convenient to make use of the identity

$$\exp\{\beta J \sigma_i \sigma_j\} = \cosh \beta J + \sigma_i \sigma_j \sinh \beta J$$

$$= \cosh \beta J (1 + v \sigma_i \sigma_j) \tag{10}$$

where $v = \tanh \beta J$. The identity (10) can be easily demonstrated by expanding the left side in a power series and using $\sigma_i^2 = 1$. The variable v, which approaches zero as T goes to infinity, can be used as an expansion parameter instead of $1/T$ and we shall construct our expansion in powers of v. Equation (9) may now be written as

$$k_B T \chi = 1 + \frac{1}{N} \frac{\sum_{i \neq j} \text{Tr } \sigma_i \sigma_j \prod_{\langle nm \rangle} (1 + v \sigma_n \sigma_m)}{\text{Tr} \prod_{\langle jm \rangle} (1 + v \sigma_j \sigma_m)} \tag{11}$$

The structure of both numerator and denominator of (11) is very similar to that of the Mayer expansion of the configuration integral that appeared in Chapter 4. We now carry out an analogous graphical expansion. We associate a line between nearest neighbors j, m with the term $v \sigma_j \sigma_m$. We also divide the top and bottom by 2^N and note that

$$2^{-N} \text{Tr } 1 = 1 = 2^{-N} \text{Tr } \sigma_i^2$$

since the trace is carried out over the 2^N states of the system. Consider now the expansion of the numerator in (11):

$$\text{Num} = 2^{-N} \sum_{i \neq j} \text{Tr } \sigma_i \sigma_j \left(1 + v \sum_{\langle nm \rangle} \sigma_n \sigma_m + v^2 \sum_{\langle nm \rangle \neq \langle rt \rangle} \sigma_n \sigma_m \sigma_r \sigma_t + \cdots \right)$$

Since $\text{Tr } \sigma_i = 0$, all terms that do not contain even powers (including zero) of any spin variable will yield zero. We do not draw a line connecting the two distinct, but otherwise unrestricted, points i, j. The first few terms of the numerator can be graphically represented as follows:

In certain types of lattices, such as the square, simple cubic, and body-centered cubic lattices, diagrams involving closed triangles cannot occur. We suppose that the lattice is of this type. We must now count the number of times that a particular diagram will appear. This number, called the lattice constant of graph g, will be denoted by (g). The sum over i, j in (11) is not restricted to nearest neighbors but $i \neq j$. Therefore, we fix i to be a specific lattice point and sum over $j \neq i$. Since v connects only nearest neighbors, we quickly obtain the lattice constants of the first three graphs appearing in the numerator

$$(\text{\textemdash}) = Nq$$

$$(\wedge) = Nq(q - 1)$$

$$(\wedge\vee) = Nq(q - 1)^2$$

where q is the coordination number of the lattice. For graph 4, the combinatorial factor is not simply $q(q - 1)^3$, as this number contains contributions from configurations in which the last point j is identical to the first point i. A more careful calculation yields

$$(\wedge\vee\wedge) \quad \begin{array}{l} = 100N \text{ on the square lattice} \\[6pt] = 726N \text{ on the simple cubic lattice} \end{array}$$

Now consider the denominator. The first nonzero contribution comes from the graph $\square$. This graph appears N times on the square lattice, $3N$ times on the simple cubic lattice. To this point, our expression for the susceptibility is

$$k_B T \chi = 1 + \frac{1}{N} \frac{(\text{\textemdash})v + (\wedge)v^2 + (\wedge\vee)v^3 + (\wedge\vee\wedge)v^4 + \cdots}{1 + (\square)v^4 + \cdots} \tag{12}$$

$$= 1 + \frac{1}{N}[(\text{\textemdash})v + (\wedge)v^2 + (\wedge\vee)v^3 + (\wedge\vee\wedge)v^4 + \cdots]$$
$$\times [1 - (\square)v^4 \cdots] \tag{13}$$

The term $(\diagup)(\square)$ is of order $N^2 v^5$, and it seems, at first sight, that the expansion will approach a limit that is dependent on N. To see that this is not the case, we must add the fifth-order contribution to the numerator. This consists of three graphs:

(a)   (b)   (c)

The factor 2 in part (b) results from the fact that the square can be attached to the line either at vertex i or vertex j. The complete term of order v^5 is therefore

$$[(\diagup\square) + 2(\diagup\square) + (\diagup\diagdown\diagup) - (\diagup)(\square)$$

The lattice constant of the disconnected graph (a) can be related to the product of the lattice constants in the last term. Since neither of the two ends of the line are allowed to touch the square, we have

$$(\diagup\square) = (\diagup)(\square) - 2(\diagup\square) - (\square)$$

The fifth-order term thus reduces to

$$[(\diagup\diagdown\diagup) - (\square)]v^5 \qquad (14)$$

Since these graphs are both connected, we see that the contribution is of order N. The cancellation of terms proportional to higher powers of N is quite general and occurs in all orders. It is clear from this example that the calculation of high-order terms in the series can be quite complicated and very sophisticated techniques have been developed to push these calculations as far as possible. Evaluating the lattice constants in (14), we obtain, through order v^5, the series

$$k_B T\chi = 1 + 4v + 12v^2 + 36v^3 + 100v^4 + 276v^5 + \cdots \qquad (15)$$

and

$$k_B T\chi = 1 + 6v + 30v^2 + 150v^3 + 726v^4 + 33510v^5 + \cdots \qquad (16)$$

on the square and simple cubic lattices. To obtain (15) and (16), we have used the following easily derived results:

$$(\diagup\diagdown\diagup) = 284N \text{ (square lattice) or } 3534N \text{ (simple cubic lattice)}$$

$$(\square) = 8N \text{ (square lattice) or } 24N \text{ (simple cubic lattice)}$$

The length of series that can be derived in any finite time clearly depends on the lattice structure, as more graphs with nonzero lattice constants exist on close-packed lattices such as the triangular and face-centered cubic lattices than on open lattices such as the simple cubic lattice. The susceptibility expansion is known through order v^{15} on the fcc lattice (McKenzie, 1975), through order v^{22} on the bcc lattice (Nickel, 1982). We do not list the series here but will use all available terms when we discuss analysis of such series in Section

5.B.(c). While the example above focuses on the susceptibility, it is clear that similar series can be (and have been) constructed for other thermodynamic functions and for other lattice models. The power of the expansion method, for lattice models, comes from the fact that the trace over spin variables is essentially trivial. The corresponding step in the Mayer expansion, the evaluation of the cluster integral, is the limiting factor that prevents the method from being as useful in the case of liquids.

(b) Low-Temperature Expansions

If the ground state of a system is known and if the excitations from this state can be classified in a simple way, it is possible to construct a series expansion that is complementary to the high-temperature series. Applications of this method have been primarily to Ising models (Domb, 1974). Models like the Heisenberg model or the XY model cannot be treated in the same way because their elementary excitations (known for both the classical and quantum Heisenberg model and not well known for the XY model) form a continuum. In the case of the Heisenberg model, a low-temperature series has been constructed by Dyson [1958] by a different approach.

Consider, once again, the Ising model in a magnetic field at $T = 0$. The ground state is the completely aligned ferromagnetic state. Excitations from this state consist of clusters of overturned spins. The cost in energy of flipping a single spin is $2qJ + 2h$, where q is the coordination number of the lattice. The partition function is, therefore,

$$Z = \left[1 + Nu^q w + \frac{N(N - q - 1)}{2} u^{2q} w^2 + \frac{qN}{2} u^{2q-2} w^2 + \cdots \right] e^{-\beta E_0}$$

where $u = e^{-2\beta J}$, $w = e^{-2\beta h}$ and where E_0 is the ground-state energy. The third and fourth terms correspond to two flipped spins. The third term arises from pairs of spins that are not nearest neighbors, the fourth from nearest-neighbor pairs of spins. Clearly, u and w are suitable variables for a power series expansion. As in the case of high-temperature series, the logarithm of Z contains only terms proportional to N. The generation of lengthy low-temperature series is a highly specialized art and we refer the reader to the review of Domb [1974] for a discussion of the details.

(c) Analysis of Series

We discuss, in this subsection, some of the simpler methods of series analysis and some of the results that have been obtained from analysis of high-temperature series. In the introduction to this topic, we showed that if a function has a simple power law singularity of the form

$$\chi(v) = (v - v_c)^{-\gamma} \tag{17}$$

the ratio of successive coefficients in the power series is

$$\frac{a_n}{a_{n-1}} = \frac{1}{v_c}\left(1 + \frac{\gamma - 1}{n}\right)$$

Let us construct the sequence

$$r_n = n\frac{a_n}{a_{n-1}} - (n - 1)\frac{a_{n-1}}{a_{n-2}}$$

This sequence should approach $1/v_c$ faster than the simple ratio a_n/a_{n-1} since the term of order $1/n$ is eliminated. If χ is of the form (17), without other singularities or analytic correction terms, then the approximants r_n should be simply $1/v_c$, where v_c is the critical value of tanh βJ in the case of the Ising model. If χ contains a term of the form (17), then we may hope that as n becomes large enough, r_n will approach a limit that we will interpret as $1/v_c$. In Table 5.1 we list these approximants for the square and simple cubic lattices using the currently available terms in the expansion. The numbers were calculated from Table I of Domb [1974].

We see that the even approximants (r_n for $n = 2, 4, 6, \ldots$) increase uniformly and the odd approximants (r_n for $n = 3, 5, 7, \ldots$) decrease and that they seem to approach a common limit. Crudely extrapolating the even and odd approximants to their intersection with the $1/n = 0$ axis, we arrive at an estimate $1/v_c = 2.4151$ corresponding to $k_B T_c/J = 2.2701$ for the square lattice. The exact Onsager result is 2.269185 and it is clear that the series ex-

TABLE 5.1

n	r_n (Square)	r_n (Simple Cubic)	S_n (Square)	S_n (Simple Cubic)
2	2	4		
3	3	5	1.7265	1.2687
4	2.1111	4.3600	1.6007	1.2188
5	2.6889	4.8136	1.7140	1.2677
6	2.2870	4.3905	1.6610	1.2245
7	2.5671	4.7368	1.7239	1.2567
8	2.3277	4.4553	1.6877	1.2276
9	2.4962	4.6966	1.7213	1.2510
10	2.3714	4.4870	1.7031	1.2288
11	2.4625	4.6716	1.7227	1.2468
12	2.3857	4.5060	1.7106	1.2287
13	2.4500	4.6556	1.7250	1.2432
14	2.3914	4.5191	1.7152	1.2280
15	2.4424	4.6445	1.7265	1.2401
16	2.3959	4.5287	1.7186	1.2270
17	2.4365	4.6363	1.7274	1.2373
18	2.3996		1.7221	
19	2.4322		1.7281	
20	2.4021		1.7227	
21	2.4292		1.7285	

pansion gives an excellent estimate of the critical temperature. The corresponding estimate of T_c for the simple cubic lattice is $k_B T_c / J = 4.515$. The method of series analysis above is very unsophisticated, and there are far more powerful methods based on Padé approximants to the series (Gaunt and Guttman, 1974). Such methods can provide extremely well-converged estimates of the critical temperatures. We note, in passing, that the oscillation of the approximants r_n for the square and simple cubic lattices is characteristic of lattices that can be divided into two sublattices with nearest neighbors of atoms on one sublattice lying on the other sublattice. In such a situation there is a competing singularity at $-v_c$ which represents the phase transition for the nearest-neighbor Ising antiferromagnet. On close-packed lattices, such as the triangular or face-centered cubic lattices, the antiferromagnetic transition occurs at a lower value of T ($T_c = 0$ on the triangular lattice). Thus the antiferromagnetic singularity is further from the origin in the complex v plane and the aforementioned oscillations in the approximants are absent.

To obtain a sequence of estimates for the exponent, γ, which characterizes the singularity, we can construct, for example, the sequence of approximants

$$S_n = 1 + n\left(\frac{a_n}{a_{n-1}} v_c - 1\right)$$

which should tend to γ as $n \to \infty$. These approximants are biased in the sense that an estimate of v_c is first required. As in the determination of T_c, far better tools exist and we use this method only as a demonstration of the utility of high-temperature series. The approximants S_n are also listed in Table 5.1 for the square and simple cubic series. Bearing in mind that we are using a rough estimate of v_c to bias the approximants, we see that the internal convergence of each sequence is remarkably good. The two-dimensional approximants increase as function of n and are certainly consistent with the exact value (see Section 5.A) of $\gamma = \frac{7}{4}$. The approximants for the simple cubic lattice are also well converged and indicate a value of γ near 1.25. For a number of years, it was thought that this result (i.e., $\gamma = \frac{5}{4}$) is exact but recent longer series (Nickel, 1982) indicate that a slightly smaller value $\gamma \approx 1.239$ is, in fact, closer to the truth.

An interesting question, at this point, is whether the nature of the singularity in χ is sensitive to details such as the type of lattice, the range of the interaction, or the magnitude of the spin. In Domb [1974], the coefficients a_n are tabulated for a number of two- and three-dimensional lattices and the interested reader may wish to carry out the straightforward analysis demonstrated above. It turns out that the exponent γ is quite insensitive to the lattice structure, whereas quantities such as T_c or v_c are not. The inclusion of a second neighbor, or longer-range, interaction also has no effect on the critical exponents, and the size of the spin also plays no role. The fact that such details are *irrelevant*, in the language of the renormalization group theory, is now understood to be a particular manifestation of "universality," a topic that we discuss in Section 5.D.

Series for other thermodynamic quantities such as the specific heat, the magnetization (low-temperature series), the second derivative of the susceptibility, $\partial^2 \chi / \partial h^2$, and a number of other quantities have been derived. Analysis of these series reveals that in general thermodynamic functions have power law singularities at the critical point with exponents which are, again, independent of lattice structure and of other details.

$$\chi(0, T) \sim A_\pm \left| T - T_c \right|^{-\gamma}$$

$$C(0, T) \sim E_\pm \left| T - T_c \right|^{-\alpha}$$

$$m(0, T) \sim B \left| T - T_c \right|^{\beta} \tag{18}$$

$$m(h, T_c) \sim \left| h \right|^{1/\delta} \text{sign} (h)$$

For the Ising model in three dimensions, estimates of the exponents were, until quite recently, consistent with the following values: $\gamma = \frac{5}{4}$, $\alpha = \frac{1}{8}$, $\beta = \frac{5}{16}$, $\delta = 5$. We quote these numbers to illustrate certain simple relations between the exponents. For example,

$$\alpha + 2\beta + \gamma = 2$$

$$\gamma = \beta(\delta - 1) \tag{19}$$

Relations of this type are called *scaling laws* and it is believed that they are exact. We discuss them further in Section 5.C.

Before turning to scaling theory in detail, we briefly review the results of series expansions for some other models. For the Heisenberg model

$$H = -J \sum_{\langle ij \rangle} \mathbf{S}_i \cdot \mathbf{S}_j$$

only high-temperature series of any substantial length have been derived and exponents such as β and δ can therefore not be determined without use of the scaling laws (19). The series for the Heisenberg model are shorter than those for the Ising model and the available critical exponents are substantially different from the corresponding Ising exponents. Again, within the margins of error of series analysis, no dependence on lattice structure or size of spin has been found. The susceptibility exponent is found to be $\gamma = 1.43 \pm 0.01$ for the spin-$\frac{1}{2}$ Heisenberg model in three dimensions and $\gamma = 1.425 \pm 0.02$ for the spin ∞ (i.e., classical spin) Heisenberg model (see Camp and van Dyke [1976] for a critical discussion).

The specific heat series are difficult to analyze and the exponent α is not very accurately known. The weight of the evidence indicates that α is small and negative, corresponding to a cusp in the specific heat, rather than a divergence. For the case of the Heisenberg model, the dependence of the critical behavior on spatial dimensionality is even more pronounced than for the Ising model. Indeed, in two dimensions, the Heisenberg model does not order at any finite temperature. This result can be proven rigorously (Mermin and Wagner, 1966).

Another model that has been studied extensively is the XY model, which has the Hamiltonian

$$H = -J \sum_{\langle ij \rangle} (S_i^x S_j^x + S_i^y S_j^y)$$

This model is thought to be a good model for the critical behavior of liquid ^{4}He at the lambda transition. As in the case of the Heisenberg model, only high-temperature series are available. The best estimates (Betts, 1974) of the critical exponents γ, α in three dimensions are $\gamma = \frac{4}{3}$, $\alpha = 0$ (logarithmic divergence). These exponents differ from both the Ising and Heisenberg exponents. As for the Heisenberg model in two dimensions, one can rigorously prove that the order parameter of the XY model is zero at any finite temperature in two dimensions. Nevertheless, there is good evidence that the XY model undergoes a transition, known as a Kosterlitz–Thouless transition, at a finite temperature in two dimensions (see Section 5.E). On the experimental side, thin films of liquid helium seem to show the same transition.

Finally, we mention the results obtained for the n-vector model (sometimes referred to as the D-vector model). This model is a generalization of the Heisenberg model to an n-dimensional spin space $S_i = (S_1, S_2, \ldots, S_n)$ and has the Hamiltonian

$$H = -J \sum_{\langle ij \rangle} \sum_{\alpha=1}^{n} S_{i\alpha} S_{j\alpha}$$

The susceptibility exponent for this model, which has the Ising, XY, and Heisenberg models as the special cases $n = 1, 2, 3$ has, in three dimensions, the approximate dependence on n

$$\gamma(n) = \frac{2n + 8}{n + 7}$$

independent of lattice (Stanley, 1974). This expression is purely phenomenological and without theoretical basis.

5.C SCALING

In Section 5.B we noted that the analysis of high- and low-temperature series suggested a number of simple relationships between critical exponents of the type (5.B.19). Experimental data on many different materials also were, within experimental error, consistent with these scaling laws (Vincentini-Missoni, 1972). In this section we pursue the matter from different points of view and show that these scaling laws indicate a particular type of structure for the free energy near the critical point.

(a) Thermodynamic Considerations

We first note that considerations of thermodynamic stability allow us to derive inequalities for the critical exponents. One of the simplest of these is due to Rushbrooke [1963]. Consider a magnetic system. The specific heats at constant field, C_H, and constant magnetization, C_M, satisfy the relationship (Problem 1.4)

$$\chi_T(C_H - C_M) = T\left(\frac{\partial M}{\partial T}\right)_H^2 \tag{1}$$

where $\chi_T = (\partial M/\partial H)_T$ is the isothermal susceptibility. Thermodynamic stability requires that χ_T, C_H, and C_M all be greater than or equal to zero. Therefore,

$$C_H > T\chi_T^{-1}\left(\frac{\partial M}{\partial T}\right)_H^2 \tag{2}$$

We now consider a system in zero field at a temperature below the critical temperature, T_c, but close enough that we may use (5.B.18). We obtain

$$(T_c - T)^{-\alpha} > \text{const. } (T_c - T)^{\gamma + 2(\beta - 1)} \tag{3}$$

which leads to the Rushbrooke inequality:

$$\alpha + 2\beta + \gamma \geq 2 \tag{4}$$

A number of similar inequalities have been derived and we refer to the book by Stanley [1971] for further details. The intriguing feature of these inequalities is that they seem to hold as equalities, and that there are only a small number of independent critical exponents.

(b) Scaling Hypothesis

To be specific, let us assume that we have a system for which the appropriate free energy is a function of two independent thermodynamic variables [e.g., $E(S, M)$, $A(T, M)$, $G(T, H)$]. At the critical point of the system, these variables have the values T_c, H_c, M_c, and so on. We introduce the relative variables

$$h = H - H_c$$

$$m = M - M_c \tag{5}$$

$$t = \frac{T - T_c}{T_c}$$

and consider the quantities

$$\chi(t, h = 0) = \left(\frac{\partial m}{\partial h}\right)_t \qquad \sim (-t)^{-\gamma} \qquad . \ t < 0$$

$$\sim t^{-\gamma} \qquad t > 0$$

$$C_h(t, 0) = -T\left(\frac{\partial^2 G}{\partial t^2}\right)_h \sim (-t)^{-\alpha'} \qquad t < 0 \tag{6}$$

$$\sim t^{-\alpha} \qquad t > 0$$

$$m(t, 0) = -\left(\frac{\partial G}{\partial h}\right)_t \sim (-t)^{\beta} \qquad t < 0$$

$$m(0, h) \sim |h|^{1/\delta} \, \text{sign} \, (h)$$

Using the Helmholtz free energy, we have for the equation of state of the system,

$$h = \frac{\partial A}{\partial m} \tag{7}$$

A number of authors (see, e.g., Griffiths [1967] for references and a more complete discussion) asked themselves what the functional form of the free energy or the equation of state must be in order to produce the correct critical exponents. We saw in Section 3.G that if the free energy is assumed to be analytic at the critical point,

$$A(t, m) = a_0 + \tfrac{1}{2}a_2 m^2 + \tfrac{1}{4}a_4 m^4 + \cdots \tag{8}$$

and $a_2 \approx at$ near $t = 0$, we have

$$h \approx am\left(t + \frac{a_4}{a}m^2 + \cdots\right)$$

for the equation of state. From this equation of state we automatically obtain the classical critical exponents $\alpha = 0$, $\beta = \frac{1}{2}$, $\gamma = 1$, $\delta = 3$. If, however, we modify the equation of state to read

$$h \approx m(t + cm^{1/\beta})$$

we can have an arbitrary value for β. This equation is still not satisfactory since if we differentiate this equation at constant t, we find

$$\frac{\partial h}{\partial m} \sim t$$

giving $\gamma = 1$ for the susceptibility exponent. The situation can be improved if, instead, we assume the equation of state

$$h \approx m(t + \text{const.} \ m^{1/\beta})^{\gamma}$$

or, more generally,

$$h = m\psi(t, m^{1/\beta})$$

where ψ is an arbitrary homogeneous function of degree γ, that is,

$$\psi(\lambda^{1/\gamma}t, \lambda^{1/\gamma}m^{1/\beta}) = \lambda\psi(t, m^{1/\beta}) \tag{9}$$

To be more systematic, let us assume that the singular part of the free energy, $G(h, t)$, near the transition is dominated by a term that changes under a change of scale according to

$$G(t, h) = \lambda G(\lambda^s t, \lambda^r h) \tag{10}$$

This form of the free energy implies that $m = -\dfrac{\partial G}{\partial h}$ will scale according to

$$m(t, h) = \lambda^{r+1}m(\lambda^s t, \lambda^r h) \tag{11}$$

while

$$\chi(t, h) = \lambda^{2r+1}\chi(\lambda^s t, \lambda^r h)$$
$$C_h(t, h) = \lambda^{2s+1}C_h(\lambda^s t, \lambda^r h) \tag{12}$$

We now consider the special cases $h = 0$, $\lambda = |t|^{-1/s}$, and $t = 0$, $\lambda = |h|^{-1/r}$ to obtain

$$m(t, 0) = (-t)^{-(r+1)/s}m(-1, 0)$$
$$m(0, h) = |h|^{-(r+1)/r}m(0, \pm1)$$
$$\chi(t, 0) = |t|^{-(2r+1)/s}\chi(\pm1, 0) \tag{13}$$
$$C_h(t, 0) = |t|^{-(2s+1)/s}C_h(\pm1, 0)$$

In (13) the coefficients of the leading power law term are the thermodynamic functions evaluated at points far from the singularity ($t = 0$, $h = 0$) and therefore are simply finite constants. We see that when critical exponents exist on both sides of the critical point [see (6)] they will have to be identical,

$$\alpha = \alpha'$$
$$\gamma = \gamma' \tag{14}$$

but the prefactor of the power law will in general be different, that is, there is no reason for $C_h(1, 0)$ or $\chi(1, 0)$ to be the same as $C_h(-1, 0)$ or $\chi(-1, 0)$. By comparing with (6), we find

$$\alpha = \frac{2s + 1}{s}$$

$$\beta = -\frac{r + 1}{s}$$

$$\gamma = \frac{2r + 1}{s} \tag{15}$$

$$\delta = -\frac{r}{r + 1}$$

We see that there are only two independent critical exponents and that the scaling relations

$$\alpha + 2\beta + \gamma = 2$$
$$\beta(\delta - 1) = \gamma$$

(16)

follow directly from (15). As we shall see, in the next section, further scaling relations appear if an assumption similar to (9) is made for the behavior of the pair correlation function near T_c.

(c) Kadanoff Block Spins

The agreement of the results of the preceding section with those of experiments (see Figure 5.6) and analyses of series is very satisfactory, but our formulation of the scaling law offers no physical justification for the basic assumption (10). An important development occurred when Kadanoff produced an intuitively appealing plausibility argument for the scaling form of the free energy. (For a very readable review of these ideas, as they were presented at the time, see Kadanoff et al. [1967].)

Let us consider an Ising model on a d-dimensional hypercubic lattice with nearest-neighbor separation a_0. The common property of all systems near a critical point is that the correlation length $\xi >> a_0$ and, exactly at T_c, $\xi = \infty$. Two parameters characterize the state of the system: $t = (T - T_c)/T_c$ and h. Consider now a small block of spins. In Figure 5.5 neighboring groups of nine spins on a square lattice have been combined into such blocks and the blocks themselves form a square lattice with nearest-neighbor spacing $3a_0$. Each block can be in one of 2^9 states, or, in general 2^n states with $n = L^d$, where L is the linear dimension of a block. In the following analysis we assume that $\xi/a_0 >> L$. If this is the case, many of these states will be suppressed because of the strong correlation over short distances.

Following Kadanoff, we now assume that each block can be characterized by an Ising spin, σ_J^B, for the Jth block and that this new dynamic variable also takes on the values ± 1, as do the original "site spins" σ_i. Heuristically,

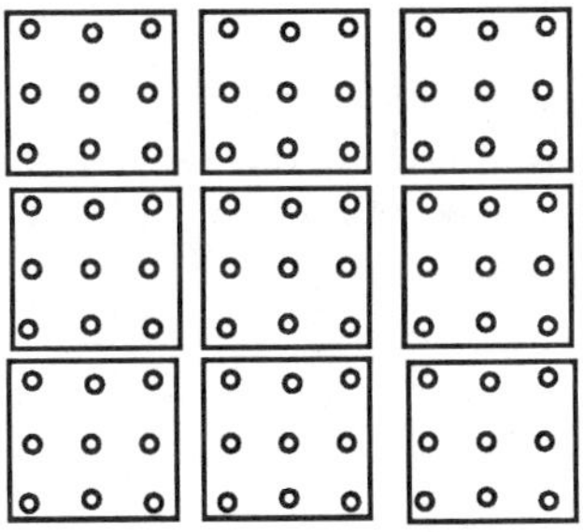

Figure 5.5 Grouping of site spins into nine-spin blocks on a square lattice.

we expect that the state of the block spin system can be described in terms of effective parameters $\bar{t}$, $\bar{h}$ which measure the distance of the block spin system from criticality. The parameters $\bar{t}$, $\bar{h}$ depend on t and h as well as on the linear dimension, L, of the block. The simplest possible relation between these variables consistent with the symmetry requirements $\bar{h} \rightarrow -\bar{h}$ when $h \rightarrow -h$ and $\bar{t} \rightarrow \bar{t}$ when $h \rightarrow -h$, as well as with the condition $\bar{t} = \bar{h} = 0$ when $t = h = 0$ is

$$\bar{h} = hL^x$$
$$\bar{t} = tL^y \tag{17}$$

where x and y are unspecified except that they must be positive so as to insure that the block spin system is further from criticality than the site spin system. The singular part of the free energy of the block spin system must be the same function of $\bar{t}$, $\bar{h}$ as the singular piece of the free energy of the site spin system is of t and h. Since there are L^d site spins per block spin, we have

$$g(t, h) = L^{-d}g(L^y t, L^x h) \tag{18}$$

which completes the derivation of the scaling form of the free energy. Since the block spin separation is L times the site spin separation, we also expect that the correlation length will be reduced by the same factor:

$$\xi(t, h) = L\xi(L^y t, L^x h) \tag{19}$$

This equation leads to new predictions. Near the critical point the correlation length diverges according to the form

$$\xi(t, 0) \sim |t|^{-\nu}$$

and we may now relate the exponent ν to the unspecified exponent y. Assuming that equations (18) and (19) hold for any value of L and letting $L = |t|^{-1/y}$ in (18) and (19), as well as $h = 0$, we obtain

$$g(t, 0) = |t|^{-d/y}g(\pm 1, 0) \tag{20}$$
$$\xi(t, 0) = |t|^{-1/y}\xi(\pm 1, 0) \tag{21}$$

The exponent d/y is related to the specific heat exponent through $d/y = 2 - \alpha = d\nu$, and we have obtained a scaling relation which explicitly involves the spatial dimensionality

$$d\nu = 2 - \alpha \tag{22}$$

called a hyperscaling equation. If we use only the scaling form of the free energy, we easily reproduce the results (9)–(16). The relation between x, y and r, s is $x = dr$, $y = ds$.

To this point we have only obtained the scaling form of the correlation length. We can also find a scaling equation for the correlation function $\Gamma(r, t, h) = \langle \sigma_0 \sigma_r \rangle - \langle \sigma_0 \rangle \langle \sigma_r \rangle$ by introducing local fields h_r, $\bar{h}_r$ and demanding that the variation of the free energy with respect to changes in the local field be the same in the site and block spin picture:

$$\Gamma(r) = \frac{\delta^2 G}{\delta h(0)\,\delta h(r)}$$

and

$$\delta^2 G = \Gamma(r)\,\delta h(0)\,\delta h(r) = L^{-2d}\Gamma\!\left(\frac{r}{L}\right)\delta\tilde{h}(0)\,\delta\tilde{h}\!\left(\frac{r}{L}\right) \tag{23}$$

Using $\tilde{h} = L^x h$, we have

$$\Gamma(r, t, h) = L^{2(x-d)}\Gamma\!\left(\frac{r}{L}, L^y t, L^x h\right) \tag{24}$$

At the critical point the correlation function $\Gamma(r, 0, 0)$ has the asymptotic form $\Gamma(r, 0, 0) \sim r^{-(d-2+\eta)}$. We let $L = r$ and find $2(x - d) = 2 - d - \eta$, or

$$2 - \eta = 2x - d \tag{25}$$

The susceptibility exponent, γ, may be obtained from (18):

$$\gamma = \frac{2x - d}{y} = (2x - d)\nu$$

$$\gamma = (2 - \eta)\nu \tag{26}$$

The derivation of further scaling relations is left to the problem section.

The experimental verification of the scaling laws is difficult. Nevertheless, all experiments of which the authors are aware are at least consistent with the exact equalities derived from the scaling form of the free energy. In Figure 5.6 we show an experimental plot of the function $h/|t|^{\gamma+\beta}$ plotted as function of the scaled magnetization $m/|t|^\beta$ for the ferromagnet $CrBr_3$. Using the scaling form of the order parameter

$$m(t, h) = |t|^\beta m\!\left(\pm 1, \frac{h}{|t|^{\beta+\gamma}}\right) = |t|^\beta \phi_\pm\!\left(\frac{h}{|t|^{\beta+\gamma}}\right) \tag{27}$$

which is easily obtained from (13), we expect that the data should fall on at most two universal curves, one for $t < 0$ corresponding to the function ϕ_-, and one for $t > 0$ corresponding to ϕ_+. We see that this is indeed the case for a range of temperatures near T_c.

A more accurate check of the scaling relations can be carried out for the two-dimensional Ising model for which $\alpha = 0$, $\beta = \frac{1}{8}$, $\gamma = \frac{7}{4}$, $\eta = \frac{1}{4}$, and $\nu = 1$ are all known exactly. Clearly, all scaling relations are found to hold. For the three-dimensional Ising model the results of series expansions have been consistent with those scaling relations not involving the correlation length exponent ν for many years. It seemed, however, that the hyperscaling relation

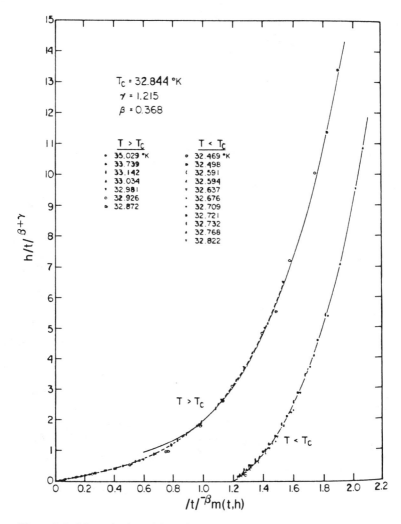

Figure 5.6 Magnetization of CrBr$_3$ in the critical region plotted in a scaled form (see the text). (From Ho and Litster [1969].)

(22) might not be satisfied. Recent work of Nickel [1982] has shown that there is no violation of hyperscaling, at least at the present level of accuracy of the series results. For a detailed discussion of experimental tests of the scaling laws, the reader is referred to the article by Vincentini-Missoni [1972].

We have emphasized Kadanoff's heuristic derivation of the scaling laws because the concept of replacing a group of dynamic variables by an effective single dynamic variable can be made into a concrete calculational tool. Indeed, it is clearly the correct way to attack problems with large correlation lengths (i.e., where dynamic variables are strongly coupled over large distances). By successively removing intermediate dynamical variables one can hope to arrive at a problem which is simple (i.e., one for which the correlation length relative to the new unit of length, La_0, is much smaller). Critical behavior is then seen to be a property of the rescaling of block spin interactions (t, h) under successive coarse graining. This is the program of the renormalization group approach that is discussed in detail in Chapter 6.

5.D UNIVERSALITY

We have seen in Section 5.B that the critical exponents of the three-dimensional Ising model are the same no matter what the underlying lattice structure is. On the other hand, they differ from the critical exponents of the two-dimensional Ising model and from those of other three-dimensional spin systems such as the Heisenberg and XY models. It is natural to ask which features of a system are important in determining the nature of the phase transition. Many possibilities spring to mind. The lattice structure has already been ruled out. Other possibilities include the range of interparticle interaction, size of spin, quantum-mechanical rather than classical spins, continuous translational freedom as in real liquids rather than lattice gases, spin space dimensionality ($n = 1, 2, 3, \ldots$, see Section 5.B), crystal field effects, and many others. A large number of investigations of different spin systems were carried out and it became clear that critical behavior was independent of a remarkable number of details.

Two parameters are clearly important. The spatial dimensionality, d, and the spin space dimensionality, n, have already been mentioned as parameters that do affect the critical exponents. On the other hand, one could ask whether a system with Hamiltonian

$$H = -J \sum_{\langle ij \rangle} \mathbf{S}_i \cdot \mathbf{S}_j - D \sum_i (S_i^z)^2 \tag{1}$$

which has spin space dimensionality $n = 3$ has the same critical exponents as the isotropic Heisenberg model ($D = 0$). It turns out that this model has the same critical behavior as the Ising model ($n = 1$) and some refinement of our concepts is required.

Empirically (i.e., from series expansions) it was found that the symmetry of the ordered phase plays a crucial role. The Hamiltonian (1) has, for $D > 0$, as its ground state the state with all spins fully aligned along the z direction. The only transformation that leaves the ground-state energy, and at $T \neq 0$ the free energy, invariant is the transformation $S_i^z \rightarrow -S_i^z$, all i. On the other hand, the Heisenberg Hamiltonian has the full three-dimensional rotational symmetry—the vector $\mathbf{m} = 1/N \sum_i \mathbf{S}_i$ can be anywhere on the surface of a three-dimensional sphere. The parameter n which we have to this point used to characterize the spin space dimensionality is now modified to denote the index n of the rotational group, $O(n)$, under which the ground state is invariant.

The fact that this index, n, is an important parameter was demonstrated by Jasnow and Wortis [1968], who studied the classical spin Hamiltonian

$$H = -J \sum_{\langle ij \rangle} (\mathbf{S}_i \cdot \mathbf{S}_j + \eta \, S_i^z S_j^z) \qquad (2)$$

The ground state of this Hamiltonian has $n = 3$ for $\eta = 0$, $n = 1$ for $\eta > 0$ and $n = 2$ for $-2 < \eta < 0$. The ground state thus changes from an XY ground state to a Heisenberg ground state and finally to an Ising ground state as the parameter η is varied. Jasnow and Wortis derived and analyzed susceptibility series for different values of η. Any finite series cannot yield completely unambiguous results, but their analysis strongly suggested that there are discontinuous changes in the exponent γ at the values of η at which the ground-state symmetry changes.

Most of the other parameters that we mentioned above were also eliminated. Some of these results can be understood qualitatively. Since the correlation length diverges at T_c, one might expect that quantum mechanics would not play a role in critical phenomena. A cluster of ξ^d spins is effectively a classical object. Similarly, the possibility of continuous rather than discrete translation should not be important. Localizing a fluid particle in a cell of size a and simply allowing it to hop to nearest-neighbor cells should not be important on a length scale of ξ. Indeed, classical fluids have, to within experimental error, the same critical exponents as the three-dimensional Ising model.

The range of interaction may, of course, be important. We now believe (Aharony, 1976) that as long as the range of the interaction is finite, or as long as the interaction decreases sufficiently rapidly as a function of separation, the critical behavior will be the same as that of a system with only nearest-neighbor interactions.

A wealth of empirical evidence led to the formulation of the *universality hypothesis* (Kadanoff, 1971) or smoothness hypothesis (Griffiths, 1970). This hypothesis simply states that the critical behavior of a system depends only on the spatial dimensionality, d, and the symmetry, n, of the ordered phase. The renormalization group has allowed us to understand in some detail this very general conclusion.

In the foregoing discussion we have considered only the symmetry group $O(n)$. Other discrete symmetries exist, of course. For example, a real spin system on a cubic lattice is in general subject to a crystal field that will align the ground-state magnetization with one of the symmetry directions of the crystal. This symmetry is neither Ising-like nor $O(n)$ with $n = 3$. Can this type of field have an effect? The answer to this question is "yes" and we will return to this point in Chapter 6. At this point we simply conclude by stating again that the concept of universality and the notion that seemingly very different systems on the microscopic level can be grouped into a small number of *universality classes* is an extremely powerful simplifying idea.

5.E TWO-DIMENSIONAL SYSTEMS WITH CONTINUOUS SYMMETRY AND THE KOSTERLITZ–THOULESS TRANSITION

To conclude this first chapter on critical phenomena we briefly discuss planar systems in which the ground state has a continuous symmetry as distinguished from, for example, the Ising model, which has only the discrete symmetry $m \rightarrow -m$. We single out these systems because in some of them there exists the possibility of an unusual finite-temperature transition to a low-temperature phase without long-range order. Such transitions are known as Kosterlitz–Thouless [1972] transitions. Physical systems which are thought to display this transition are certain planar magnets, films of liquid helium (the two-dimensional version of the lambda transition), thin superconducting films, liquid-crystal monolayers, and in some cases gases adsorbed on crystal surfaces. We will return to discuss some of these situations in more detail but first begin by showing, heuristically, that spatial dimensionality 2 separates simple critical behavior (as discussed in the rest of this chapter and in Chapter 6) from a nonordering situation.

We use, as an example, a two-dimensional XY model (Section 5.B) with spins modeled by classical vectors. In Chapter 7 we show that Bose condensation in an ideal Bose gas cannot occur for $d \leq 2$ and in Chapter 8 the reader is asked to demonstrate (Problem 8.6) the same result for the Heisenberg model in the spin wave approximation.

Consider the Hamiltonian

$$H = -J \sum_{\langle ij \rangle} (S_{ix} S_{jx} + S_{iy} S_{jy}) = -JS^2 \sum_{\langle ij \rangle} \cos (\phi_i - \phi_j) \tag{1}$$

where the spins $\mathbf{S}_i$ are classical vectors of magnitude S constrained to lie in the S_x–S_y plane. These spins can therefore be specified by their orientation $\phi_i (0 \leq \phi_i < 2\pi)$ with respect to the S_x axis. For the present we assume that the sites i lie on a d-dimensional hypercubic lattice. We note that the Hamiltonian (1) has a continuous symmetry: the transformation $\phi_i \rightarrow \phi_i + \phi_0$, for all i, leaves the Hamiltonian invariant. In Chapter 7 we show that the same symmetry exists in the BCS theory of superconductivity and in the interacting Bose gas.

The ground state of (1) is the fully aligned state $\phi_i = \phi$ for all i, where ϕ can be any angle in the range 0 to 2π. We now assume that at low enough temperature $|\phi_i - \phi_j| << 2\pi$ for i, j nearest neighbors and approximate (1) by the expression

$$H = -\frac{qNJS^2}{2} + \frac{1}{2} JS^2 \sum_{\langle ij \rangle} (\phi_i - \phi_j)^2 \tag{2}$$

$$= E_0 + \frac{JS^2}{4} \sum_{\mathbf{r}, \mathbf{a}} [\phi(\mathbf{r} + \mathbf{a}) - \phi(\mathbf{r})]^2 \tag{3}$$

where E_0 is the ground-state energy and the sum over $\mathbf{a}$ runs over all nearest neighbors of site $\mathbf{r}$. If $\phi(\mathbf{r})$ is a slowly varying function of $\mathbf{r}$, we may further approximate (3) by a continuum model. Replacing the finite differences in (3) by derivatives and the sum over lattice sites by an integral, we obtain the expression

$$H = E_0 + \frac{JS^2}{2a^{d-2}} \int d^d r \, [\nabla\phi(\mathbf{r}) \cdot \nabla\phi(\mathbf{r})] \tag{4}$$

where a is the nearest-neighbor distance. The second term in (4) is the classical form of the spin wave energy (see Chapter 8).

The constraint that ϕ must be in the range 0 to 2π is inconvenient and we relax this condition and allow ϕ to range from $-\infty$ to ∞. It is then a trivial matter to calculate the partition function and other thermodynamic properties of the system. In particular, we wish to calculate the correlation function

$$g(r) = \langle \exp\{i[\phi(\mathbf{r}) - \phi(0)]\} \rangle \tag{5}$$

which, in a phase with conventional long-range order, will approach a constant as $r \to \infty$. In the ground state, of course, $g(r) = 1$. Using periodic boundary conditions and writing

$$\phi(\mathbf{r}) = \frac{1}{\sqrt{N}} \sum_{\mathbf{k}} \phi_{\mathbf{k}} e^{i\mathbf{k}\cdot\mathbf{r}} \tag{6}$$

we obtain

$$H = E_0 + \frac{JS^2 a^2}{2} \sum_{\mathbf{k}} k^2 \phi_{\mathbf{k}} \phi_{-\mathbf{k}} \tag{7}$$

$$= E_0 + JS^2 a^2 \sum_{\mathbf{k}}{}' k^2 (\alpha_{\mathbf{k}}^2 + \gamma_{\mathbf{k}}^2) \tag{8}$$

where $\phi_{\mathbf{k}} = \alpha_{\mathbf{k}} + i\gamma_{\mathbf{k}} = (\phi_{-\mathbf{k}})^*$ and where $\sum'$ indicates that we have combined the two terms in (7) for $\mathbf{k}$ and $-\mathbf{k}$ and are summing over half the Brillouin zone. The expectation value (5) can now be easily evaluated (Problem 5.6). The result is

$$g(r) = \exp\left\{-\frac{k_B T}{NJS^2 a^2} \sum_{\mathbf{k}} \frac{1 - \cos(\mathbf{k}\cdot\mathbf{r})}{k^2}\right\} \tag{9}$$

We transform the sum over k in (9) to an integral in the usual way:

$$\frac{1}{N}\sum_{\mathbf{k}} \longrightarrow \left(\frac{a}{2\pi}\right)^d \int d^d k$$

and arrive at the expression

$$g(r) = \exp\left\{-\frac{k_B T a^{d-2}}{(2\pi)^d JS^2} \int d^d k \frac{1 - \cos(\mathbf{k}\cdot\mathbf{r})}{k^2}\right\} \tag{10}$$

If we now take $d = 2$, ignore the geometry of the Brillouin zone, and carry out the integration in polar coordinates, we have, using

$$\int_0^{2\pi} d\theta \cos{(kr \cos{\theta})} = 2\pi J_0(kr)$$

where J_0 is the zeroth-order Bessel function,

$$g(r) = \exp\left\{-\frac{k_B T}{2JS^2\pi} \int_0^{\pi/a} dk \frac{1 - J_0(kr)}{k}\right\} \tag{11}$$

where the upper limit, π/a, is roughly the distance to the zone boundary. Substituting $x = kr$, we obtain for the integral in (11),

$$\int_0^{\pi r/a} dx \frac{1 - J_0(x)}{x}$$

For large r/a the dominant contribution comes from the region $x \gg 1$ in which we can ignore the Bessel function and we finally obtain

$$g(r) \approx \exp\left\{-\frac{k_B T}{2JS^2\pi} \ln{\frac{\pi r}{a}}\right\} = \left(\frac{\pi r}{a}\right)^{-k_B T/2JS^2\pi} = \left(\frac{\pi r}{a}\right)^{-\eta(T)} \tag{12}$$

We see, therefore, that the correlation function falls off algebraically at all finite temperatures and that there is no long-range order in the system. If, in (10), we take $d = 3$, we easily see that $g(r)$ approaches a constant as r becomes large (Problem 5.6). In physical terms, the absence of long-range order at finite temperatures is due to the excitation of long-wavelength low-energy spin waves which are weighted, in (10), by a phase space factor k^{d-1} that becomes more and more important for low dimensionality.

The argument above can be made rigorous (Mermin and Wagner, 1966) and holds as well for the Heisenberg model, for superfluid films and superconductors (Hohenberg, 1967) and, indeed, for any two-dimensional system with short-range interactions in which the ordered phase has a continuous symmetry. The continuous symmetry then implies that at least one branch of the spectrum of elementary excitations has the property that the energy approaches zero continuously as the wavelength becomes large (Goldstone boson; see, e.g., Anderson [1984] for a discussion).

We note also that equation (12) is reminiscent of a system with a finite-temperature phase transition precisely at its critical point. Recall that we generally write, for the correlation function

$$g(r) \approx \frac{\exp{\{-r/\xi(T)\}}}{r^{d-2+\eta}} \qquad T \neq T_c$$

$$\approx \frac{1}{r^{d-2+\eta}} \qquad T = T_c \tag{13}$$

In these planar magnets we therefore find a temperature-dependent "critical" exponent $\eta(T)$. In our spin wave approximation the system seems to be at a critical point at all temperatures. This result is clearly unphysical. At high enough temperature we expect the correlation function to fall off exponentially. We now argue, following Kosterlitz and Thouless [1972], that there exists another set of excitations which take the system from its low-temperature "critical" phase described by the spin wave approximation to a simple high-temperature disordered phase.

These excitations were identified by Kosterlitz and Thouless to be *vortices* which at low temperatures occur in tightly bound pairs that unbind at a critical temperature. To see that such a mechanism could give rise to a transition, consider first an isolated vortex which is displayed in Figure 5.7. We label the orientation of a spin at position r, θ by $\phi(r, \theta)$. In the continuum approximation $\phi(r, \theta) = n\theta$, where n is the strength of the vortex. Thus

$$\oint dl \cdot \nabla\phi = 2\pi n \quad \text{and} \quad \nabla\phi = \frac{n}{r}a_\theta.$$

The energy of an isolated vortex is easily calculated:

$$E = \frac{JS^2}{2} \int d^2r \, \nabla\phi(\mathbf{r}) \cdot \nabla\phi(\mathbf{r}) = \pi JS^2 n^2 \int_a^L dr \frac{1}{r} = \pi JS^2 n^2 \ln \frac{L}{a} \qquad (14)$$

where L is the linear dimension of the system and a the lattice constant. Thus the energy of an isolated vortex is infinite in the thermodynamic limit. The entropy associated with a single vortex is given by $S = k_B \ln (L/a)^2$ and the change in free energy due to formation of a vortex is

$$\Delta G = (\pi JS^2 n^2 - 2k_B T) \ln \frac{L}{a} \qquad (15)$$

This quantity is positive for $k_B T < \pi JS^2/2$ and isolated vortices will therefore not occur for temperatures lower than this value.

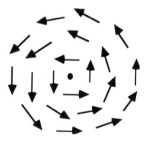

Figure 5.7 Schematic picture of vortex of unit strength in planar magnet.

Consider next a pair of vortices separated by a distance r. We note that the ground-state configuration of the spin system is given by

$$\delta E(\{\phi\}) = \delta \int d^2r \frac{JS^2}{2} [\nabla\phi(\mathbf{r})]^2 = 0 \qquad (16)$$

which yields

$$\nabla^2 \phi(\mathbf{r}) = 0 \qquad (17)$$

This Laplace equation is supplemented by the condition

$$\oint_C \nabla\phi \cdot d\mathbf{l} = 2\pi n_1 \qquad (18)$$

for a contour C which only encloses vortex 1, of strength n_1, and a similar condition for contours enclosing vortex 2 only.

Equation (18) is reminiscent of Ampère's law for the magnetic induction due to a current distribution:

$$\oint_C \mathbf{B} \cdot d\mathbf{l} = \mu I$$

where I is the current enclosed by the contour C. In this analogy we can identify the equivalent magnetic field with $\nabla\phi$. In SI units the correspondence is $I_1 = 2\pi J S^2 n_1$ and $\mu = 1/JS^2$, where μ is the "permittivity" and I_1 the equivalent current associated with a vortex of strength n_1. Using this analogy, one then obtains, for the interaction energy of a pair of vortices of strength n_1 and n_2,

$$E_{\text{pair}}(\mathbf{r}_1, \mathbf{r}_2) = -2\pi J S^2 n_1 n_2 \ln \left| \frac{\mathbf{r}_1 - \mathbf{r}_2}{a} \right| \qquad (19)$$

where we have set the energy of nearest-neighbor vortices equal to zero. This logarithmic dependence on separation of the interaction energy also occurs in the case of "two-dimensional" charged particles, or more accurately, lines of charge. Thus we also have an analogy between excitations of the planar magnet and the two-dimensional Coulomb gas. We note that (19) implies that the state of minimum energy for oppositely "charged" vortices is the tightly bound configuration in which they are nearest neighbors.

Since the size of the system does not appear in the expression (19), but does appear in the expression for the entropy, we see that the low-temperature state of the system will consist of an equilibrium density of bound vortex pairs. This equilibrium density is determined by pair–pair interactions which we have not considered. At higher temperatures, the vortex unbinding mechanism (15) will then destroy this condensed phase.

We also note that the arguments above yield no information on the nature of the vortex unbinding transition. They merely show that two qualitatively different states of the system can exist in different temperature ranges. It is possible to determine further properties of the transition using renormalization group methods. We refer the reader to the article by José et al. [1977].

The foregoing discussion applies directly to the case of superfluid films. In a superfluid (see also Section 7.B) the quantity $\nabla\phi$ is proportional to the velocity of the film relative to the substrate on which it is adsorbed and the vortices quite literally represent circulation of material.

The case of melting of two-dimensional crystals is considerably more complicated. We first note that it is important, in physisorbed materials, to distinguish between lattice gases and floating monolayers. An example of a lattice gas (helium adsorbed onto the basal plane of graphite) is discussed in Chapter 6. In such a system the adsorbed layer does not have a continuous translational symmetry. To a first approximation, the atoms occupy discrete sites on the substrate and thermal excitation results in hopping of atoms between eligible sites. Such lattice gases disorder in the conventional manner.

A floating monolayer, on the other hand, is not strongly perturbed by the periodic component of the interaction between adsorbate and substrate. The ground-state configuration is, in an ideal case, determined entirely by the interparticle interaction of the adsorbate and the entire layer can be displaced uniformly by any amount parallel to the substrate surface without cost in energy. For such floating monolayers one can show (Mermin, 1968) that the two-dimensional crystal does not have long-range positional order at any nonzero temperature. The experimental manifestation of this result is that in a diffraction experiment one would not observe true Bragg peaks (δ-function peaks centered on the reciprocal lattice vectors of the two-dimensional crystal) but rather peaks whose intensity falls off as a power law in the vicinity of the reciprocal lattice vectors. In terms of real space correlations, the analog of the spin–spin correlation function (5) is the function

$$g_G(\mathbf{R}) = \langle e^{i\mathbf{G} \cdot [\mathbf{u}(\mathbf{R}) - \mathbf{u}(0)]} \rangle \tag{20}$$

where the positions of the atoms are given by $\mathbf{r} = \mathbf{R} + \mathbf{u}(\mathbf{R})$ and where $\mathbf{G}$ is a reciprocal lattice vector of the ground-state crystal. One can show (see e.g., Nelson, 1983) that, in the harmonic approximation,

$$g_G(\mathbf{R}) \sim |\mathbf{R}|^{-\eta_G(T)} \tag{21}$$

where the "critical" exponent $\eta_G(T)$ depends linearly on the temperature and quadratically on the magnitude of the reciprocal lattice vector $\mathbf{G}$. Thus we have essentially the same situation as in the spin wave theory of the planar magnet (12).

There is, however, an added feature to the crystallization problem in two dimensions. It *is* possible for long-range *orientational* order to exist at finite temperature. We consider the case of a triangular lattice ground state and let

$$g_\theta(\mathbf{r}) = \langle e^{6i[\theta(\mathbf{r}) - \theta(0)]} \rangle \tag{22}$$

where $\theta(\mathbf{r})$ is the angle of a nearest-neighbor bond between two atoms, one of which is at position $\mathbf{r}$. The function $g_\theta(\mathbf{r})$ is then a measure of orientational order. One can show that the long-wavelength phonons which destroy long-range positional order do not destroy orientational long-range order and that

$$\lim_{r \to \infty} g_\theta(\mathbf{r}) = \text{const.}$$

for low-enough temperature.

Although the case of melting of a two-dimensional crystal is considerably more complicated than the disordering of a planar magnet, an analogous picture of the process can be constructed. The topological defects analogous to the vortices in the magnet are dislocations. These interact via a logarithmic potential (19) as do the vortices but the corresponding "charges" are the Burgers vectors of the dislocations and thus vector rather than scalar quantities. A dislocation mediated theory of melting has been constructed by Halperin and Nelson and by Young (see Nelson [1983] for a review and for the original references).

We now briefly discuss the experimental (computer and laboratory) situation. In the case of ^{4}He films the Kosterlitz–Thouless theory (with subsequent elaborations) predicts a universal discontinuity in the superfluid density $\rho_s(T_c)/T_c$. The critical temperature can be varied by changing the thickness of the film and such experiments have been carried out, for example, by Bishop and Reppy [1978] and Rudnick [1978] and the results are consistent with the Kosterlitz–Thouless predictions. Computer experiments on planar magnets (Tobochnik and Chester, 1979; see also Saito and Müller–Krumbhaar [1984] for an extensive review) also are consistent with both the low-temperature predictions of spin wave theory and the vortex unbinding mechanism.

The situation as far as two-dimensional melting is concerned is more controversial. Melting behavior consistent with Kosterlitz-Thouless theory has been observed in colloid suspensions (Murray and Van Winkle 1987) while in other cases the melting transition seems to be a conventional first-order transition. Laboratory experiments are complicated by substrate effects and by long relaxation times; the latter also plague computer experiments. It seems clear, however, that the nature of the transition depends on microscopic parameters of the system in question and is therefore not a universal feature of two-dimensional melting. Rather than report on the results of specific calculations or experiments on these fascinating systems, we refer the reader to the reviews by Abraham [1986], Saito and Müller-Krumbhaar [1984], and Nelson [1983].

PROBLEMS

5.1. *Approximate Solution of the Ising Model on the Square Lattice.* Consider the modified transfer matrix

$$\mathbf{V} = (2 \sinh 2K)^{M/2} \exp\left\{ K^* \sum_{j=1}^{M} \sigma_{jX} + K \sum_{j=1}^{M} \sigma_{jZ}\sigma_{j+1,Z} \right\}$$

obtained from (5.A.15) if one ignores the fact that the operators $\mathbf{V}_1$ and $\mathbf{V}_2$ do not commute. The largest eigenvalue of this transfer matrix can be found by the methods of Sections 5.A(b) and 5.A(c) with some reduction in complexity. Calculate the free energy and show that the specific heat diverges logarithmically at the critical point.

5.2. *Internal Energy and Specific Heat of the Two-Dimensional Ising Model.* **(a)** Supply the missing steps between (5.A.52) and (5.A.53). *Hint:* The incomplete elliptic integrals $F(\phi, q)$ and $E(\phi, q)$ are defined as follows:

$$F(\phi, q) = \int_0^\phi dx \frac{1}{\sqrt{1 - q^2 \sin^2 x}}$$

$$E(\phi, q) = \int_0^\phi dx \sqrt{1 - q^2 \sin^2 x}$$

First show that

$$\frac{\partial F(\phi, q)}{\partial q} = \frac{1}{1 - q^2} \left[\frac{E(\phi, q) - (1 - q^2)F(\phi, q)}{q} - \frac{q \sin \phi \cos \phi}{\sqrt{1 - q^2 \sin^2 \phi}} \right]$$

and hence

$$\frac{dK_1(q)}{dq} = \frac{E_1(q)}{q(1 - q^2)} - \frac{K_1(q)}{q}$$

(b) Show that (5.A.53) implies the logarithmic singularity of the specific heat (5.A.54).

5.3 *High-Temperature Series for the Susceptibility of the Heisenberg Model.* Consider the spin-$\frac{1}{2}$ Heisenberg model on the simple cubic lattice:

$$H = -J \sum_{\langle ij \rangle} \mathbf{S}_i \cdot \mathbf{S}_j - h \sum_i S_{iz}$$

(a) Construct the high-temperature series for the susceptibility per spin

$$\chi(0, T) = \frac{\partial}{\partial h} \langle S_{iz} \rangle = \beta \sum_j \frac{\text{Tr } S_{iz} S_{jz} \exp \{\beta J \sum_{\langle nm \rangle} \mathbf{S}_n \cdot \mathbf{S}_m\}}{\text{Tr } \exp \{\beta J \sum_{\langle nm \rangle} \mathbf{S}_n \cdot \mathbf{S}_m\}}$$

up to, and including, the term of order J^2.

(b) Analyze this short series by writing

$$\chi(0, T) = \frac{1}{T} \left(\frac{a_1 - a_2/T}{1 - a_3/T} \right)$$

which is a simple Padé approximant (see Gaunt and Guttman, 1974). Find T_c.

(c) Compare with the critical temperature obtained from a two-term expansion and with the best estimate $k_B T_c / J\hbar^2 = 0.84$ obtained from longer series.

5.4. *Analysis of High-Temperature Series.* Analyze the first 10 terms of the high-temperature series for the zero-field susceptibility of the Ising ferromagnet on the triangular lattice and obtain estimates of T_c and γ. The coefficients of the series can be found in Domb [1974].

5.5. *Scaling.* For $T = T_c$ and h small we expect that the correlation length ξ will have the scaling form

$$\xi(h, 0) \sim |h|^{-\nu_H}$$

and that the pair correlation function will have the approximate form

$$g(r, h, 0) \sim \frac{e^{-r/\xi}}{r^{d-2+\eta_H}}$$

(a) Use Landau–Ginzburg theory (Chapter 3) to derive the classical values of the critical exponents ν_H and η_H.

(b) We also expect that the susceptibility will diverge as $|h| \to 0$ on the critical isotherm with an exponent γ_H. Express γ_H in terms of ν_H and η_H.

(c) Using the scaling form (5.C.10), find a relation between γ_H, α_H, and δ where α_H is the specific heat exponent for the critical isotherm.

5.6. *Correlation Function in the Spin Wave Approximation.*

(a) Complete the calculation of the correlation function $g(r)$ defined in (5.E.5) to obtain equation (5.E.9).

(b) Show that in three dimensions the function $g(r)$ approaches a constant as r approaches infinity.

6

Critical Phenomena II:
The Renormalization Group

In this chapter we introduce the renormalization group approach to critical phenomena. In contrast to Chapter 5, we do not proceed historically, but begin in Section 6.A with a renormalization treatment of a simple exactly solvable model, the familiar Ising chain (Section 3.F). We proceed in Section 6.B to discuss properties of fixed points, the relation to scaling, and the notion of universality. Section 6.C continues with the cumulant approach to the position space renormalization group, and the critical exponents are calculated approximately, but in a nontrivial fashion, for a two-dimensional model. In Section 6.D we describe the application of other position space renormalization methods to phase transitions. Section 6.E describes the momentum space approach and the ϵ expansion is developed.

6.A RENORMALIZATION GROUP FOR THE ISING CHAIN

Consider again the Ising model for a one-dimensional chain with periodic boundary conditions (Section 3.F). The Hamiltonian is

$$\tilde{H} = -J \sum_{i=1}^{n} \sigma_i \sigma_{i+1} - \tilde{h} \sum_{i=1}^{N} \sigma_i \tag{1}$$

with $\sigma_i = \pm 1$ and $\sigma_{N+1} = \sigma_1$. We define the dimensionless Hamiltonian

$$H = -\beta\tilde{H} = K \sum_{i=1}^{N} \sigma_i \sigma_{i+1} + h \sum_{i=1}^{N} \sigma_i \tag{2}$$

with $K = \beta J$, $h = \beta \bar{h}$. The partition function is given by

$$Z_c = \text{Tr } e^H = \sum_{\{\sigma_i\}=\pm 1} \exp \left\{ \sum_{i=1}^{N} [K\sigma_i\sigma_{i+1} + \tfrac{1}{2}h(\sigma_i + \sigma_{i+1})] \right\} \qquad (3)$$

We now carry out the sum over the degrees of freedom in two steps,

$$\sum_{\{\sigma_i\}=\pm 1} e^H = \sum_{\sigma_2=\pm 1} \sum_{\sigma_4=\pm 1} \cdots \sum_{\sigma_N} \left[\sum_{\sigma_1=\pm 1} \sum_{\sigma_3=\pm 1} \cdots \sum_{\sigma_{N-1}=\pm 1} e^H \right] \qquad (4)$$

The sums inside the brackets are easy to carry out. Each spin with an odd index is connected by the nearest-neighbor interaction only to spins with an even index. Thus the terms in H which involve σ_1 are simply

$$K\sigma_1(\sigma_N + \sigma_2) + h\sigma_1$$

and carrying out the trace over σ_1, we find that

$$\sum_{\sigma_1=\pm 1} e^{K\sigma_1(\sigma_N+\sigma_2)+h\sigma_1} = 2 \cosh [K(\sigma_N + \sigma_2) + h]$$

Using the property $\sigma^{2n} = 1$, $\sigma^{2n+1} = \sigma$ of Ising spins, we write

$$2e^{h(\sigma_N+\sigma_2)/2} \cosh [K(\sigma_N + \sigma_2) + h]$$

$$= \exp \{2g + K'\sigma_N\sigma_2 + \tfrac{1}{2}h'(\sigma_N + \sigma_2)\} \qquad (5)$$

where

$$K' = \frac{1}{4} \ln \frac{\cosh (2K + h) \cosh (2K - h)}{\cosh^2 h} \qquad (6)$$

$$h' = h + \frac{1}{2} \ln \frac{\cosh (2K + h)}{\cosh (2K - h)} \qquad (7)$$

and

$$g = \tfrac{1}{8} \ln [16 \cosh (2K + h) \cosh (2K - h) \cosh^2 h)] \qquad (8)$$

All other sums inside the brackets in (4) yield identical results and we have

$$\sum_{\sigma_1,\sigma_3...\sigma_{N-1}=\pm 1} e^H = \exp \{Ng(K, h) + K' \sum_i \sigma_{2i}\sigma_{2i+2} + h' \sum_i \sigma_{2i}\} \qquad (9)$$

where the sum over spins in the exponential is over the remaining even-numbered sites. We notice that the sum over the even spins constitutes a problem of exactly the same type as the calculation of the original partition function. The remaining spins of the thinned-out chain interact with their nearest neighbors through a "renormalized" coupling constant K' and with a renormalized magnetic field. We therefore have the relation

$$Z_c(N, K, h) = e^{Ng(K,h)} Z_c(\tfrac{1}{2}N, K', h') \qquad (10)$$

The situation is shown schematically as

It is clear that the process may be continued indefinitely. Notice that from (10) we may obtain a formula for the free energy

$$-\beta G(N, K, h) = \ln Z_c(N, K, h)$$
$$= Ng(K, h) + \ln Z(\tfrac{1}{2}N, K', h') \tag{11}$$

or

$$-\frac{\beta G}{N} = \mu(K, h) = g(K, h) + \tfrac{1}{2}g(K', h') + \tfrac{1}{4}g(K'', h'') + \cdots \tag{12}$$
$$= \sum_{j=0}^{\infty} (\tfrac{1}{2})^j g(K_j, h_j)$$

The important feature of this equation is that the form of the function g is the same at each stage of the iteration since the renormalized Hamiltonian always has the same form. To discuss the convergence of the sum (12), we must understand the "flow" of the coupling constants K, h. Let us first consider the case $h = 0$. Then $h_j = 0$ for all j. From (6) we obtain

$$K' = \tfrac{1}{2} \ln \cosh 2K \le K \tag{13}$$

The equality $K' = K$ holds at the two special points $K = 0$ and $K = \infty$. These are called the *fixed points* of the renormalization transformation. For any finite K the successive thinning out of degrees of freedom produces a Hamiltonian for which the remaining spins are more weakly coupled. The flow in coupling constant space is thus toward a Hamiltonian which consists of noninteracting degrees of freedom. This fixed point, which also can be thought of as an infinite temperature fixed point is *stable*. Conversely, the other fixed point at $K = \infty$, or $T = 0$, is unstable. A dimensionless Hamiltonian which deviates from $K = \infty$ will under renormalization flow toward $K = 0$. It is now clear that the sum (12) will converge for any finite coupling constant K. From (8) we have

$$g(K, 0) = \tfrac{1}{2} \ln 2 + \tfrac{1}{4} \ln (\cosh 2K) \tag{14}$$

As K becomes smaller, the second term in (14) approaches zero, and if we neglect the contribution of this term to the sum in (12) for $j > n$, we obtain

$$\mu(K, 0) = \sum_{j=0}^{n} g(K_j, 0)(\tfrac{1}{2})^j + 2^{-(n+1)} \ln 2 \tag{15}$$

The last term is simply the entropy/k_B per particle of the remaining $N/2^{n+1}$ spins, which are effectively noninteracting.

It is interesting to examine the flow of the coupling constants for nonzero h. From (7) we see that

$$\frac{\partial h'}{\partial h} > 1$$

for all finite K. Thus a small magnetic field will under iteration become larger, and since K becomes smaller, the flow will be toward the line $K = 0$. This flow is shown schematically in Figure 6.1 for a number of different starting points in the K–h plane.

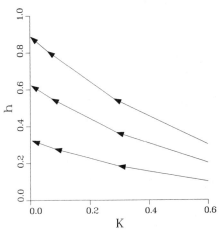

Figure 6.1 Renormalization flow for one-dimensional Ising model.

In Section 3.F we calculated the correlation length and found that $\xi = 0$ at $K = 0$ and $\xi = \infty$ at $K = \infty$. The flow of the coupling constants can be understood in terms of Kadanoff's scaling picture and the behavior of the correlation length. Crudely speaking, we have in our renormalization replaced a pair of spins by a block spin. The block spins are separated by twice the site spin separation. In Kadanoff's scaling picture, one expects the correlation length for block spins to be smaller than the correlation length for site spins, unless the system is critical ($\xi = \infty$), or noninteracting ($\xi = 0$). The two fixed points can thus be understood as corresponding to Hamiltonians for which the correlation lengths are invariant under rescaling. One Hamiltonian ($K = 0$) is trivial, one ($K = \infty$) is critical in this picture. In a higher-dimensional system, with a finite temperature phase transition, one therefore expects at least three fixed points. Two of these will be trivial (infinite and zero temperature) and one will be the critical fixed point.

Before we go on to discuss the properties of fixed points in more detail, we summarize the important results of this section. First, we have developed a new way of evaluating the partition function by successively thinning out degrees of freedom. Because the form of the Hamiltonian remained invariant under this "scale transformation," we were able to evaluate the free energy by a simple iterative scheme. The Kadanoff picture of decreasing correlation lengths shows up in this scheme as a flow toward Hamiltonians with successively smaller coupling constants. This procedure of thinning out degrees of freedom has been termed renormalization group by Wilson because two successive such operations have the property

$$R_b(R_b(\{K\})) = R_{2b}(K)$$

where $R_b(K) = K'$ describes the effect on coupling constants of replacing b^d spins by one block spin. There is no inverse operation R_b^{-1} and therefore the word "group" is a misnomer.

6.B. FIXED POINTS

We now turn to a more general discussion of the renormalization group approach. In Section 6.A we derived recursion relations for the pair of dimensionless coupling constants (K, h) of the one-dimensional Ising model. We now wish to consider a system that is specified, on a d-dimensional lattice, in terms of a set of coupling constants $\{K\} = (K_1, K_2, \ldots, K_n)$. Here K_1 might correspond to nearest neighbor interactions, K_2 to second neighbor, K_3 to a magnetic field, and so on. We suppose that this set of coupling constants is complete, in the sense that a renormalization transformation which replaces b^d degrees of freedom by one, results in a Hamiltonian with exactly the same type of interactions between the remaining degrees of freedom. We describe the system in terms of a dimensionless Hamiltonian

$$H = -\beta \tilde{H} = \sum_{\alpha=1}^{n} K_\alpha \psi_\alpha(\sigma_i) \tag{1}$$

where, for example,

$$\psi_1 = \sum_{\langle ij \rangle} \sigma_i \sigma_j \qquad \psi_2 = \sum_{ijnnn} \sigma_i \sigma_j \qquad \psi_3 = \sum_i \sigma_i \cdots$$

A renormalization transformation will then produce a new Hamiltonian,

$$H' = \sum_{\alpha=1}^{n} K'_\alpha \psi_\alpha(\sigma_l) + Ng(\{K\}) \tag{2}$$

where the remaining degrees of freedom $\{\sigma_l\}$ have the same algebraic properties as the original ones and the functional form of the ψ_α's is unchanged by the transformation. In (2) a term $g(\{K\})$ has been included because, as we saw in (6.A.8), there will in general be a spin-independent term as a result of the partial trace. Assuming that the thinning-out operation can be carried out, we have the relations

$$K'_\alpha = R_\alpha(K_1, K_2, \ldots, K_n) \tag{3}$$

and

$$\mathop{\mathrm{Tr}}_{\{\sigma_i\}} e^H = e^{Ng(\{K\})} \mathop{\mathrm{Tr}}_{\{\sigma_l\}} e^{H'(\{K'\})} \tag{4}$$

or with

$$\text{Tr } e^H = e^{Nf(\{K\})}$$
$$\text{Tr } e^{H'} = e^{Nf(\{K'\})/b^d}$$

(5)

we obtain

$$f(\{K\}) = g(\{K\}) + b^{-d}f(\{K'\})$$

(6)

Equation (6.A.11) is a special case of this formula. We leave the recursion relation for the free energy aside for the moment and concentrate on the relation (3) for the coupling constants. We have already seen in Section 6.A that the fixed points of this recursion relation correspond to either noninteracting or critical Hamiltonians. However, as we shall presently see, there will in general be critical points which are not fixed points. Let us imagine a two-dimensional space of coupling constants K_1, K_2 with a critical point at K_{1c}, K_{2c}. We note that this point is in general a point on a line of critical points. To see this, imagine a number of different systems with different ratios J_2/J_1 of second-nearest to nearest-neighbor interactions. The critical temperature T_c will depend on this ratio. Thus as J_2/J_1 varies, the point

$$(K_{1c}, K_{2c}) = \left(\frac{J_1}{k_B T_c}, \frac{J_2}{k_B T_c} \right)$$

describes a curve in the K_1–K_2 plane. Each point on the curve corresponds to the critical point of a particular model in the family of Hamiltonians. This situation is depicted schematically in Figure 6.2.

The dotted line in Figure 6.2 is the path that a particular system follows

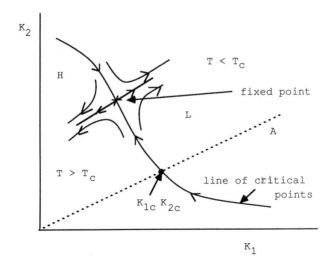

Figure 6.2 Lines with arrows indicate direction of flow in recursion relation. Dashed line indicates possible states for a system with a given value of $K_2/K_1 = J_2/J_1$.

in coupling constant space as the temperature is lowered from $T = \infty$ ($K_1 = K_2 = 0$) to $T = 0$ with $K_2/K_1 = J_2/J_1$ held fixed. We now attempt to relate the properties of the system, as it describes this path, to the flow of the coupling constants under a renormalization transformation. The flows have a number of simple properties. First, it is clear that the flows cannot approach the line of critical points. This is because the correlation length ξ is infinite on this curve and is finite everywhere else. We have argued that as the degrees of freedom are thinned out, the correlation length relative to the new spacing can only decrease. The states on the right side of the line of critical points correspond to a low-temperature phase and will be ordered. The states on the high-temperature side will be disordered. This aspect cannot be changed by the thinning out of the degrees of freedom, and the flow will therefore not cross the critical line. We conclude that the flow for $T > T_c$ and $T < T_c$ must be as shown in Figure 6.2. In region $H(T > T_c)$ the flow will be toward a $T = \infty$ noninteracting fixed point $K_1 = K_2 = 0$. In region $L(T < T_c)$ the flow will be toward a zero temperature (ground state) fixed point. Conversely, the flow from points K_{1c}, K_{2c} on the critical line must remain on this line since $\xi = \infty$. It is possible that all points on the critical line are stationary (i.e., are fixed points of the renormalization group transformation), but this turns out to be an exceptional case. More generally, one finds a finite number of isolated fixed points. Let us therefore assume that K_1^*, K_2^* is a fixed point on the critical line,

$$K_1^* = R_1(K_1^*, K_2^*)$$
$$K_2^* = R_2(K_1^*, K_2^*) \tag{7}$$

and that the flow along the critical line approaches this point. Consider now the flow near K_1^*, K_2^*. Let $\delta K_1 = K_1 - K_1^*$, $\delta K_2 = K_2 - K_2^*$. To first order,

$$K_1' = R_1(K_1^* + \delta K_1, K_2^* + \delta K_2)$$
$$= K_1^* + \delta K_1 \frac{\partial R_1}{\partial K_1}\bigg|_{K_1^* K_2^*} + \delta K_2 \frac{\partial R_1}{\partial K_2}\bigg|_{K_1^* K_2^*} \tag{8}$$

A similar expression obtains for K_2'. We write these expressions in matrix form,

$$\delta K_1' = M_{11}\, \delta K_1 + M_{12}\, \delta K_2$$
$$\delta K_2' = M_{21}\, \delta K_1 + M_{22}\, \delta K_2 \tag{9}$$

and find an appropriate coordinate system for describing the flow by solving the left eigenvalue problem

$$\sum_{i=1,2} \varphi_{\alpha i} M_{ij} = \lambda_\alpha \varphi_{\alpha j} = b^{y_\alpha} \varphi_{\alpha j} \tag{10}$$

using the group property of the renormalization transformation $\lambda_\alpha(b)\lambda_\alpha(b) = \lambda_\alpha(b^2)$. Therefore,

$$\lambda_\alpha = b^{y_\alpha} \tag{11}$$

defines y_α. Consider now the new variables ($\alpha = 1, 2$)

$$U_\alpha = \delta K_1 \varphi_{\alpha 1} + \delta K_2 \varphi_{\alpha 2} \tag{12}$$

We apply the linearized renormalization transformation (9) and use the fact that φ_α is a left eigenvector to obtain

$$\begin{aligned} U'_\alpha &= \delta K'_1 \varphi_{\alpha 1} + \delta K'_2 \varphi_{\alpha 2} \\ &= \lambda_\alpha U_\alpha = b^{y_\alpha} U_\alpha \end{aligned} \tag{13}$$

Geometrically, U_α corresponds to a projection of the deviation (δK_1, δK_2) from the fixed point on the basis vector φ_α The U'_α's are called *scaling fields* for reasons that will become clear.

We can make one further statement regarding the flow near the fixed point. One of the exponents, say y_2, must be negative, the other, y_1 positive. The reason for this is the assumption that flows which originate in the critical line must tend toward the fixed point. Thus one of the basis vectors, φ_2, must be tangential to the critical surface at the fixed point. The other vector must point out of the critical surface. Since the matrix M is generally not symmetric, φ_1 and φ_2 are not necessarily orthogonal, but this has no bearing on the discussion that follows. The exceptional case, $y_2 = 0$, corresponds to a line of fixed points rather than the case of an isolated fixed point.

Let us now return to the recursion relation (6) for the free energy. Since the first term $g(\{K\})$ arises from the removal of short distance fluctuations, which play no role in the phase transition, $g(\{K\})$ is expected to be an analytic function of the coupling constants. Therefore, the singular part of the free energy obeys the relation

$$f_s(\{K\}) = b^{-d} f_s(\{K'\}) \tag{14}$$

Let us now suppose that the point $\{K\}$ is close enough to $\{K^*\}$ that we may use the linearized recursion relations (11). Re-expressing K, K' in terms of U_1, U_2, we have

$$f_s(U_1, U_2) = b^{-d} f_s(b^{y_1} U_1, b^{y_2} U_2) \tag{15}$$

that is, a scaling form of the free energy. We now assume that a change in temperature at constant field corresponds to a change in U_1 at constant U_2. The connection with the critical exponents defined in Chapter 5 can now made by realizing that if $T \neq T_c$, $U_1 \neq 0$. Conversely, if $T = T_c$, $U_1 = 0$ as the system point must lie on the critical line. Defining

$$t = U_1 = \frac{T - T_c}{T_c}$$

we obtain

$$f_s(t, U_2) = b^{-d} f_s(b^{y_1} t, b^{y_2} U_2) \tag{16}$$

A relation of this type must hold for any rescaling parameter b and we may therefore let $b = |t|^{-1/y_1}$ and obtain

$$f_s(t, U_2) = |t|^{d/y_1} f_s(t/|t|, |t|^{-y_2/y_1} U_2) \tag{17}$$

This equation demonstrates two important features of critical points. First, the role of the fixed point is clarified: the critical exponents are determined by the eigenvalues of the linearized recursion relations at the fixed point. Since the specific heat is proportional to the second derivative of f with respect to t, we have $f_s \propto t^{2-\alpha}$, where α is the specific heat critical exponent. We thus have $d/y_1 = 2 - \alpha$ and $y_1 = \ln \lambda_1/\ln b$. Next we see the concept of universality emerging from the theory. We argued above that $y_2 < 0$. As $t \to 0$ the term

$$|t|^{-y_2/y_1} U_2 \to 0$$

and the asymptotic behavior of the free energy is independent of U_2. In other words, all systems whose Hamiltonians flow under renormalization to the same critical fixed point have the same critical exponents. These are the most important qualitative results that follow from the renormalization group approach.

To obtain a complete description, we now generalize our analysis to higher-dimensional spaces. The fixed point in the n-dimensional space of coupling constants is given by

$$K_j^* = R_j(K_1^*, \ldots, K_n^*)$$

The matrix M becomes an $n \times n$ matrix with

$$\delta K_j' = \sum_l M_{jl} \, \delta K_l$$

There are now n eigenvalues corresponding to the solution to the eigenvalue problem

$$\boldsymbol{\varphi}_\alpha M = b^{y_\alpha} \boldsymbol{\varphi}_\alpha \tag{18}$$

and the generalization of (15) is

$$f_s(U_1, \ldots, U_n) = b^{-d} f_s(b^{y_1} U_1, \ldots, b^{y_n} U_n) \tag{19}$$

Ordinary critical points are characterized (Section 5.C) by two independent exponents and we therefore expect that two of the y's, say y_1 and y_2, are positive, the rest negative. In the generalized Ising model, we expect that $U_1 \propto (T - T_c)/T_c$, $U_2 \propto h$ and that all other scaling fields (corresponding, for example, to $J_2/J_1 - J_2^*/J_1^*$ with J_2 the second-neighbor interaction) will play no role in the asymptotic critical behavior.

It is, of course, possible that the critical surface will contain several fixed points with different domains of attraction. The critical surface of the anisotropic Heisenberg model (see Section 5.D) is presumably such a critical surface. In this case, there is one fixed point, the Heisenberg fixed point, which is unstable with respect to a third scaling field proportional to the anisotropy parameter η (5.D.2). For any nonzero η the flow in the critical surface is either toward the Ising or XY fixed point. On the basis of the analysis above, we see that it is possible to observe the Heisenberg critical exponents only if this scaling field or anisotropy parameter η is exactly zero. In any other situation we would observe either XY or Ising exponents as long as we were sufficiently close to the critical point. This statement, which follows very simply from (19), is consistent with the series expansion results of Jasnow and Wortis [1968].

We mention also that there is a conventional terminology for the different types of scaling fields. A scaling field U_α with exponent $y_\alpha > 0$ is termed *relevant;* if $y_\alpha < 0$ it is called *irrelevant* and in the special case $y_\alpha = 0$ it is *marginal.*

Before closing this section we note that we have made certain tacit assumptions. First, we have implicitly assumed that the renormalization transformation is analytic everywhere in coupling constant space and in particular at the fixed point. Moreover, we have assumed that a system can be characterized in terms of a finite number of coupling constants. That this is the case is by no means obvious and in practice it is usually a specific truncation approximation that guarantees that the number of coupling constants remains finite and that the recursion relations are analytic. We shall have more to say about these points later.

We have also assumed, in a rather cavalier manner, that the free energy $f(K)$ which obeys the relation (6) has a singular piece which obeys the simple relation

$$f_s(\{K\}) = b^{-d} f_s(\{K'\})$$

We could use (6) to obtain an infinite series for the free energy in terms of the analytic function $g(\{K\})$

$$f(\{K\}) = \sum_{j=1}^{\infty} b^{-jd} g(\{K^{(j)}\})$$

and attempt to show how the singular piece emerges from this sum. This has been done by Niemeijer and van Leeuwen [1976] and the interested reader is referred to this article.

Finally, we have assumed that the linear approximation is valid even far from the fixed point as long as the system is close to the critical surface. This assumption may also be removed by systematically including higher-order terms in (8) (Niemeijer and van Leeuwen, 1976).

In the next section, we turn toward some concrete examples of renormalization group calculations for systems which, unlike the one-dimensional Ising model, do display a phase transition.

6.C POSITION SPACE RENORMALIZATION GROUP:
THE CUMULANT APPROACH

In Section 6.A we constructed a simple renormalization transformation for the one-dimensional Ising model. We now carry out the analogous calculation for the two-dimensional model on the triangular lattice (Niemeijer and van Leeuwen, 1976). This calculation will illustrate some of the important features discussed in Section 6.B and will also highlight a number of difficulties that arise in such calculations. The dimensionless Hamiltonian is

$$H = \sum_{ij} K_{ij}\sigma_i\sigma_j + h\sum_i \sigma_i \tag{1}$$

where $\sigma_i = \pm 1$, with the spins occupying the sites of a triangular lattice, and the sum no longer restricted to nearest-neighbor sites.

We see from Figure 6.3 that the lattice can be divided into triangular blocks each containing three spins. These triangles in turn form a triangular lattice with separation $\sqrt{3} \times$ the original separation. Our procedure will be to map the original system onto a system of the same form as (1) but with block spins $\mu_I = \pm 1$ representing the state of the three spins σ_{1I}, σ_{2I}, σ_{3I} composing block I. We formulate this mapping process in terms of a projection operator

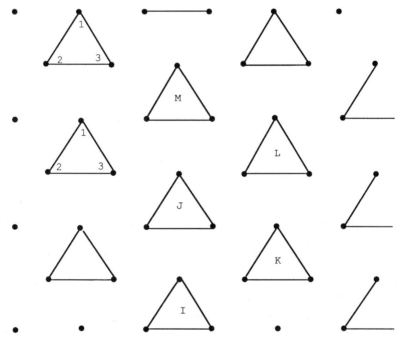

Figure 6.3 Partitioning of a triangular lattice into blocks of three spins. Numbers indicate labeling within each block; capital letters label blocks.

$$P(\mu_I, \sigma_{1I}, \sigma_{2I}, \sigma_{3I}) \equiv P(\mu_I, \{\sigma_I\}) \tag{2}$$

satisfying

$$e^{Ng(K, h) + H'(\{\mu\}, K', h')} = \text{Tr}_{\{\sigma\}}\left(\prod_I P(\mu_I, \{\sigma\})\right)e^{H(\{\sigma\}, K, h)} \tag{3}$$

If we demand that the projection operator P satisfies the relation

$$\text{Tr}_{\{\mu\}} P(\mu_I, \{\sigma_I\}) = 1 \tag{4}$$

then

$$\text{Tr}_{\{\mu\}} e^{Ng + H'} = \text{Tr}_{\{\sigma\}} e^{H} \tag{5}$$

and the free energy will be exactly preserved by the transformation, provided that we can carry out the operation (3). A possible candidate for the projection operator P is

$$P(\mu, \{\sigma\}) = \delta_{\mu\varphi(\{\sigma\})} \tag{6}$$

where $\varphi(\{\sigma\}) = \frac{1}{2}(\sigma_1 + \sigma_2 + \sigma_3 - \sigma_1\sigma_2\sigma_3)$ and $\delta_{\mu,\varphi}$ is the Kronecker symbol. Note that $\varphi(\{\sigma\}) = 1$ whenever two or more spins on the triangle are 1 and $\varphi(\{\sigma\}) = -1$ whenever two or more spins are -1. The projection operator thus assigns the block spin μ the value $+1$ or -1 according to a *majority rule*.

At this point our formulation is still exact, but the problem of evaluating (3) is intractable without some approximation. Before carrying out an approximate calculation, we note that the question of whether (3) and (6) define an analytic recursion relation (see Section 6.B) constitutes a difficult mathematical problem. This problem was addressed by Griffiths and Pearce [1978], who showed that this renormalization transformation is in general *not* analytic. However, approximate versions of (3) with (6) do produce analytic recursion relations, as do more sophisticated renormalization transformations.

Many different approximate treatments of (3) are possible. We outline first the cumulant approach of Niemeijer and van Leeuwen [1976], which is perhaps the simplest technically. We divide the Hamiltonian into two parts,

$$H(\{\sigma\}, K, h) = H_0(\{\sigma\}, K, h) + V(\{\sigma\}, K, h) \tag{7}$$

H_0 represents the part of the Hamiltonian that does not involve couplings between spins in different blocks, and V contains the coupling between blocks. Thus

$$H_0 = \sum_I K_1(\sigma_{1I}\sigma_{2I} + \sigma_{1I}\sigma_{3I} + \sigma_{2I}\sigma_{3I}) + h\sum_I(\sigma_{1I} + \sigma_{2I} + \sigma_{3I}) \tag{8}$$

and

$$V = \sum_{I,J} K_n \sum_{\alpha,\beta} \sigma_{\alpha I} \sigma_{\beta J} \qquad (9)$$

where K_n is the nth nearest-neighbor coupling constant and the labels α, β run over the appropriate labels on blocks I, J. From Figure 6.3 we see that in the case of nearest-neighbor interaction, adjacent blocks interact through two types of coupling terms.

$$V_{IJ} = K_1(\sigma_{1I}\sigma_{2J} + \sigma_{1I}\sigma_{3J})$$
$$V_{IK} = K_1(\sigma_{1I}\sigma_{2K} + \sigma_{3I}\sigma_{2K}) \qquad (10)$$

We may now write (3) in the form

$$\mathrm{Tr}_{\{\sigma\}}\left(\prod_I P(\mu_I, \{\sigma_I\})\right)e^H = \mathrm{Tr}_{\{\sigma\}}\left(\prod_I P(\mu_I, \{\sigma_I\})\right)e^{H_0}e^V = Z_0\langle e^V\rangle \qquad (11)$$

where

$$Z_0 = \mathrm{Tr}_{\{\sigma\}}\left(\prod_I P(\mu_I, \{\sigma\})\right)e^{H_0} \qquad (12)$$

and

$$\langle A\rangle = \frac{1}{Z_0}\mathrm{Tr}_{\{\sigma\}}\left(\prod_I P(\mu_I, \{\sigma\})\right)A e^{H_0} \qquad (13)$$

This formula is still exact. We now approximate $\langle e^V\rangle$ by a truncated cumulant expansion

$$\langle e^V\rangle = \langle 1 + V + \frac{V^2}{2!} + \cdots + \frac{V^n}{n!} + \cdots\rangle$$

$$= \exp\left\{\langle V\rangle + \frac{1}{2!}(\langle V^2\rangle - \langle V\rangle^2) + \frac{1}{3!}(\langle V^3\rangle - 3\langle V\rangle\langle V^2\rangle + 2\langle V\rangle^3) + \cdots\right\}$$

$$= \exp\left\{C_1 + C_2 + C_3 + \cdots\right\} \qquad (14)$$

The jth cumulant approximation then corresponds to retaining the first j cumulants $C_1 \cdots C_j$.

(a) First-Order Approximation

Let us assume that only nearest-neighbor interactions are present and set $K_1 = K$, $K_2 = K_3 = \cdots = 0$. The trace (12) over the decoupled blocks is easy to carry out. Writing

$$Z_0 = \prod_I e^{A+B\mu_I} \qquad (15)$$

we obtain

$$e^{A+B} = e^{3K+3h} + 3e^{-K+h}$$

$$e^{A-B} = e^{3K-3h} + 3e^{-K-h}$$

or

$$A = \tfrac{1}{2} \ln \left[(e^{3K+3h} + 3e^{-K+h})(e^{3K-3h} + 3e^{-K-h}) \right]$$

$$B = \frac{1}{2} \ln \frac{e^{3K+3h} + 3e^{-K+h}}{e^{3K-3h} + 3e^{-K-h}} \tag{16}$$

Expectation values of the type $\langle \sigma_{\alpha I} \sigma_{\beta J} \rangle$ can be factored since they are to be evaluated with respect to the Hamiltonian H_0 which contains no coupling between blocks I and J. For this reason the expectation value $\langle V \rangle$ is also easy to obtain:

$$\langle V \rangle = K \sum_{\langle \alpha I, \, \beta J \rangle} \langle \sigma_{\alpha I} \rangle \langle \sigma_{\beta J} \rangle \tag{17}$$

The expectation value $\langle \sigma_{\alpha I} \rangle$ is by symmetry independent of α and can be written as

$$\langle \sigma_{\alpha I} \rangle = C + D\mu_I \tag{18}$$

with

$$C = \frac{1}{2} \left(\frac{e^{3K+3h} + e^{-K+h}}{e^{3K+3h} + 3e^{-K+h}} - \frac{e^{3K-3h} + e^{-K-h}}{e^{3K-3h} + 3e^{-K-h}} \right)$$

$$D = \frac{1}{2} \left(\frac{e^{3K+3h} + e^{-K+h}}{e^{3K+3h} + 3e^{-K+h}} + \frac{e^{3K-3h} + e^{-K-h}}{e^{3K-3h} + 3e^{-K-h}} \right) \tag{19}$$

Combining these results, we have

$$Ng(K, h) + H'(\{\mu_I,\} K', h')$$

$$= \frac{1}{3} NA(K, h) + B\sum_I \mu_I + 2K \sum_{\langle IJ \rangle} (C + D\mu_I)(C + D\mu_J) \tag{20}$$

where the functions A, B, C, D are given by (16–19). Reexpressing the renormalized Hamiltonian H' in the original form (1),

$$H' = K' \sum_{\langle IJ \rangle} \mu_I \mu_J + h' \sum_I \mu_I \tag{21}$$

we obtain the recursion relations

$$K' = 2KD^2(K, h)$$

$$h' = B(K, h) + 12KC(K, h)D(K, h) \tag{22}$$

$$g(K, h) = \tfrac{1}{3} A(K, h) + 2KC^2(K, h)$$

The flow in the K–h plane from the recursion relations (22) is indicated in Figure 6.4. Since the phase transition to the ferromagnetic state takes place at

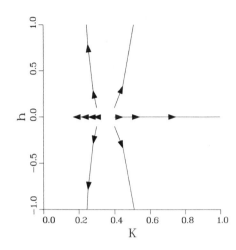

Figure 6.4 Recursion flow for the two-dimensional Ising model in first-order cumulant approximation.

$h = 0$, we look for a fixed point $(K^*, h^*) = (K_c, 0)$ where $K_c = J/k_B T_c$. For $h = 0$, $B = C = 0$ and our recursion relations (22) reduce to

$$K' = 2K\left(\frac{e^{3K} + e^{-K}}{e^{3K} + 3e^{-K}}\right)^2 \tag{23}$$

$$h' = h = 0 \tag{24}$$

The recursion relation (23) has very simple limiting behavior. For $K \ll 1$ $K' \approx (2K)(\frac{1}{2})^2 = \frac{1}{2}K$, while for $K \gg 1$ $K' \approx 2K$ and we see that the flow reverses itself at some finite value of K. For small K the flow is toward the noninteracting high-temperature fixed point, for large K toward the $K = \infty$ ground-state fixed point. The critical point is given by

$$\frac{e^{3K^*} + e^{-K^*}}{e^{3K^*} + 3e^{-K^*}} = \frac{1}{\sqrt{2}} \tag{25}$$

This equation is easily solved analytically, by making the substitution $x = \exp\{4K^*\}$ and solving for x, to yield $K^* = J/k_B T_c = 0.3356$. The exact result for this model is $K^* = 0.27465$ and the mean field result is $K^* = 0.1667$. We see that the critical temperature in the first-order approximation does not agree particularly well with the exact result. Nevertheless, this treatment of the model, because of the very structure of the renormalization theory, does produce nontrivial (i.e., non-mean-field-like) critical exponents. By symmetry,

$$\left.\frac{\partial h'}{\partial K}\right|_{K^*, h^*} = \left.\frac{\partial K'}{\partial h}\right|_{K^*, h^*} = 0$$

so that we quickly obtain

$$b^{y_t} = \left.\frac{\partial K'}{\partial K}\right|_{K^*, h^*} \qquad b^{y_h} = \left.\frac{\partial h'}{\partial h}\right|_{K^*, h^*} \tag{26}$$

with $b = \sqrt{3}$. Evaluating the derivatives at the fixed point we find $y_t = 0.882$, $y_h = 2.034$. The specific heat exponent $\alpha = 2 - d/y_t = -0.267$ and the magnetization exponent is $\beta = (d - y_h)/y_t = -0.039$ in the first-order approximation. Again these results are not particularly accurate ($\alpha = 0$, $\beta = 0.125$ in the exact theory), but we have developed a theory that captures the essence of critical behavior and can be improved systematically.

(b) Second-Order Approximation

The next order approximation in the hierarchy (14) consists of retaining the terms

$$\langle V \rangle + \tfrac{1}{2}(\langle V^2 \rangle - \langle V \rangle^2)$$

The calculation is then more cumbersome, but we outline it here since it will illustrate some features of our qualitative discussion of Section 6.B and also highlight some of the technical problems in renormalization calculations.

Consider again Figure 6.3, in which some typical cells have been drawn. Cells I, J are nearest neighbors, while cells I, L are second neighbors and I, M third neighbors in the triangular lattice of cells. It is clear that the second-order cumulant $\langle V^2 \rangle - \langle V \rangle^2$ contains interactions of longer range than nearest neighbor. For example, the third nearest-neighbor cells I and M are coupled by terms of the form

$$K^2 \langle \sigma_{1I}(\sigma_{2J} + \sigma_{3J})\sigma_{1J}(\sigma_{2M} + \sigma_{3M})\rangle - K^2 \langle \sigma_{1I}(\sigma_{2J} + \sigma_{3J})\rangle\langle \sigma_{1J}(\sigma_{2M} + \sigma_{3M})\rangle$$

$$= 4K^2\{\langle \sigma_{1I}\rangle\langle \sigma_{1J}\sigma_{2J}\rangle\langle \sigma_{1M}\rangle - \langle \sigma_{1J}\rangle\langle \sigma_{2J}\rangle^2\langle \sigma_{2M}\rangle\} \tag{27}$$

Let us consider the simple case $h = 0$ for which (27) can be easily evaluated. We need one new expectation value

$$\langle \sigma_{1J}\sigma_{2J}\rangle = \frac{e^{3K} - e^{-K}}{e^{3K} + 3e^{-K}} = E(K) \tag{28}$$

which is independent of μ_J. The contribution (27) will appear twice in the expansion of $\langle V^2 \rangle$ canceling the $\tfrac{1}{2}$ in the cumulant expansion (14) and we see that the renormalized Hamiltonian will contain a third nearest-neighbor coupling term $K_3'\mu_I\mu_M$. Noting that for $h = 0$, $B = C = 0$, we find that

$$K_3' = 4K^2[D^2(K)E(K) - D^4(K)] \tag{29}$$

where $D(K)$ and $E(K)$ are given by (19) and (28), respectively. In the next iteration the second cumulant will generate still longer-range interactions through terms of order $(K_3')^2$. Therefore, we must find some way of truncating the system of recursion relations.

The first approach of Niemeijer and van Leeuwen [1976] ordered the coupling constants into a hierarchy according to the power of the nearest-neighbor coupling constant at which they are first generated. Thus the unique term of order 1 is the nearest-neighbor coupling constant; the terms of order 2 are the second and third neighbor constants K_2 and K_3. In the second cumulant approximation the second and third neighbor interactions are included only in $\langle V \rangle$, while the first neighbor interaction is retained in $\langle V \rangle$ and $\langle V^2 \rangle - \langle V \rangle^2$. With this choice a fixed number of coupling constants appear at every iteration. This classification of interactions is rather arbitrary, and we will in Section 6.D(a) discuss a better method, the finite cluster approximation, which also was developed by Niemeijer and van Leeuwen [1976]. However, we note that all position space calculations on models which are not exactly solvable do require some ad hoc approximation procedure, to avoid dealing with the infinite number of coupling constants implied by (3). The rationalization for such a truncation is that at the fixed point there appear to be only a small number of relevant scaling fields (Section 6.B). One hopes that by a proper choice of a finite set of coupling constants, one can obtain an accurate representation of the relevant scaling fields.

The second-order recursion relations are

$$K_1' = 2K_1 D^2 + 4(D^2 + D^2 E - 2D^4)K_1^2 + 3D^2 K_2 + 2D^2 K_3$$

$$K_2' = K_1^2(7D^2 E + D^2 - 8D^4) + K_3 D^2 \qquad (30)$$

$$K_3' = 4K_1^2(D^2 E - D^4)$$

where D and E are given by (19) and (28). The fixed point is located at

$$K_1^* = 0.27887 \qquad K_2^* = -0.01425 \qquad K_3^* = -0.01523$$

The critical point in the nearest-neighbor model can be located by finding the intersection of the critical surface and the K_1 axis. One finds $K_{1c} = 0.2575$ in better agreement with the exact result ($K_c = 0.27465$) than the first-order approximation.

The linearized recursion relations at $\{K^*\}$ yield the matrix M (see Section 6.B). Numerically,

$$M = \begin{bmatrix} 1.8313 & 1.3446 & 0.8964 \\ -0.0052 & 0 & 0.4482 \\ -0.0781 & 0 & 0 \end{bmatrix}$$

with eigenvalues $\lambda_1 = 1.7728$, $\lambda_2 = 0.1948$, and $\lambda_3 = -0.1364$. The relevant scaling field corresponds to λ_1 and has exponent $y_T = y_1 = 1.042$, which is close to the exact value $y_T = 1.0$. The specific heat exponent is 0.081 (i.e., quite small). In the exact solution there is only a logarithmic singularity in the specific heat ($\alpha = 0$).

One can now ask whether still better results can be obtained with higher-order cumulants, larger cells, or weighting functions other than the majority rule. The experience so far has been rather discouraging. However, there are ways other than the cumulant method of implementing the position space renormalization method which yield much better convergence. We describe some of these methods in the next section.

6.D OTHER POSITION SPACE RENORMALIZATION GROUP METHODS

(a) Finite Lattice Methods

The finite lattice approach or cluster approximation represents one of the most useful renormalization group techniques. The basic notion is that the recursion relations for an infinite system can be modeled by exact recursion relations for a small system. We will not attempt any comprehensive review of the method here, and we refer the interested reader to the review article by Niemeijer and van Leeuwen [1976]. We illustrate here the procedure by considering the smallest cluster that can model the Ising ferromagnet on the triangular lattice. A further example involving a system of more direct physical interest is discussed in the next subsection.

Consider the system made up of a pair of nearest-neighbor cells of three spins (see Figure 6.5). We will use the "majority rule" projection (6.C.6), just as we did in our discussion of the cumulant approximation. It is now no longer necessary to approximate $\langle e^V \rangle$; we can evaluate this expression simply by carrying out the trace over the 2^6 configurations of the six spins. For simplicity we only consider $h = 0$, in which case the renormalized Hamiltonian takes the form

$$e^{g(K)+H'(K',\mu)} = e^{g+K'\mu_I\mu_J}$$

$$= \operatorname*{Tr}_\sigma P(\mu, \{\sigma\}) \exp\left\{K \sum_{\langle ij \rangle} \sigma_i \sigma_j\right\} \tag{1}$$

The quantities g, K' can be obtained by noting that

$$e^{g+K'} = \operatorname*{Tr}_\sigma P(\mu_I = +1, \{\sigma\})P(\mu_J = +1, \{\sigma\})e^H$$

$$e^{g-K'} = \operatorname*{Tr}_\sigma P(\mu_I = +1, \{\sigma\})P(\mu_J = -1, \{\sigma\})e^H \tag{2}$$

Because of the restrictions placed on the trace by the projection operator 16 terms contribute in each case. We find

$$e^{g+K'} = e^{8K} + 3e^{4K} + 2e^{2K} + 3 + 6e^{-2K} + e^{-4K}$$

$$e^{g-K'} = 2e^{4K} + 2e^{2K} + 4 + 6e^{-2K} + 2e^{-4K} \tag{3}$$

or

$$K' = \frac{1}{2} \ln \frac{e^{8K} + 3e^{4K} + 2e^{2K} + 3 + 6e^{-2K} + e^{-4K}}{2e^{4K} + 2e^{2K} + 4 + 6e^{-2K} + 2e^{-4K}} \tag{4}$$

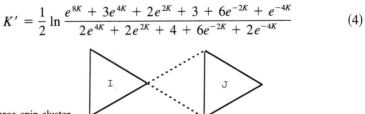

Figure 6.5 Three-spin cluster.

This recursion relation yields a fixed point $K^* = K_c = 0.3653$ and an exponent $y_t = 0.7922$. The results of this simple calculation are not impressive, but Niemeijer and van Leeuwen [1976] have used larger and more symmetric clusters to obtain very good convergence for both the critical temperature and the critical exponents for the Ising ferromagnet on the triangular lattice. It is not apparent from the simple example above how longer-range coupling constants enter into the calculation. A little reflection will convince the reader that all couplings consistent with the symmetry of the Hamiltonian and the cluster will eventually be generated under iteration. Thus, for the Ising model on the square lattice in zero magnetic field a 16-spin cluster divided into four-spin cells allows the first and second nearest-neighbor interactions K_1 and K_2 and a four-spin interaction of the form $K_4' \mu_1 \mu_2 \mu_3 \mu_4$. The calculation $K_1, K_2, K_4 \to K_1', K_2', K_4'$ has been carried out by Nauenberg and Nienhuis [1974], who obtained excellent agreement with the exact results of Onsager. Since there is an even number of spins in each block, it is necessary to modify the majority rule to handle tie votes.

At this point, the reader may wonder whether the renormalization group approach, aside from the fact that it automatically produces a scaling free energy and universality, is limited as a calculational tool to models which are already well understood. In the next section we discuss a model which is relevant to experimental situations, and where mean field theory gives misleading results.

(b) Adsorbed Monolayers: The Ising Antiferromagnet

A number of physisorbed systems have been studied intensely in recent years, since they provide realizations of a rich variety of phase transitions, some of which are peculiar to two dimensions. We discuss here one such system, namely helium on the surface of graphite. The graphite surface has a honeycomb structure, as indicated in Figure 6.6.

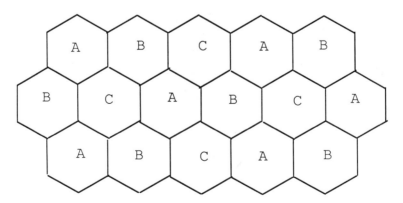

Figure 6.6   Honeycomb structure of a graphite monolayer.

The graphite–helium interaction gives rise to preferred adsorption sites directly above the honeycomb centers. In order to move from one preferred adsorption site to the next, a helium atom will have to pass over a potential barrier. We will here make the idealization that these barriers are large enough that we can treat the system as a two-dimensional lattice gas with the adsorption sites being either filled ($n_i = 1$) or empty $n_i = 0$. If there is a pairwise He–He interaction $V(r_{ij})$, we obtain a Hamiltonian of the form

$$H = \tfrac{1}{2} \sum_{ij} V(r_{ij}) n_i n_j \tag{5}$$

The He–He interaction can be approximated by a Lennard-Jones potential (4.A.3), with a minimum somewhere between the nearest-neighbor distance (2.46 Å) and the second nearest honeycomb to honeycomb distance ($\sqrt{3} \times 2.46 = 4.26$ Å). We neglect the second- and higher-neighbor interactions and work with the idealized Hamiltonian

$$H = V_0 \sum_{\langle ij \rangle} n_i n_j \tag{6}$$

with $V_0 > 0$ and the sum extending over nearest-neighbor pairs of hexagon centers. It is convenient to work in the grand canonical ensemble (in an experiment the adsorbed He atoms are in equilibrium with He vapor). We therefore add a term

$$-\mu \sum_i n_i$$

to the Hamiltonian. To express the Hamiltonian in the Ising form, we make the transformation

$$n_i = \tfrac{1}{2}(1 + \sigma_i) \tag{7}$$

with $\sigma_i = \pm 1$. Then

$$H = J \sum_{\langle ij \rangle} \sigma_i \sigma_j - h \sum_i \sigma_i + c \tag{8}$$

where $J = V_0/4$, $h = \tfrac{1}{2}(\mu - 3V_0)$, and $c = \tfrac{1}{4} N(3V_0 - 2\mu)$, that is, we have the equivalent problem of an Ising antiferromagnet in a magnetic field. Zero magnetization corresponds to exactly half the lattice sites being occupied. In the special case $h = 0$ the model has been solved by Houtappel [1950] and Husimi and Syozi [1950]. They found that the system remains disordered for all nonzero temperatures and that there are no phase transitions. The physical reason for this absence of phase transitions is the very high degeneracy of the ground state. To see this we note that the triangular lattice may be divided into three sublattices (labeled A, B, and C in Figure 6.6). All sites on one sublattice have nearest neighbors on the two other sublattices but none on its own kind. Some of the degenerate ground-state configurations have $\sigma_i = +1$ on one of the sublattices (say, A) and $\sigma_i = -1$ on another (e.g., B). Because of the antiferromagnetic coupling this lowers the energy relative to the completely disordered configuration. However, once the assignment of the A and B spins have been made the C spins are completely "frustrated" (i.e., it does not matter what their orientation is). The degeneracy of the ground state is therefore greater than $2^{N/3}$ and there will be a residual entropy even at zero temperature.

On the other hand, if $h \neq 0$ in (8), the degeneracy of the ground state will be broken and we expect a phase diagram of the type sketched in Figure 6.7.

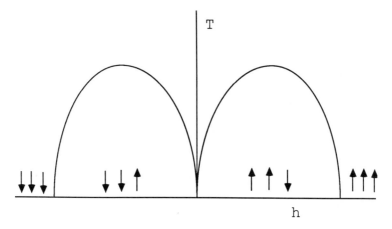

Figure 6.7   Schematic phase diagram for the two-dimensional Ising antiferromagnet on the triangular lattice.

We now briefly outline a renormalization group treatment of the Hamiltonian (8) due to Schick et al. [1977]. In dealing with models in which there is an underlying symmetry, it is important to preserve this symmetry under renormalization. The reason for this is that although formulas such as (6.C.3) and (6.C.11) are exact and hold for any choice of blocking, they cannot be evaluated without approximation. In an approximate calculation a disregard of symmetry may result in a renormalized Hamiltonian which belongs to a different universality class than the original Hamiltonian.

In the present problem we wish to retain the equivalence of the three sublattices under renormalization. The $\sqrt{3} \times \sqrt{3}$ ordered state, corresponding to a coverage of $\frac{1}{3}$, can have either the A, the B, or the C sites predominantly occupied, and we wish to preserve this feature under renormalization.

We note in passing that the ordered state of this system, the case where the coverage is exactly $\frac{1}{3}$, can be described by a two component order parameter. In the disordered phase the density on each of the sublattices is $\frac{1}{3}$ and the degree to which one sublattice is preferred is given by the three numbers $\langle n_A \rangle - \frac{1}{3}$, $\langle n_B \rangle - \frac{1}{3}$, $\langle n_C \rangle - \frac{1}{3}$. At fixed density these three numbers must add to zero, and there are only two independent variables which describe the system. We now introduce the two-component order parameter $\mathbf{m} = (m_x, m_y)$

with

$$\langle n_A \rangle = \tfrac{1}{3} + \tfrac{2}{3} m_y$$
$$\langle n_B \rangle = \tfrac{1}{3} + (3)^{-1/2} m_x - \tfrac{1}{3} m_y \qquad (9)$$
$$\langle n_C \rangle = \tfrac{1}{3} - (3)^{-1/2} m_x - \tfrac{1}{3} m_y$$

In terms of this order parameter the ground state with all particles on the A sublattice is the state $\mathbf{m} = (0, 1)$, the state with all particles on the B sublattice is $(3^{1/2}/2, -1/2)$ and the state with all particles on the C sublattice is $(-3^{1/2}, -\tfrac{1}{2})$. The situation is depicted in Figure 6.8. The discrete threefold symmetry of the system (corresponding to rotation of the order parameter by $2\pi/3$) is often referred to as the three-state Potts symmetry, and we expect the phase transitions of this system to be the same as for the three-state Potts [1952] model. The Landau free energy of the Potts model (or our antiferromagnet) contains cubic terms signifying a first order transition. In this case the Landau theory is qualitatively incorrect—it can be proven rigorously (Baxter, 1973) that the transition is continuous in two dimensions. In three dimensions the Landau result seems to be correct.

In view of the discussion above it is clear that a renormalization transformation should preserve the identity of the three sublattices. The simplest blocking scheme consistent with this criterion is shown in Figure 6.9. The three interpenetrating triangles are three blocks of equivalent sites, and we can

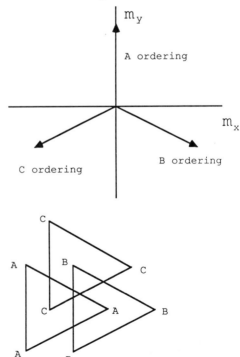

Figure 6.8   Three-state Potts symmetry.

Figure 6.9   Blocking scheme preserving the threefold symmetry.

associate a block spin (e.g., σ_A) with the three site spins on the original lattice. At this point we may choose an approximation scheme (cluster method, cumulant method, etc.). The resulting renormalized Hamiltonian should have the proper threefold symmetry. We shall not carry out this calculation here. The details of a finite cluster calculation may be found in the article by Schick et al. [1977].

(c) Monte Carlo Renormalization

In the preceding sections we have developed the renormalization group formalism and described a few methods of carrying out renormalization group calculations. None of these methods were extremely successful in producing accurate values of the critical coupling constants or exponents. Their value today lies more in the insights they provide (e.g., when there are a number of competing fixed points) than in the accuracy of the specific results. We now turn to a renormalization group scheme which is capable of producing, at least in the case of classical spins, highly accurate results for the critical exponents.

The Monte Carlo renormalization group method was invented by Ma [1976b] and further developed by Swendsen and co-workers (Swendsen, 1979, 1984a,b; Swendsen and Krinsky, 1979; Blöte and Swendsen, 1979; Pawley et al. 1984). Consider an Ising system on a d-dimensional lattice. We shall write the Hamiltonian in the compact form

$$H = \sum_\alpha K_\alpha \psi_\alpha(\sigma_i) \tag{1}$$

where the index α refers to a generic type of coupling (6.B.1). At this point we make no specific restrictions on the type of couplings that we will include in H. We note that the expectation values and correlation functions of the spin operators are given by

$$\langle \psi_\alpha \rangle = \frac{\partial \ln Z}{\partial K_\alpha} \tag{2}$$

where $Z = \text{Tr } e^H$ and

$$\langle \psi_\alpha \psi_\beta \rangle - \langle \psi_\alpha \rangle \langle \psi_\beta \rangle = \frac{\partial \langle \psi_\alpha \rangle}{\partial K_\beta} \equiv S_{\alpha\beta} \tag{3}$$

Consider now a large finite system with the Hamiltonian (1) and (usually) periodic boundary conditions. We have already indicated how such a system can be simulated by the Monte Carlo method [Section 4.C(a)]. The expectation values (2) and (3) can be obtained to arbitrary accuracy (for the finite system) by sampling for a sufficiently long time.

We now divide the finite system into blocks. For example, on a d-dimensional cubic lattice one may choose blocks of 2^d spins and the $N/2^d$ such blocks will once again form a cubic lattice. We also define a rule (perhaps the majority rule) which assigns values to the block spins on the basis of the configuration of the original spins. As we progress through a Monte Carlo run we can thus keep track of the configurations of both the block and site spins. Indeed, if the original lattice is large enough, we may accumulate information on several generations of spins (N_0 site spins, $N_1 = N_0/2^d$ block spins, $N_2 = N_1/2^d$ second-generation block spins, etc.).

We suppose that the Hamiltonian for the nth generation of the block spins can be written in the form

$$H^{(n)} = \sum_\alpha K_\alpha^{(n)} \psi_\alpha(\sigma_i^{(n)}) \tag{4}$$

where the functions ψ_α are the same at all levels. Our goal is to obtain a recursion relation $K_\alpha^{(n)} = R_\alpha(K_1^{(n-1)}, K_2^{(n-1)}, \ldots)$. We write

$$\exp\{H^{(n)}\} = \operatorname*{Tr}_{\sigma^{(n-1)}} P(\sigma^{(n)}, \sigma^{(n-1)}) \exp\{H^{(n-1)}\} \tag{5}$$

We define $S_{\alpha\beta}^n$ to be the correlation function (3) computed at the nth level of blocking, and denote by $S_{\alpha\beta}^{n,n-1}$ the correlation function connecting the two blocking levels n and $n-1$:

$$S_{\beta\alpha}^{n,n-1} \equiv \frac{\partial \langle \psi_\beta^n \rangle}{\partial K_\alpha^{n-1}} = \frac{\partial}{\partial K_\alpha^{n-1}} \frac{\operatorname*{Tr}_{\sigma^n} \psi_\beta^n \operatorname*{Tr}_{\sigma^{n-1}} P(\sigma^n, \sigma^{n-1}) \exp\{H^{n-1}\}}{\operatorname*{Tr}_{\sigma^n} \operatorname*{Tr}_{\sigma^{n-1}} P(\sigma^n, \sigma^{n-1}) \exp\{H^{n-1}\}} \tag{6}$$

Differentiating the right-hand side of (6) and using $\operatorname*{Tr}_{\sigma^n} P(\sigma^n, \sigma^{n-1}) = 1$ we obtain

$$S_{\beta\alpha}^{n,n-1} = \frac{\operatorname*{Tr}_{\sigma^n} P(\sigma^n, \sigma^{n-1}) \psi_\beta^n \psi_\alpha^{n-1} \exp\{H^{n-1}\}}{\operatorname*{Tr} \exp\{H^{n-1}\}}$$

$$- \frac{\operatorname*{Tr}_{\sigma^n} \psi_\beta^n \exp\{H^n\}}{\operatorname*{Tr}_{\sigma^n} \exp\{H^n\}} \frac{\operatorname*{Tr}_{\sigma^{n-1}} \psi_\beta^{n-1} \exp\{H^{n-1}\}}{\operatorname*{Tr}_{\sigma^{n-1}} \exp\{H^{n-1}\}} \tag{7}$$

Using the chain rule of differentiation, we find that

$$S_{\beta\alpha}^{n,n-1} \equiv \frac{\partial \langle \psi_\beta^n \rangle}{\partial K_\alpha^{n-1}} = \sum_\gamma \frac{\partial \langle \psi_\beta^n \rangle}{\partial K_\gamma^n} \frac{\partial K_\gamma^n}{\partial K_\alpha^{n-1}} \tag{8}$$

or

$$S_{\beta\alpha}^{n,n-1} = \sum_\gamma S_{\beta\gamma}^n \frac{\partial K_\gamma^n}{\partial K_\alpha^{n-1}} \tag{9}$$

Equation (9) is the basic equation of the Monte Carlo renormalization group approach. The matrix elements $S_{\beta\gamma}^n$ and $S_{\beta\alpha}^{n,n-1}$ can be calculated by averaging over the configurations generated in a Monte Carlo run. They depend only on the coupling constants of the site spin Hamiltonian and on the projection operator P which defines the blocking. If the site spin Hamiltonian is on a critical surface, the recursion relations

$$\frac{\partial K_\gamma^n}{\partial K_\alpha^{n-1}}$$

will, as n becomes large, approach the linearized recursion relations at the stable fixed point, that is,

$$T_{\gamma\alpha}^{n,n-1} \equiv \frac{\partial K_\gamma^n}{\partial K_\alpha^{n-1}} \to M_{\gamma\alpha} \qquad (10)$$

In matrix notation,

$$M = [S^n]^{-1}[S^{n,n-1}] \qquad (11)$$

The eigenvalues of this matrix yield the critical exponents.

To this point we have been rather vague about a number of important technical points. As we have noted previously, the space of coupling constants is infinite dimensional. In a finite-size spin system we can accommodate only a finite number of independent spin operators and coupling constants. In the Monte Carlo approach outlined here, the number of coupling constants that can be used is determined by the number of block spins in the final generation. It is straightforward, however, to include more coupling constants by going to larger site spin systems while using the same number of generations or by blocking fewer times.

In (11) we have implicitly assumed that we will, with a finite number of iterations, come sufficiently close to the fixed point that the matrix T has effectively become the same as the matrix M. In practice the relevant eigenvalues of the matrix $T^{n,n-1}$ rather rapidly approach stationary values as functions of n for a small range of coupling constants near the critical surface, and there is therefore good reason to believe that the procedure is sensible. For further technical details, and the results of some of the more important applications of the Monte Carlo renormalization group method, we refer to the references quoted earlier in this section.

6.E ϵ EXPANSION

We now develop the renormalization group from a different point of view. Instead of dealing with very specific models, such as the Ising model on a particular lattice, we wish to exploit the notion of universality and retain in our Hamiltonian only what we believe to be the most essential features. Our development will follow Wilson's original formulation of the renormalization group (Wilson 1971, Wilson and Kogut, 1974). To start with, consider an n-component spin system on a d-dimensional cubic lattice. The Hamiltonian is taken to be

$$\tilde{H} = -\sum_{r,\delta} J(\delta) S_r \cdot S_{r+\delta} - \sum_r \tilde{h} S_r^\alpha \tag{1}$$

where $J(\delta)$ is an interaction energy, $\tilde{h}$ a magnetic field pointing in the α direction, and r the points on the d-dimensional lattice. The spin variables are taken to be continuous (classical) n-component vectors

$$S_r = (S_r^x, S_r^y, \dots)$$

with

$$\sum_{\gamma=1}^n (S_r^\gamma)^2 = 1$$

It is often convenient to remove the restriction that the spins have a fixed magnitude and to replace, in the calculation of the partition function, this restriction by a weighting function

$$Z = \left(\prod_{r,\gamma} \int_{|S_r|=1} dS_r^\gamma \right) e^{-\beta \tilde{H}} \longrightarrow \left(\prod_{r,\gamma} \int_{-\infty}^\infty dS_r^\gamma \, W(|S_r|) \right) e^{-\beta \tilde{H}} \tag{2}$$

If we choose

$$W(|S_r|) = \delta \left(\sum_\gamma (S_r^\gamma)^2 - 1 \right)$$

the replacement (2) will be an identity. It is, however, possible to simplify the calculation by replacing the δ function by a continuous function such as

$$W(S_r) = e^{-(1/2)b|S_r|^2 - U|S_r|^4} \tag{3}$$

With $b < 0$, $U = -b/4$, this function has a maximum at $|S| = 1$ and if $|b|$ is large W will decrease rapidly as $|S|$ deviates from unity.

Instead of treating the Hamiltonian in position space, we go to the momentum space representation and write

$$S_r = N^{-1/2} \sum_q S_q e^{iq \cdot r} \tag{4}$$

where the q vectors are restricted to the first Brillouin zone of the simple cubic d-dimensional lattice $(-\pi/a < q_i < \pi/a, \; i = 1, 2, \dots, d)$. Substituting into the Hamiltonian, we get

$$H = -\beta \tilde{H} = \sum_q K(q) S_q \cdot S_{-q} + N^{1/2} h_0 S_0^\alpha \tag{5}$$

where

$$K(q) = \beta \sum_\delta J(\delta) e^{-iq \cdot \delta} \tag{6}$$

and $h_0 = \beta\tilde{h}$. Consider a d-dimensional cubic lattice with nearest-neighbor spacing a and nearest-neighbor interactions only. With $\beta J(\delta) = \frac{1}{2}K_0$, we have, in the limit of long wavelength (small q),

$$K(q) = K_0 \sum_{j=1}^{d} \cos q_j a \approx dk_0 - \frac{1}{2}K_0 a^2 q^2 \tag{7}$$

We now argue that we only need to retain terms to order q^2, since the important fluctuations near the critical point are the long-wavelength ones, and the essential physics will be contained in the leading term. The partition function now becomes

$$Z(K_0, h) = \prod_{q,\gamma} \int dS_q^\gamma W(\{S_q^\gamma\}) e^{H(\{S_q^\gamma\})} \tag{8}$$

with

$$H(\{S_q^\alpha\}) = \sum_{q,\gamma} (dK_0 - \frac{1}{2}K_0 a^2 q^2)S_q^\gamma S_{-q}^\gamma + (N)^{1/2}h_0 S_0^\alpha \tag{9}$$

(a) The Gaussian Model

We first consider a modified version of the weighting function W, which may be somewhat unphysical, but which allows us to illustrate the momentum space renormalization procedure in a transparent way. If we take the weighting function to be a Gaussian

$$W(|S_r|) = e^{-(1/2)b|S_r|^2} \tag{10}$$

with $b > 0$ we obtain a distribution which is peaked around $|S_r| = 0$ rather than $|S_r| = 1$. The functional

$$W(\{S_q\})e^{H(\{S_q\})}$$

now contains only quadratic terms in S_q and can therefore be handled quite simply:

$$Z = \left(\prod_{q,\alpha} \int dS_q^\alpha\right) \exp\left\{-\frac{1}{2}b \sum_q \mathbf{S}_q \cdot \mathbf{S}_{-q}\right\} e^{H(\{S_q\})}$$

$$= \left(\prod_{q,\alpha} \int dS_q^\alpha\right) \exp\left\{-\sum_q \left(\frac{b}{2} - dK_0 + \frac{1}{2}K_0 a^2 q^2\right)\mathbf{S}_q \cdot \mathbf{S}_{-q} + (N)^{1/2}h_0 S_0^\alpha\right\} \tag{11}$$

We now rescale the spin variable so as to make the coefficient of q^2 in the exponential equal to $\frac{1}{2}$ and obtain

$$Z \propto \left(\prod_{q,\alpha} \int dS_q^\alpha\right) \exp\left\{-\frac{1}{2}\sum_q (r + q^2)\,\mathbf{S}_q \cdot \mathbf{S}_{-q} + (N)^{1/2}S_0^\alpha\right\} \tag{12}$$

where

$$r = \frac{b - 2dk_0}{K_0 a^2} \qquad h = \frac{h_0}{(K_0 a^2)^{1/2}}$$

We next define a renormalization transformation in the following way:

1. Carry out the functional integration over all $S_{\mathbf{q}}$ with $q > q_l = \pi/(la)$, where $l > 1$ is a parameter. In the position space approach, this corresponds to the operation of coarse graining or choosing a block spin. In carrying out the integration over $S_{\mathbf{q}}$ with $q > \pi/(la)$, the minimum-length scale is changed from a to la; correlations at distances shorter than la can no longer be resolved. The integration produces the result (with A a constant)

$$Z = A \left(\prod_{q < q_{l, \alpha}} \int dS_{\mathbf{q}}^{\alpha} \right) \exp \left\{ -\sum_{q < q_l} \frac{1}{2} \mathbf{S}_{\mathbf{q}} \cdot \mathbf{S}_{-\mathbf{q}}(r + q^2) + N^{1/2} h S_0^{\alpha} \right\} \quad (13)$$

2. Rescale lengths to the original scale (i.e., let $\mathbf{q} = \mathbf{q}'/l$ with $q' \leq \pi/a$). With this transformation the exponent in (13) becomes

$$H' = -\frac{1}{2} \sum_{\mathbf{q}'} \left[r + \left(\frac{q'}{l} \right)^2 \right] l^{-d} \mathbf{S}_{\mathbf{q}'/l} \cdot \mathbf{S}_{-\mathbf{q}'/l} + N^{1/2} h S_0^{\alpha} \quad (14)$$

The factor l^{-d} compensates for the extra degrees of freedom that have been introduced by the expansion (at constant density of points in $\mathbf{q}$ space) of the remaining part of the Brillouin zone. We note that in these operations we have approximated the hypercubic Brillouin zone $(-\pi/a < q_\alpha < \pi/a)$ by a hyperspherical zone. This is in keeping with our philosophy of ignoring nonessential features such as details of the lattice structure.

3. We require that aside from additive constants, H' must have the identical form as H. We have stipulated that the coefficient of the term $\frac{1}{2} q^2 \mathbf{S}_{\mathbf{q}} \cdot \mathbf{S}_{-\mathbf{q}}$ in H should be unity. the corresponding term in (14) is $\frac{1}{2} q'^2 l^{-(d+2)} \mathbf{S}_{\mathbf{q}'/l} \cdot \mathbf{S}_{-\mathbf{q}'/l}$ and in order to satisfy the condition that this term have a fixed coefficient, independent of l, we make the spin rescaling transformation

$$\mathbf{S}_{\mathbf{q}'/l} = \zeta(l) \mathbf{S}_{\mathbf{q}'}$$
$$\zeta(l) = l^{1+d/2} \quad (15)$$

This spin rescaling operation is similar in spirit to the requirement that the block spin in the position space renormalization group (Sections 6.B and 6.C) be a variable of the same type as the site spin (see Pfeuty and Toulouse [1977], p. 66, for further discussion of this point).

The final form of H' is therefore

$$H' = -\frac{1}{2} \sum_{\mathbf{q}'} (rl^2 + q'^2) \, \mathbf{S}_{\mathbf{q}'} \cdot \mathbf{S}_{-\mathbf{q}'} + N^{1/2} h l^{1+d/2} S_0^{\alpha}$$

$$= -\frac{1}{2} \sum_{\mathbf{q}} (r' + q^2) \mathbf{S}_{\mathbf{q}} \cdot \mathbf{S}_{-\mathbf{q}} + N^{1/2} h' S_0^{\alpha} \quad (16)$$

with

$$r' = rl^2 \qquad h' = l^{1+d/2}h \tag{17}$$

Equations (17) are the recursion relations for the Gaussian model. We see that there are three fixed points for $h = 0$: $r = +\infty$ corresponding to $T = \infty$; $r = -\infty$ corresponding to $T = 0$, and $r = 0$, which we shall see is the critical fixed point. Assuming that r is a temperature like variable we have the scaling form of the singular part of the free energy (5.C.18)

$$g(t, h) = l^{-d}g(tl^2, hl^{1+d/2}) \tag{18}$$

The critical exponents are easily found to be

$$\alpha = 2 - \frac{1}{2}d \qquad \beta = \frac{d}{4} - \frac{1}{2} \qquad \gamma = 1$$

and we see that these exponents become identical with the Landau exponents for $d = 4$.

Consider now the spatial correlation function $\Gamma(\mathbf{r}) = \langle S_0^\alpha S_r^\alpha \rangle$

$$\Gamma(\mathbf{r}) = \frac{1}{N} \sum_x \langle S_x^\alpha S_{x+r}^\alpha \rangle = \frac{1}{N^2} \sum_{qq'x} \langle S_q^\alpha S_{q'}^\alpha \rangle e^{i\mathbf{q}\cdot\mathbf{x}} e^{i\mathbf{q'}\cdot(\mathbf{x}+\mathbf{r})} \tag{19}$$

$$= \frac{1}{N} \sum_q \langle S_q^\alpha S_{-q}^\alpha \rangle e^{-i\mathbf{q}\cdot\mathbf{r}} = \frac{1}{N} \sum_q \Gamma(\mathbf{q})e^{-i\mathbf{q}\cdot\mathbf{r}}$$

where

$$\Gamma(\mathbf{q}) = \frac{\int dS_q^\alpha dS_{-q}^\alpha S_q^\alpha S_{-q}^\alpha \exp\{-(r + q^2)S_q^\alpha S_{-q}^\alpha\}}{\int dS_q^\alpha dS_{-q}^\alpha \exp\{-(r + q^2)S_q^\alpha S_{-q}^\alpha\}} \tag{20}$$

To obtain (20) we have taken $h = 0$ and carried out the integral over all spin variables except S_q^α, S_{-q}^α. The factors $\frac{1}{2}$ in the exponentials have disappeared because the term $S_q^\alpha S_{-q}^\alpha$ appears twice in the Hamiltonian. Separating the spin variables into real and imaginary parts,

$$S_q^\alpha = x + iy \qquad S_{-q}^\alpha = x - iy$$

the integration can be carried out to yield

$$\Gamma(q) = \frac{\int_{-\infty}^{\infty} dx \int_{-\infty}^{\infty} dy \ (x^2 + y^2) \exp\{-(r + q^2)(x^2 + y^2)\}}{\int_{-\infty}^{\infty} dx \int_{-\infty}^{\infty} dy \ \exp\{-(r + q^2)(x^2 + y^2)\}}$$

$$= \frac{1}{r + q^2} \tag{21}$$

The spatial correlation function is therefore

$$\Gamma(\mathbf{x}) = \left(\frac{a}{2\pi}\right)^d \int d^d q \frac{1}{r + q^2} e^{i\mathbf{q}\cdot\mathbf{x}} \tag{22}$$

The value of the integral depends on the dimensionality d, but for $r \to 0$ will be dominated by the contribution from the pole at $q = ir^{1/2}$ and we thus obtain

$$\Gamma(x) \sim e^{-|x|r^{1/2}} \equiv e^{-|x|/\xi} \tag{23}$$

with $\xi = r^{-1/2}$. Since $\xi \propto |t|^{-\nu}$ we have, in any dimension, $\nu = \frac{1}{2}$, which is the Landau–Ginzburg result.

It is interesting to carry out the renormalization procedure described above for the correlation function $\Gamma(q)$ for $q < \pi/(la)$. We modify the Hamiltonian slightly by adding a q-dependent magnetic field

$$H = -\frac{1}{2} \sum_q (r + q^2) \mathbf{S_q} \cdot \mathbf{S_{-q}} + \sum_q \mathbf{h_q} \cdot \mathbf{S_{-q}} \tag{24}$$

Then

$$\langle \mathbf{S_q} \rangle = \frac{\partial \ln Z}{\partial \mathbf{h_{-q}}}$$

and

$$\Gamma(\mathbf{q}, r, \mathbf{h}) = \langle \mathbf{S_q} \cdot \mathbf{S_{-q}} \rangle - \langle \mathbf{S_q} \rangle \cdot \langle \mathbf{S_{-q}} \rangle = \frac{\partial^2 \ln Z}{\partial \mathbf{h_q} \, \partial \mathbf{h_{-q}}} \tag{25}$$

We now integrate over the spin variables with $q > \pi/(la)$ and use (15) with $\mathbf{q} = \mathbf{q}'/l$. Then

$$H' = -\frac{1}{2} \sum_q (rl^2 + q'^2) \mathbf{S_{q'}} \, \mathbf{S_{-q'}} + \sum_{q'} \mathbf{h_{q'/l}} \zeta l^{-d} \mathbf{S_{q'}}$$
$$= -\frac{1}{2} \sum_q (r' + q^2) \mathbf{S_q} \cdot \mathbf{S_{-q}} + \sum_q \mathbf{h'_q} \cdot \mathbf{S_{-q}} \tag{26}$$

with $r' = rl^2$, $\mathbf{h'_q} = \mathbf{h_{q/l}} \zeta l^{-d}$. The correlation function $\Gamma(\mathbf{q}, r', \mathbf{h}')$ for the block spin system is given by

$$\Gamma(\mathbf{q}, r', \mathbf{h}') = \frac{\partial^2 \ln Z'(r', \mathbf{h}')}{\partial \mathbf{h'_q} \partial \mathbf{h'_{-q}}} = \zeta^{-2} l^{2d} \frac{\partial^2 \ln Z'(r', \mathbf{h}')}{\partial \mathbf{h_{q/l}} \partial \mathbf{h_{-q/l}}} \tag{27}$$

Finally, we let $q \to lq$, to obtain

$$\Gamma(l\mathbf{q}, r', \mathbf{h}') = \zeta^{-2} l^{2d} \frac{\partial^2 \ln Z'(r', \mathbf{h}')}{\partial \mathbf{h_q} \partial \mathbf{h_{-q}}} = \zeta^{-2} l^d \Gamma(\mathbf{q}, r, \mathbf{h}) \tag{28}$$

where we have used the fact (18) that $\ln Z' = l^{-d} \ln Z$. We next let $h \to 0$ to obtain the scaling relation for the correlation function

$$\Gamma(l\mathbf{q}, r') = l^{-2} \Gamma(\mathbf{q}, r) \tag{29}$$

At the fixed point $r = 0$, and letting $l = q^{-1}$ we find that

$$\Gamma(\mathbf{q}, 0) = q^{-2} \Gamma(\hat{q}, 0)$$

Conventionally, one writes $\Gamma(q, 0) \propto q^{-2+\eta}$ and we therefore find as in (21), that in the Gaussian model $\eta = 0$.

(b) The S^4 Model

We now turn to the more general weighting function (3) and consider the effective Hamiltonian

$$H = -\frac{1}{2} \sum_{\mathbf{q}} (r + q^2)\mathbf{S_q} \cdot \mathbf{S_{-q}} - \frac{U}{N} \sum_{\mathbf{q_1q_2q_3}} (\mathbf{S_{q_1}} \cdot \mathbf{S_{q_2}})(\mathbf{S_{q_3}} \cdot \mathbf{S_{-q_1-q_2-q_3}})$$

$$-\frac{W}{N^2} \sum_{\mathbf{q_1}\cdots\mathbf{q_5}} (\mathbf{S_{q_1}} \cdot \mathbf{S_{q_2}})(\mathbf{S_{q_3}} \cdot \mathbf{S_{q_4}})(\mathbf{S_{q_5}} \cdot \mathbf{S_{-q_1-q_2-q_3-q_4-q_5}}) + \cdots + hN^{1/2}S_0^\alpha$$

$$(30)$$

where sixth- and higher-order terms have been added because we expect them to be generated by the renormalization transformation. The Hamiltonian (30) is commonly referred to as the Landau–Ginzburg–Wilson Hamiltonian because of its similarity to the Landau–Ginzburg free-energy functional (Section 3.J). Our renormalization procedures will be the same as in the preceding section, but since the integrals over the fourth- and higher-order spin terms are difficult, we construct a cumulant expansion for the partition function. We write

$$H = H_0 + H_1 \tag{31}$$

where

$$H_0 = -\frac{1}{2} \sum_{\mathbf{q}} (r + q^2)\mathbf{S_q} \cdot \mathbf{S_{-q}} + hN^{1/2}S_0^\alpha \tag{32}$$

and

$$H_1 = -\frac{U}{N} \sum_{\mathbf{q_1}..\mathbf{q_4}} \left(\mathbf{S_{q_1}} \cdot \mathbf{S_{q_2}}\right)\left(\mathbf{S_{q_3}} \cdot \mathbf{S_{q_4}}\right)\Delta(\mathbf{q_1} + \mathbf{q_2} + \mathbf{q_3} + \mathbf{q_4}) + \cdots \tag{33}$$

In (33), $\Delta(\mathbf{q_1} + \mathbf{q_2} + \mathbf{q_3} + \mathbf{q_4})$ is the Kronecker delta. Proceeding as in Section 6.C(6.C.14), we have

$$e^{H'} \propto \left(\prod_{\alpha,q>q_l}\int dS_\mathbf{q}^\alpha\right)e^{H_0+H_1} = \left(\prod_{\alpha,q>q_l}\int dS_\mathbf{q}^\alpha\right)\left(1 + H_1 + \tfrac{1}{2}H_1^2 + \cdots\right)e^{H_0}$$

$$= \left(\prod_{\alpha,q>q_l}\int dS_\mathbf{q}^\alpha\right)e^{H_0} \exp\{\langle H_1\rangle + \tfrac{1}{2}(\langle H_1^2\rangle - \langle H_1\rangle^2) + \cdots\} \tag{34}$$

where

$$\langle A \rangle = \frac{\prod_{\alpha,q>q_l} \int dS_\mathbf{q}^\alpha A e^{H_0}}{\prod_{\alpha,q>q_l} \int dS_\mathbf{q}^\alpha e^{H_0}} \tag{35}$$

Clearly, the first factor in (34) produces simply the renormalized Gaussian Hamiltonian. Let us calculate the contribution of the four spin terms to $\langle H_1 \rangle$:

$$-\frac{U}{N} \underset{q_1 \ldots q_4}{\Sigma} \frac{(\underset{q>q_l, \alpha}{\Pi} \int dS_q^\alpha) \exp\{-\frac{1}{2}\underset{q}{\Sigma}(r + q^2)S_q \cdot S_{-q}\}(S_{q_1} \cdot S_{q_2})(S_{q_3} \cdot S_{q_4})\Delta(\overset{4}{\underset{1}{\Sigma}} q_i)}{(\underset{q>q_l, \alpha}{\Pi} \int dS_q^\alpha) \exp\{-\frac{1}{2}\underset{q}{\Sigma}(r + q^2)(S_q \cdot S_{-q})\}}$$

(36)

The contributions to (36) fall into the following main categories:

1. q_1, q_2, q_3, q_4 are all in the region $q > q_l$. The contribution from such terms to the numerator will simply be a constant, since all variables are integrated over. These terms will contribute to the free energy but not to the renormalized Hamiltonian.

2. One, or three, of the vectors q_i lie in the region $q > q_l$. The contribution from such terms is zero by symmetry.

3. Two of the vectors q_i are larger than q_l in magnitude. There are two distinct possibilities:

   (a) $q_1 = -q_2$, $q_3 = -q_4$. Let the latter two vectors be the ones which are larger than q_l. The expectation value of such a term is then

   $$n\Gamma(q_3) \underset{q_1, \alpha}{\Sigma} S_{q_1}^\alpha S_{-q_1}^\alpha$$

   where $\Gamma(q)$ is the Gaussian correlation function calculated above and n is the number of components of the spin vectors.

   (b) $q_1 = -q_3$, $q_2 = -q_4$ (or $q_1 = -q_4$, $q_2 = -q_3$). If we select the first possibility with the two first vectors smaller than q_l, we obtain contributions of the form

   $$\underset{\alpha}{\Sigma} S_{q_1}^\alpha S_{q_3}^\alpha \langle S_{q_2}^\alpha S_{q_4}^\alpha \rangle \Delta(q_1 + q_2 + q_3 + q_4) = S_{q_1} \cdot S_{-q_1} \Gamma(q_2)$$

Adding up terms, we find that

$$\langle H_1 \rangle = -\frac{U}{N} \underset{q_1 < q_l}{\Sigma} (2n + 4)S_{q_1} \cdot S_{-q_1} \underset{q > q_l}{\Sigma} \Gamma(q)$$

$$-\frac{U}{N} \underset{q_1 \cdots q_4 < q_l}{\Sigma} (S_{q_1} \cdot S_{q_2})(S_{q_3} \cdot S_{q_4})\Delta(q_1 + q_2 + q_3 + q_4)$$

(37)

For the time being, we ignore the sixth-order term and analyze the Hamiltonian

$$H' = - \underset{q < q_l}{\Sigma} \left[\frac{1}{2}(r + q^2) + \frac{U}{N} \underset{q_1 > q_l}{\Sigma}(2n + 4)\Gamma(q_1)\right] S_q \cdot S_{-q} + N^{1/2} h S_0^\alpha$$

$$-\frac{U}{N} \underset{q_1 \cdots q_4 < q_l}{\Sigma} \Delta(q_1 + q_2 + q_3 + q_4)(S_{q_1} \cdot S_{q_2})(S_{q_3} \cdot S_{q_4})$$

(38)

We follow the rescaling procedure (13)–(17) to obtain the recursion relations

$$r' = rl^2 + 4(n + 2)l^2 \frac{U}{N} \sum_{q' > \pi/(la)} \Gamma(q') \tag{39a}$$

$$U' = Ul^{-3d} \zeta^4 = Ul^{4-d} \tag{39b}$$

$$h' = hl^{1+\frac{1}{2}d} \tag{39c}$$

Without evaluating the remaining sum in (39a), we see from (39b) that $U' < U$ if $d > 4$. In this case the Gaussian fixed point is stable with respect to the addition of a fourth-order term. Conversely, for $d < 4$, $U' > U$ and the fourth-order term grows, indicating that another fixed point with nonzero values or r^*, U^*, determines the critical exponents of the system. The structure of (39b) is very suggestive. Suppose that we carried out the cumulant expansion to order U^2. We then expect (39b) to become

$$U' = Ul^{4-d} + \psi(l, r)U^2 \tag{40}$$

where $\psi(l, r)$ is some function. This equation can be used to determine U^*:

$$U^* = -\frac{l^{4-d} - 1}{\psi(l, r^*)} = -\frac{l^\epsilon - 1}{\psi(l, r^*)} \approx -\frac{\ln(l)}{\psi(l, r^*)} \epsilon$$

where $\epsilon = 4 - d$. The parameter ϵ is a natural expansion parameter for this problem. The celebrated ϵ expansion of Wilson and Fisher [1972] is based on this notion. One classifies the coupling constants U, W, . . . in (30) according to the leading power of ϵ with which they appear in the fixed-point equation and systematically constructs the fixed point to order ϵ^0, ϵ, ϵ^2, and so on. The critical exponents are then found to be given by a power series in ϵ which one can attempt to evaluate for $\epsilon = 1$ ($d = 3$). It turns out, as we shall see below, that the sixth-order coupling constant $W^* \propto \epsilon^3$, and higher-order coupling constants in turn are proportional to still-higher powers of ϵ at the fixed point. We now carry out the renormalization calculation to order ϵ.

(c) Critical Exponents to Order ϵ

We must now consider the second-order term in the cumulant expansion (34). To simplify the formalism we will take the rescaling parameter l to be $1 + \delta$, where δ is an infinitesimal. We must also consider the the first-order contribution from the term of sixth order in the spin variables

$$-\frac{W}{N^2} \sum_{q_1 \ldots q_6} (\mathbf{S}_{q_1} \cdot \mathbf{S}_{q_2})(\mathbf{S}_{q_3} \cdot \mathbf{S}_{q_4})(\mathbf{S}_{q_5} \cdot \mathbf{S}_{q_6})\Delta(\mathbf{q}_1 + \mathbf{q}_2 + \mathbf{q}_3 + \mathbf{q}_4 + \mathbf{q}_5 + \mathbf{q}_6)$$

in addition to second-order contributions from

$$V = -\frac{U}{N} \sum_{q_1 \ldots q_4} (\mathbf{S}_{q_1} \cdot \mathbf{S}_{q_2})(\mathbf{S}_{q_3} \cdot \mathbf{S}_{q_4})\Delta(\mathbf{q}_1 + \mathbf{q}_2 + \mathbf{q}_3 + \mathbf{q}_4)$$

The recursion relation for the coupling constant W is quickly obtained:

$$W' = l^{-5d}\zeta^6 W + O(U^2)$$
$$= l^{-2d+6}W + O(U^2) \tag{41}$$

Since the term of order U^2 is of order ϵ^2 at the fixed point, it would seem that $W^* \propto \epsilon^2$. We will show, however, that the term of order U^2 in (41) vanishes and that the leading term is at least of order U^3. To see this we must consider the contribution of order U^2 to the renormalized Hamiltonian, which is of sixth order in the renormalized spin variables. In general,

$$\tfrac{1}{2}(\langle V^2 \rangle - \langle V \rangle^2)$$

$$= \frac{U^2}{2N^2} \sum_{q_1 \ldots q_8} \langle (S_{q_1} \cdot S_{q_2})(S_{q_3} \cdot S_{q_4})(S_{q_5} \cdot S_{q_6})(S_{q_7} \cdot S_{q_8}) \rangle \Delta(q_1 \ldots q_4)\Delta(q_5 \ldots q_8)$$

$$- \tfrac{1}{2}\langle V \rangle^2 \tag{42}$$

Because of the term $-\tfrac{1}{2}\langle V \rangle^2$ in (42) we require that some of the wave vectors $q_1 \ldots q_4$ be paired with vectors from the set $q_5 \ldots q_8$. Thus, in sixth order in spin variables we must have one q, say q_4, from $q_1 \ldots q_4$, in the outer shell $\pi/[a(1+\delta)] < q_4 < \pi/a$. Similarly, one vector, say q_8, must be in the same shell. From condition 2 following (36) we have $\langle V \rangle = 0$ for this assignment of wave vectors and we obtain a contribution of the form

$$\frac{U^2}{2N^2} \sum_{q_1 \ldots q_7} (S_{q_1} \cdot S_{q_2})(S_{q_3} \cdot S_{q_7})(S_{q_5} \cdot S_{q_6})\Gamma(q_4)$$

$$\Delta(q_1 + q_2 + q_3 + q_4)\Delta(q_5 + q_6 + q_7 - q_4) \tag{43}$$

The product of Kronecker deltas can be put into the form

$$\Delta(q_1 + q_2 + q_3 + q_5 + q_6 + q_7)\Delta(q_5 + q_6 + q_7 - q_4)$$

Since q_4 is in a shell of width $\pi\delta/a$, the second Kronecker delta restricts the sum of $q_5 + q_6 + q_7$ to a shell of width δ. As $\delta \to 0$ these terms vanish and we have no contribution of order U^2. Thus it is safe to ignore W as long as we are only interested in fixed points and exponents to order ϵ. We may also ignore the contribution of order U^2 to r' as we already have $r' = O(U) = O(\epsilon)$ [see (39)]. Our task is therefore to evaluate the contribution to the renormalized four-spin term from (42).

As in the evaluation of the sixth-order terms, we may, because of the presence of the term $-\tfrac{1}{2}\langle V \rangle^2$, only consider contributions in which two of the momenta $q_1 \ldots q_4$ and of $q_5 \ldots q_8$ lie in the outer shell. There are two main cases, which we illustrate below.

1. The two vectors in the outer shell from each set come from the same scalar product. A typical term has $q_3 = -q_7$ and $q_4 = -q_8$ in the outer shell. There are eight such terms, which add up to

$$\frac{4U^2}{N^2} \sum_{\substack{q_1,q_2,q_5,q_6<q_l \\ q_3,q_4>q_l}} (\mathbf{S}_{\mathbf{q}_1} \cdot \mathbf{S}_{\mathbf{q}_2})(\mathbf{S}_{\mathbf{q}_5} \cdot \mathbf{S}_{\mathbf{q}_6}) \sum_{\alpha\beta} \langle S_{\mathbf{q}_3}^\alpha S_{-\mathbf{q}_3}{}^\beta S_{\mathbf{q}_4}^\alpha S_{-\mathbf{q}_4}^\beta \rangle$$

$$\times \Delta(\mathbf{q}_1 + \mathbf{q}_2 + \mathbf{q}_3 + \mathbf{q}_4)\Delta(\mathbf{q}_5 + \mathbf{q}_6 - \mathbf{q}_3 - \mathbf{q}_4)$$

$$= \frac{4nU^2}{N} \sum_{q_1,q_2,q_5,q_6<q_l} (\mathbf{S}_{\mathbf{q}_1} \cdot \mathbf{S}_{\mathbf{q}_2})(\mathbf{S}_{\mathbf{q}_5} \cdot \mathbf{S}_{\mathbf{q}_6})\Delta(\mathbf{q}_1 + \mathbf{q}_2 + \mathbf{q}_5 + \mathbf{q}_6) \tag{44}$$

$$\times \frac{1}{N} \sum_{q_3, q_4>q_l} \Gamma(q_3)\Gamma(q_4)\Delta(\mathbf{q}_5 + \mathbf{q}_6 - \mathbf{q}_3 - \mathbf{q}_4)$$

2. The remaining 64 pairings of two vectors from the sets $\mathbf{q}_1 \ldots \mathbf{q}_4$ and $\mathbf{q}_5 \ldots \mathbf{q}_8$ all yield the same expectation values, given for the case $\mathbf{q}_2 = -\mathbf{q}_6$ and $\mathbf{q}_4 = -\mathbf{q}_8$ by

$$\frac{32U^2}{N^2} \sum_{\substack{q_1,q_3,q_5,q_7<q_l \\ q_2,q_4>q_l,\alpha,\beta,\gamma,\nu}} S_{\mathbf{q}_1}^\alpha S_{\mathbf{q}_3}^\beta S_{\mathbf{q}_5}^\gamma S_{\mathbf{q}_7}^\nu \langle S_{\mathbf{q}_2}^\alpha S_{-\mathbf{q}_2}^\gamma S_{\mathbf{q}_4}^\beta S_{-\mathbf{q}_4}^\nu \rangle$$

$$\times \Delta(\mathbf{q}_1 + \mathbf{q}_2 + \mathbf{q}_3 + \mathbf{q}_4)\Delta(\mathbf{q}_5 + \mathbf{q}_6 + \mathbf{q}_7 + \mathbf{q}_8)$$

$$= \frac{32U^2}{N^2} \sum_{\substack{q_1,q_3,q_5,q_7<q_l \\ q_2,q_4>q_l,\alpha,\beta}} S_{\mathbf{q}_1}^\alpha S_{\mathbf{q}_5}^\alpha S_{\mathbf{q}_3}^\beta S_{\mathbf{q}_7}^\beta \Gamma(q_2)\Gamma(q_4)$$

$$\times \Delta(\mathbf{q}_1 + \mathbf{q}_2 + \mathbf{q}_3 + \mathbf{q}_4)\Delta(\mathbf{q}_5 - \mathbf{q}_2 + \mathbf{q}_7 - \mathbf{q}_4)$$

$$= \frac{32U^2}{N} \sum_{q_1, q_3, q_5, q_7<q_l} (\mathbf{S}_{\mathbf{q}_1} \cdot \mathbf{S}_{\mathbf{q}_5})(\mathbf{S}_{\mathbf{q}_3} \cdot \mathbf{S}_{\mathbf{q}_7})\Delta(\mathbf{q}_1 + \mathbf{q}_3 + \mathbf{q}_5 + \mathbf{q}_7)$$

$$\frac{1}{N} \sum_{q_2, q_4>q_l} \Gamma(q_2)\Gamma(q_4)\Delta(\mathbf{q}_2 + \mathbf{q}_4 - \mathbf{q}_5 - \mathbf{q}_7) \tag{45}$$

Cases such as when $\mathbf{q}_3 = -\mathbf{q}_4 = \mathbf{q}_7 = -\mathbf{q}_8$ all are in the outer shell have vanishingly small phase space and can be ignored.

Thus the renormalized fourth-order term in the Hamiltonian becomes, after rescaling,

$$\frac{U}{N} l^{4-d} \sum_{q_1, q_2, q_3, q_4} (\mathbf{S}_{\mathbf{q}_1} \cdot \mathbf{S}_{\mathbf{q}_2})(\mathbf{S}_{\mathbf{q}_3} \cdot \mathbf{S}_{\mathbf{q}_4})\Delta(\mathbf{q}_1 + \mathbf{q}_2 + \mathbf{q}_3 + \mathbf{q}_4)$$

$$\times \left[1 - \frac{U}{N}4(n + 8) \sum_{q,q'>q_l} \Gamma(q)\Gamma(q')\Delta(\mathbf{q}_3 + \mathbf{q}_4 - \mathbf{q} - \mathbf{q}') \right] \tag{46}$$

For the second-order term we have

$$r' = rl^2 + 4(n + 2)Ul^2\frac{1}{N} \sum_{q > q_l} \Gamma(q) \tag{47}$$

We now approximate the sum appearing in (47):

$$\frac{1}{N} \sum_{q>q_l} \Gamma(q) \approx \left(\frac{a}{2\pi}\right)^d \int_{\pi/a(1+\delta)}^{\pi/a>q>} d^d q \frac{1}{r+q^2} \tag{48}$$

Since $\delta << 1$ we can use (2.A.12) for the surface area of a d-dimensional sphere. Letting $C = 2^{1-d}\pi^{(1/2)d}/(\frac{1}{2}d - 1)!$ and choosing $\pi/a = 1$, we find

$$r' = rl^2 + 4(n+2)Ul^2\frac{C\delta}{1+r} \tag{49}$$

The integrals appearing in the four-spin term (46) are of the form

$$\frac{1}{N} \sum_{q,q'>q_l} \Gamma(q)\Gamma(q')\Delta(\mathbf{q}_3 + \mathbf{q}_4 - \mathbf{q} - \mathbf{q}') \tag{50}$$

which we approximate by

$$\frac{1}{N} \sum_{q>q_l} \Gamma^2(q) = \frac{C\delta}{(1+r)^2} \tag{51}$$

Substituting, we finally obtain

$$U' = l^{4-d}\left[U - \frac{4(n+8)U^2 C\delta}{(1+r)^2}\right] \tag{52}$$

The recursion relations (49) and (52) can be written in the form

$$r' - r = \delta\frac{\partial r}{\partial \delta} = [(1+\delta)^2 - 1]r + \frac{4(n+2)C\delta}{1+r}U \tag{53}$$

$$U' - U = \delta\frac{\partial U}{\partial \delta} = [(1+\delta)^{4-d} - 1]U + \frac{4(n+8)C\delta}{(1+r)^2}U^2 \tag{54}$$

At the fixed point $r = r' = r*$, $U = U' = U*$, and expanding to lowest order in δ we find that

$$U* = \frac{\epsilon(1+r*)^2}{4(n+8)C} \qquad r* = -\frac{2(n+2)C}{1+r*}U* \tag{55}$$

Retaining only the first-order term in ϵ, we obtain

$$U* = \frac{\epsilon}{4(n+8)C} \qquad r* = -\frac{(n+2)\epsilon}{2(n+8)} \tag{56}$$

Let us now consider the linearized recursion relations at this fixed point. From (49), (52), and (56) we have, again to first order in ϵ and δ,

$$\frac{\partial r'}{\partial r} = l^2 - \frac{4(n+2)l^2 C\delta U*}{(1+r*)^2} = l^2 - \frac{(n+2)\delta\epsilon}{n+8} = 1 + 2\delta - \delta\epsilon\frac{n+2}{n+8}$$

$$\approx (1+\delta)^{2-\epsilon(n+2)/(n+8)} \tag{57a}$$

$$\frac{\partial r'}{\partial U} = \frac{4(n + 2)C\delta}{1 + r*} \tag{57b}$$

$$\frac{\partial U'}{\partial r} = O(\epsilon^2) \tag{57c}$$

$$\frac{\partial U'}{\partial U} = (1 + \delta)^\epsilon - 8(n + 8)U*C\delta = (1 + \delta)^\epsilon - 2\epsilon\delta \approx (1 + \delta)^{-\epsilon} \tag{57d}$$

The matrix M which determines the eigenvalues, and thus the critical exponents, takes the form

$$M = \begin{bmatrix} l^{2 - \epsilon(n+2)/(n+8)} & 4(n + 2)C \\ 0 & l^{-\epsilon} \end{bmatrix} \tag{58}$$

and we obtain

$$y_1 = 2 - \epsilon\frac{n + 2}{n + 8} \qquad y_2 = -\epsilon \tag{59}$$

To this point we have ignored the magnetic field. The field-dependent term in the Hamiltonian is simply rescaled under renormalization as in (17):

$$h'_\alpha = l^{1+d/2}h_\alpha$$

Thus

$$y_h = 1 + \tfrac{1}{2}d = 3 - \tfrac{1}{2}\epsilon \tag{60}$$

The usual critical exponents are determined from the scaling form of the free energy. In terms of the scaling fields u_1 and u_2 which are the eigenvectors of M (58), we have for the free energy per spin,

$$g(u_1, u_2, h) = l^{-d}g(u_1 l^{y_1}, u_2 l^{y_2}, h l^{y_h}) \tag{61}$$

which leads to the critical exponents (Section 5.C)

$$\alpha = 2 - \frac{d}{y_1} = \epsilon\left[\frac{1}{2} - \frac{n + 2}{n + 8}\right] + O(\epsilon^2) \tag{62a}$$

$$\beta = \frac{d - y_h}{y_1} = \frac{1}{2} - \frac{3\epsilon}{2(n + 8)} + O(\epsilon^2) \tag{62b}$$

$$\gamma = \frac{2y_h - d}{y_1} = 1 + \frac{(n + 2)\epsilon}{2(n + 8)} + O(\epsilon^2) \tag{62c}$$

$$\delta = \frac{y_h}{d - y_h} = 3 + \epsilon \tag{62d}$$

The susceptibility exponent γ for the Ising model ($n = 1$) in three dimensions to order ϵ is therefore $\tfrac{7}{6}$. For the XY model ($n = 2$) $\gamma = \tfrac{6}{5}$ and for the

Heisenberg model ($n = 3$) we obtain $\frac{27}{22}$. These estimates are not very accurate, but do display the correct qualitative trend to larger γ as n increases. The best series expansion estimates for γ are 1.24, 1.33, and 1.43 for the Ising, XY, and Heisenberg models.

We also recover, at least to order ϵ, the property of universality of critical exponents. Since the exponent y_2 is negative, the scaling field u_2 is irrelevant and thus does not affect the critical exponents. We note, however, that the four-spin interaction in (30) does not break the n-dimensional rotational symmetry of the Hamiltonian. To determine which of the fixed points (Ising, XY, or Heisenberg) governs the critical behavior of the anisotropic Heisenberg model, one has to introduce a symmetry breaking term into (30) and examine the stability of the Heisenberg fixed point with respect to such fields (Problem 6.5).

Finally, we briefly discuss the flows in coupling constant space. An examination of the linearized recursion relations near the nontrivial fixed point shows that the coupling constants approach the fixed point along the line $(r - r^*)/(U - U^*) = -2(n + 2)/C$. This line connects the unphysical (for $\epsilon > 0$) Gaussian fixed point with the nontrivial fixed point. The coupling constants move away from the fixed point along the line $U^* = U$. Thus the flows behave qualitatively as sketched in Figure 6.10. As ϵ decreases, the physical fixed point approaches the Gaussian fixed point along the critical surface, and at $\epsilon = 0$ the two fixed points merge. For $\epsilon < 0$ ($d > 4$), $U^*(\epsilon) < 0$ and the nontrivial fixed point becomes unstable, and thus unphysical. For all $d > 4$ the critical properties of these spin systems are determined by the Gaussian fixed point. The magnetic field direction is not depicted in the diagram. It may be thought of as a third axis orthogonal to both the r and U axes.

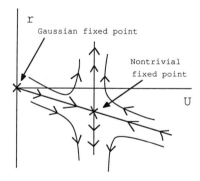

Figure 6.10 Recursion flow for the S^4 model when $d < 4$ ($\epsilon > 0$).

(d) Conclusion

The preceding derivation has been rather lengthy. This is in part due to our assumption that the reader may not be familiar with Feynman graph techniques. More elegant derivations of the foregoing results can be found in the books of Ma [1976a] or Pfeuty and Toulouse [1977] and the reviews of Wilson and Kogut [1974] and Fisher [1974]. With considerable effort the epsilon expansion can be extended to higher orders—presently the state of the art is ϵ^5 (see Gorishny et al., 1984; Le Guillou and Zinn-Justin, 1987). Although the epsilon expansion is an asymptotic rather than a convergent series, powerful summation techniques can be used to evaluate the critical exponents in three and even two dimensions. These estimates are found to be in excellent agreement with high-temperature series results in $d = 3$ and in remarkably good agreement with the exactly known Ising model exponents in $d = 2$ (Le Guillou and Zinn-Justin, 1987).

Other methods of calculating the critical exponents of the Landau–Ginzburg–Wilson Hamiltonian (30) have also been developed. In particular, field-theoretic methods (Brezin et al., 1976) have proven to be very effective and the results of Baker et al. [1976] for the Ising model in three dimensions were in fact the first convincing demonstration that the renormalization group could yield critical exponents of the same accuracy as high-temperature series.

The most important result, in our opinion, to come from the renormalization group approach is the elaboration of the notion of universality. One now has a tool that can be used to determine which features of a microscopic Hamiltonian are important in determining critical behavior. For example, it is quite straightforward to incorporate symmetry breaking terms (cubic or uniaxial anisotropy), or dipolar and isotropic long-range interactions, into the Landau–Ginzburg–Wilson Hamiltonian and to determine the flow on the critical surface to low order in ϵ. It is generally believed that the stability of fixed points for $d < 4$ does not depend sensitively on ϵ and that reliable conclusions can be drawn from low-order calculations. Thus an almost global flow diagram for the critical surface can be constructed, and the asymptotic behavior of a system close to its critical point is then determined by the fixed point into whose basin of attraction its critical point falls. For a discussion of this mapping out of the critical surface, the reader is referred to Fisher [1974] or Aharony [1976].

Before leaving this subject we note that the concepts and techniques of the renormalization group have found much wider applicability than we have demonstrated in this chapter. Since its discovery the renormalization group has been applied to critical dynamics, chaotic dynamics, the Kondo problem, conductivity in disordered materials, and to many other problems (see also Chapter 9). It has now become a standard tool of condensed matter theory, and for this reason we have devoted considerable space to it.

PROBLEMS

6.1. *Renormalization of the Ising Chain.* Carry out the iteration procedure indicated by (6A.13)–(6.15) to obtain K_j, $g(K_j, 0)$ to fourth order starting with $K = 1$. Use the results to obtain an approximate expression for the dimensionless free energy per spin μ. Your result should be quite close to the exact value in Section 3.F ($\mu = 1.127$).

6.2. *First-Order Cumulant Approximation.* Find the critical exponents of the two-dimensional spin-$\frac{1}{2}$ Ising model on the (a) square, (b) triangular, and (c) honeycomb lattices in the first-order cumulant approximation using blocks of spins as in Figure 6.11. Along the way you will also find the critical temperatures for these lattices. The exact values are $K_{c,\square} \approx 0.441$, $K_{c,\triangle} \approx 0.275$, $K_{c,\bigcirc} \approx 0.658$.

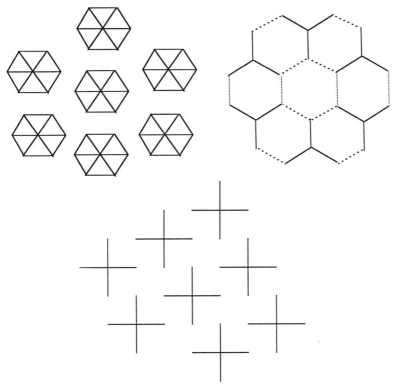

Figure 6.11   Selection of blocks for Problem 6.2.

6.3. *Second-Order Cumulant Approximation.* Use the truncation procedure outlined below (6.C.29) to obtain the recursion relations (6.C.30) for the second-order cumulant approximation to the Ising model on a triangular lattice.

6.4. *Migdal–Kadanoff Transformation.* Consider the Ising model on the square lattice and construct a renormalization transformation according to the following two-step process:

Step 1. Shift one-half of the horizontal bonds by one lattice spacing to obtain the modified interactions shown in Figure 6.12. It is now easy to perform the trace over the spins on the sites denoted by open circles to produce the anisotropic renormalized Hamiltonian shown in Figure 6.13.

Step 2. Shift one-half of the vertical bonds by one lattice spacing to obtain Figure 6.14. Once again carry out the trace over the variables at the open circles to obtain the new renormalized Hamiltonian indicated in Figure 6.15. Symmetrize the Hamiltonian by defining

$$K' = \tfrac{1}{2}(2\gamma' + 2\gamma) = \gamma'(K) + \gamma(K)$$

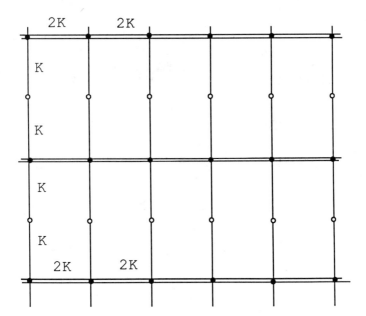

Figure 6.12   First step in Migdal–Kadanoff procedure.

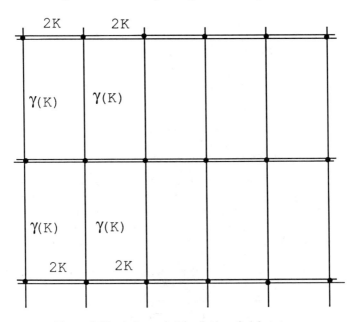

Figure 6.13   Anisotropic Ising lattice after first step.

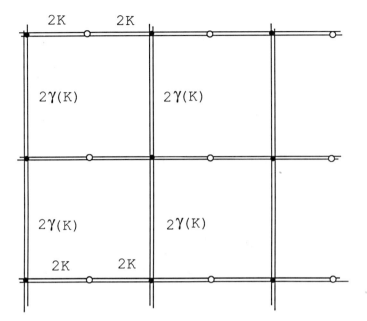

Figure 6.14   Second step in Migdal–Kadanoff transformation.

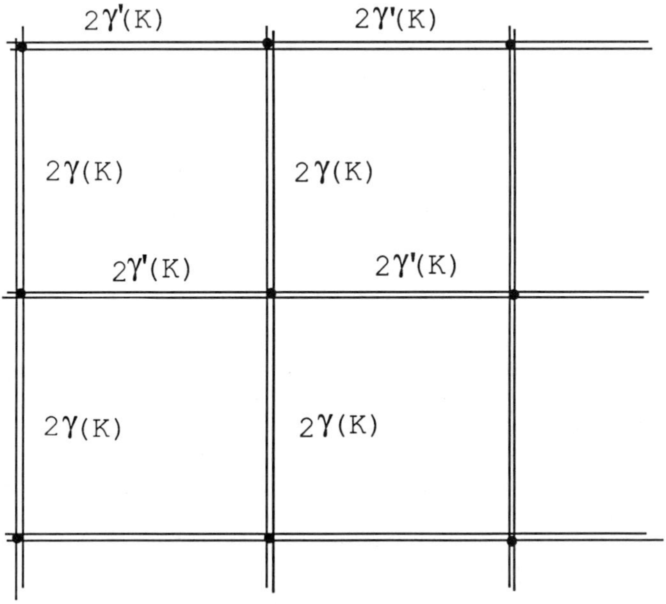

Figure 6.15   Renormalized Ising lattice after second step.

The rescaling parameter is $b = 2$ in this case. Solve for the fixed point of the transformation and obtain the leading thermal exponent y_1. Compare with the exact critical temperature and specific heat exponent.

6.5. ϵ *Expansion for the Anisotropic Heisenberg Model.* Apply the methods of Section 6.E to the anisotropic n-vector model with Landau–Ginzburg–Wilson Hamiltonian

$$H = -\frac{1}{2}\sum_q \sum_{\alpha=1}^{n-1} (r_s + q^2) S_q^\alpha S_{-q}^\alpha - \frac{1}{2}\sum_q (r_n + q^2) S_q^n S_{-q}^n$$

$$+ \frac{V_1}{N} \sum_{\{q_j\}} \sum_{\alpha,\,\beta<n} S_{q_1}^\alpha S_{q_2}^\alpha S_{q_3}^\beta S_{-q_1-q_2-q_3}^\beta$$

$$+ \frac{2V_2}{N} \sum_{\{q_j\}} \sum_{\alpha<n} S_{q_1}^\alpha S_{q_2}^\alpha S_{q_3}^n S_{-q_1-q_2-q_3}^n$$

$$+ \frac{V_3}{N} \sum_{\{q_j\}} S_{q_1}^n S_{q_2}^n S_{q_3}^n S_{-q_1-q_2-q_3}^n$$

The isotropic n-vector model is clearly the special $r_s = r_n$, $V_1 = V_2 = V_3$. The Ising model corresponds to $r_s = \infty$, $V_1 = V_2 = 0$ and the $m = (n-1)$-vector model to $r_n = \infty$, $V_2 = V_3 = 0$.

(a) Show that the recursion relations to order ϵ are given by

$$r_s' = l^2\left[r_s + \frac{4(n+1)c\delta}{1+r_s}V_1 + \frac{4c\delta}{1+r_n}V_2\right]$$

$$r_n' = l^2\left[r_n + \frac{4(n-1)c\delta}{1+r_s}V_2 + \frac{12c\delta}{1+r_n}V_3\right]$$

$$V_1' = l^\epsilon\left[V_1 - \frac{4(n+7)c\delta}{(1+r_s)^2}V_1^2 - \frac{4c\delta}{(1+r_n)^2}V_2^2\right]$$

$$V_2' = l^\epsilon\left[V_2 - \frac{16c\delta}{(1+r_s)(1+r_n)}V_2^2 - \frac{4(n+1)c\delta}{(1+r_s)^2}V_1V_2 - \frac{12c\delta}{(1+r_n)^2}V_2V_3\right]$$

$$V_3' = l^\epsilon\left[V_3 - \frac{36c\delta}{(1+r_n)^2}V_3^2 - \frac{4(n-1)c\delta}{(1+r_s)^2}V_2^2\right]$$

where $l = 1 + \delta$ and where the approximations (6.E.48) and (6.E.51) have been made in integrating correlation functions over the outer shell.

(b) Show that the linearized recursion relations (6.E.57) and (6.E.58) are recovered for the Heisenberg, XY, and Ising cases.

(c) Consider the general linearized recursion relations at the *isotropic* n-vector fixed point. In particular, show that to order ϵ the matrix M [analogous to (6.E.58)] has $M_{j1} = M_{j2} = 0$ for $j = 3, 4, 5$. Thus two of the right eigenvectors are of the form

$$\begin{bmatrix} a \\ b \\ 0 \\ 0 \\ 0 \end{bmatrix}$$

Show that the corresponding exponents are given by

$$y_1 = 2 - \frac{n + 2}{n + 8}\epsilon \qquad y_2 = 2 - \frac{2}{n + 8}\epsilon$$

For small anisotropy

$$a = \frac{r_n - r_s}{r_s}$$

we may express the singular part of the zero-field free energy near the critical point in the scaling form

$$g_s(t, a, 0) = l^{-d}g_s(tl^{y_1}, al^{y_2}, 0) = |t|^{d/y_1}\psi(a/|t|^{\varphi})$$

where

$$\varphi = \frac{y_2}{y_1} = 1 + \frac{n}{2(n + 8)}\epsilon$$

is called the "crossover" exponent. Note that the stable fixed point in this case is the Ising fixed point, not the isotropic fixed point.

(d) Consider an anisotropic Heisenberg magnet with $a = 10^{-3}$. For what range of temperatures could one expect to measure the asymptotic Ising critical exponent?

(e) Carry out the analysis of part (c) (i.e., for $n = 3$) at the Ising and XY fixed points and show that the exponent y_2 is negative (i.e., both fixed points are stable).

7

Quantum Fluids

In this chapter we begin, in Section 7.A, by discussing in detail one of the most striking consequences of quantum statistics, the condensation of a noninteracting Bose gas. We next turn our attention (Section 7.B) to an interacting Bose system and to the phenomenon of superfluidity. Our treatment of this subject is limited mainly to low-temperature properties and is primarily qualitative. The first part of Section 7.C is devoted to the Bardeen, Cooper, Schrieffer [1957] (BCS) theory of superconductivity, in which fermion pairs undergo a transition that is similar to Bose condensation. We also consider the macroscopic theory of superconductivity due to Ginzburg and Landau. This approach is also applicable to the theory of superfluidity and briefly discussed in that context. We encountered the Landau–Ginzburg formalism previously in Sections 3.J and 4.E. In the present context we use it to describe some of the important physical properties of superconductors and to indicate why the mean field approach of the BCS theory works so effectively. A useful general reference for much of the material of this chapter is Part 2 of the book by Landau and Lifshitz [1980].

7.A BOSE CONDENSATION

Consider a system of noninteracting bosons confined to a cubical box of volume $V = L^3$. We use periodic boundary conditions in the solution of the single-particle Schrödinger equation:

$$\psi(\mathbf{r} + L\hat{a}_i) = \psi(\mathbf{r})$$

where $\hat{a}_i$ is the unit vector in the ith direction.[1] With these boundary conditions the eigenfunctions of the Schrödinger equation are

$$\psi_{\mathbf{k}}(\mathbf{r}) = \frac{1}{\sqrt{V}}\, e^{i\mathbf{k}\cdot\mathbf{r}} \tag{1}$$

where $\mathbf{k} = 2\pi/L(n_1, n_2, n_3)$ with $n_i = 0, \pm 1, \pm 2, \ldots$. The single-particle energies are given by

$$\epsilon(\mathbf{k}) = \frac{\hbar^2 k^2}{2m} \tag{2}$$

The logarithm of the grand canonical partition (2.D.10) is

$$\ln Z_G = -\sum_{\mathbf{k}} \ln\left(1 - \exp\{-\beta[\epsilon(\mathbf{k}) - \mu]\}\right) \tag{3}$$

where μ is the chemical potential. The mean occupation number of the state with wave vector $\mathbf{k}$ is

$$\langle n_{\mathbf{k}}\rangle = \frac{1}{\exp\{\beta[\epsilon(\mathbf{k}) - \mu]\} - 1} \tag{4}$$

For a large system one may attempt to evaluate sums over the closely spaced but discrete values of $\mathbf{k}$ by converting the sum to an integral. Consider first the mean number of particles:

$$\langle N \rangle = \sum_{\mathbf{k}} \frac{1}{\exp\{\beta[\epsilon(\mathbf{k}) - \mu]\} - 1} \stackrel{?}{=} \frac{V}{(2\pi)^3} \int \frac{d^3 k}{\exp\{\beta[\epsilon(\mathbf{k}) - \mu]\} - 1} \tag{5}$$

The integral, in spherical coordinates, is a special case of the Bose–Einstein integral

$$g_\nu(z) = \frac{1}{\Gamma(\nu)} \int_0^\infty dx \frac{x^{\nu-1}}{z^{-1}e^x - 1} = \sum_{j=1}^\infty \frac{z^j}{j^\nu} \tag{6}$$

where $\Gamma(\nu)$ is the gamma function. If the replacement of the sum by the integral is legitimate, we obtain, using $\Gamma(\frac{3}{2}) = \frac{1}{2}(\pi)^{1/2}$,

$$\frac{\langle N \rangle}{V} \stackrel{?}{=} \left(\frac{mk_B T}{2\pi \hbar^2}\right)^{3/2} g_{3/2}(z) \tag{7}$$

where $z = e^{\beta\mu}$ is the fugacity. Since occupation numbers cannot be negative or infinite, we must have $\mu < 0$ or $0 \le z < 1$. The function $g_{3/2}(z)$ is finite in this interval and is depicted in Figure 7.1. The limiting value is

---

[1] A discussion of other boundary conditions may be found in Pathria [1983] and references therein.

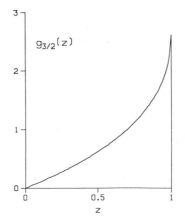

Figure 7.1 Function $g_{3/2}(z)$.

$g_{3/2}(1) = \zeta(\frac{3}{2}) = 2.612\ldots$, where $\zeta(x)$ is the Riemann zeta function. It is clear that the approximation of replacing the sum over **k** by an integral is not always valid since (7) yields a solution for z in terms of the density in the allowed region $(z < 1)$ only for temperatures greater than

$$k_B T_c = \frac{2\pi\hbar^2}{m}\left[\frac{\langle N\rangle}{\zeta(\frac{3}{2})V}\right]^{2/3} \tag{8}$$

Below this temperature it is impossible to satisfy (7) and we use the subscript c to denote that this is the critical temperature of the system. In the replacement of the sum over **k** by an integral, we have implicity assumed that the function $\langle n_k\rangle$ is smoothly varying. However, when μ comes close enough to zero [i.e., $\mu \sim O(1/V)$], the ground state of the system will become macroscopically occupied [i.e., $\langle n_0\rangle/V \sim O(1)$]. Since the spacing between energy levels is $\sim O(V^{-2/3})$, all the higher-energy states will have occupation of at most $\langle n_k\rangle/V \sim O(V^{-1/3})$. We must therefore single out the ground state in the sum over states in (5), while the occupation numbers of all the other states can be summed, in the thermodynamic limit, by replacing the sum by an integral as we did previously. For $T > T_c$ there is no macroscopic occupation of the single-particle ground state and we have

$$\frac{\langle N\rangle}{V} = n = \left(\frac{mk_B T}{2\pi\hbar^2}\right)^{3/2} g_{3/2}(z) \qquad T > T_c \tag{9}$$

while for $T < T_c$ we consider the gas to be a mixture of two phases,

$$n = n_n(T) + n_0(T) \tag{10}$$

The density of particles occupying the $k \neq 0$ states is

$$n_n(T) = \left(\frac{mk_B T}{2\pi\hbar^2}\right)^{3/2} \zeta(\frac{3}{2}) \tag{11}$$

The density of particles in the ground state adjusts itself to make up for the deficit:

$$n_0(T) = n\left[1 - \left(\frac{T}{T_c}\right)^{3/2}\right] \tag{12}$$

The temperature dependence of $n_0(T)$ and of the chemical potential $\mu(T)$ is shown in Figures 7.2 and 7.3. The singular behavior of $n_0(T)$ and $\mu(T)$ at $T = T_c$ is reflected in nonanalyticities in all other thermodynamic functions. If we convert (3) into an integral, with due attention to the possible macroscopic occupation of the ground state, we find, using (6), that

$$\ln Z_G = -\ln(1 - z) + \left(\frac{mk_B T}{2\pi\hbar^2}\right)^{3/2} V g_{5/2}(z) \tag{13}$$

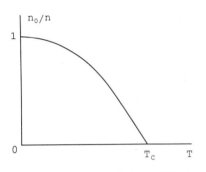

Figure 7.2 Fraction of the particles in the $k = 0$ state as a function of temperature.

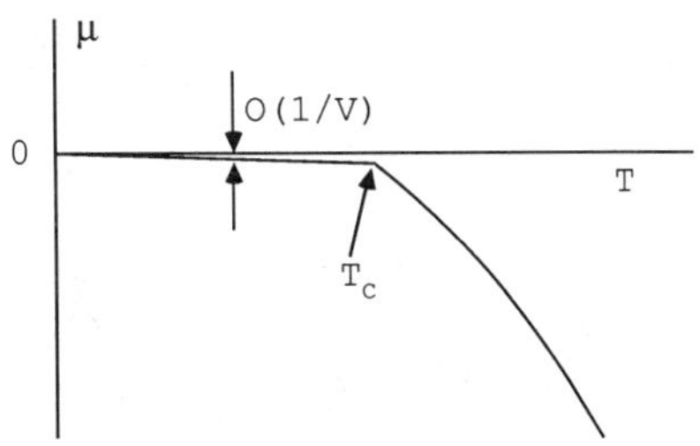

Figure 7.3 Temperature dependence of the chemical potential of an ideal Bose gas.

Since $(1 - z) \sim O(1)$ above T_c and $(1 - z) \sim O(1/V)$ below, we have

$$\frac{1}{V}\ln(1 - z) \to 0 \text{ as } V \to \infty \tag{14}$$

and find, for the pressure,

$$P = \left(\frac{m}{2\pi\hbar^2}\right)^{3/2}(k_B T)^{5/2} g_{5/2}(z) \qquad T > T_c$$

$$= \left(\frac{m}{2\pi\hbar^2}\right)^{3/2}(k_B T)^{5/2} \zeta\left(\tfrac{5}{2}\right) \qquad T \le T_c \tag{15}$$

where $\zeta\left(\tfrac{5}{2}\right) = g_{5/2}(1) = 1.341$. By combining (15) with (8) and eliminating the temperature, we find that the phase transition line in a PV diagram will be $(v = V/\langle N \rangle)$

$$P_c v_c^{5/3} = \frac{\zeta\left(\tfrac{5}{2}\right)}{[\zeta\left(\tfrac{3}{2}\right)]^{5/3}} \frac{2\pi\hbar^2}{m} \tag{16}$$

In Figure 7.4 we plot some of the isotherms and the line of phase transitions in the P–v plane.

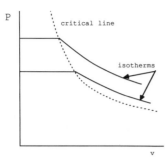

Figure 7.4 *PV* plane isotherms (solid curves) and critical line (dotted curve).

The entropy may be obtained from

$$S = \left(\frac{\partial(k_B T \ln Z_G)}{\partial T}\right)_{\mu,v} = \left(\frac{\partial PV}{\partial T}\right)_{\mu,v} \tag{17}$$

Below T_c we have, from (15), in the thermodynamic limit,

$$S = \frac{5}{2} k_B V \left(\frac{mk_B T}{2\pi \hbar^2}\right)^{3/2} \zeta\left(\frac{5}{2}\right) = \frac{5}{2} k_B n_n V \frac{\zeta\left(\frac{5}{2}\right)}{\zeta\left(\frac{3}{2}\right)} \tag{18}$$

where we have used (11) in the last step. This result supports our interpretation, given above, of two coexisting phases below T_c with densities n_n and n_0, respectively. The normal component has an entropy per particle

$$\frac{5}{2} k_B \frac{\zeta\left(\frac{5}{2}\right)}{\zeta\left(\frac{3}{2}\right)} \tag{19}$$

while the particles condensed into the ground state carry no entropy. The flat parts of the isotherms in Figure 7.4 are thus seen to be coexistence curves. As the normal particles condense there is a latent heat per particle

$$L = T\Delta S = \frac{5}{2} k_B T_c \frac{\zeta\left(\frac{5}{2}\right)}{\zeta\left(\frac{3}{2}\right)} \tag{20}$$

This equation for the latent heat may also be obtained from the Clausius–Clapeyron equation

$$\frac{dP_c(T)}{dT} = \frac{5}{2} k_B \left(\frac{mk_B T}{2\pi \hbar^2}\right)^{3/2} \zeta\left(\frac{5}{2}\right)$$

$$= \frac{5}{2} k_B \frac{\zeta\left(\frac{5}{2}\right)}{\zeta\left(\frac{3}{2}\right)} \frac{1}{v_c} = \frac{L}{T \Delta V} \tag{21}$$

The interpretation is, once again, in terms of two-fluid coexistence, one part with specific volume v_c, the other with zero specific volume (the condensate).

If, on the other hand, one considers the system at constant volume with a fixed total number of particles, the appropriate specific heat is

$$C_{V,N} = T\left(\frac{\partial S}{\partial T}\right)_{V,N}$$

We leave it as an exercise (Problem 7.1) for the reader to show that $C_{V,N}$ is continuous at the transition and has a discontinuity in its first derivative (Figure 7.5).

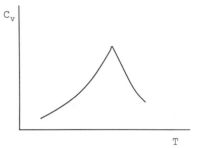

Figure 7.5 Specific heat of the ideal Bose gas.

For these reasons Bose–Einstein condensation is sometimes called a first-order transition (latent heat!) and sometimes a third-order transition (discontinuous derivative of $C_{V,N}$), depending on the point of view.

We have emphasized the phase-transition aspect of Bose condensation since there are a number of striking similarities with the phenomena of superfluidity and superconductivity. To exploit the analogy, it is useful to comment on the nature of the order parameter of the low-temperature phase. For the properties discussed to this point, the two-fluid description (10) is adequate. This description is also used in phenomenological theories of superfluidity and superconductivity, but then there are, as we shall see, important aspects of the low-temperature phases which require description in terms of a two-component order parameter.

In the case of a Bose condensate, the natural choice for a two-component order parameter is the wave function $\psi = |\psi| e^{i\phi}$ of the macroscopically occupied state. The aspect of Bose condensation which formally has an analog in superconductivity and superfluidity is that of *off-diagonal long-range order* (ODLRO) (see Yang, 1962; Penrose and Onsager, 1956; Penrose, 1951). In the case of the ideal Bose gas, in the momentum space representation, the expectation value $\langle b_{\mathbf{k}'}^{+}, b_{\mathbf{k}} \rangle = \delta_{\mathbf{k}\mathbf{k}'} \langle n_{\mathbf{k}} \rangle$ and is given by (4). Here $b_{\mathbf{k}}^{+}$, $b_{\mathbf{k}}$ are the creation and annihilation operators for a particle in state $\mathbf{k}$ with $\chi^{+}(\mathbf{r})$, $\chi(\mathbf{r})$ the corresponding field operators (see the Appendix). The corresponding quantity in the coordinate representation is the single-particle (reduced) density matrix $\rho(\mathbf{r}, \mathbf{r}')$ given by

$$\rho(\mathbf{r}, \mathbf{r}') = \langle \chi^{+}(\mathbf{r})\chi(\mathbf{r}') \rangle = \frac{1}{V} \sum_{\mathbf{k}} \langle n_{\mathbf{k}} \rangle e^{i\mathbf{k} \cdot (\mathbf{r}-\mathbf{r}')} \tag{22}$$

Above T_c all the occupation numbers are of order unity and $\rho(\mathbf{r}, \mathbf{r}') \to 0$ as $|\mathbf{r} - \mathbf{r}'| \to \infty$. Below the transition we have

$$\lim_{|\mathbf{r}-\mathbf{r}'|\to\infty} \rho(\mathbf{r}, \mathbf{r}') = n_0$$

and we may say that the existence of the condensate implies this nonzero limit or this ODLRO. Similarly, if the condensation were to be into a state with wave vector $\mathbf{k} \neq 0$, we would have

$$\lim_{|\mathbf{r}-\mathbf{r}'|\to\infty} \rho(\mathbf{r}, \mathbf{r}') = \langle n_\mathbf{k}\rangle e^{i\mathbf{k}\cdot(\mathbf{r}-\mathbf{r}')}$$

Generalizing, if the condensation is into a state with wave function ψ, we would have

$$\rho(\mathbf{r}, \mathbf{r}') \longrightarrow n_0 \psi^*(\mathbf{r})\psi(\mathbf{r}') \tag{23}$$

It was conjectured by Yang [1962] that the long-range order predicted by (22) and (23) is retained in the case of an interacting superfluid phase despite the fact that single-particle momentum states are no longer eigenstates of the Hamiltonian. The concept of off-diagonal long-range order is also seen to fit well with the concepts of the Landau–Ginzburg theory (see Section 3.J and later in this chapter).

We now briefly discuss the ideal Bose gas in arbitrary spatial dimension d. We assume that the particles occupy a hypercube of volume L^d ($d = 1, 2, 3,$ . . .) and, as in three dimensions, impose periodic boundary conditions on the single-particle wave functions. The generalization of equation (7) is immediate. Ignoring the possible macroscopic occupation of the ground state, we find that

$$\frac{N}{V} = \left(\frac{mk_\mathrm{B}T}{2\pi\hbar^2}\right)^{d/2} g_{d/2}(z) \tag{24}$$

Examining the power series (6) for $g_\nu(z)$, we see immediately that $g_\nu(z) \to \infty$ as $z \to 1$ for $\nu \leq 1$. Thus for $d = 2$ or less, there exists a solution $z < 1$ of equation (24) at any finite temperature. The ground state, as a consequence, is macroscopically occupied only at $T = 0$ and there is no finite-temperature Bose condensation for $d \leq 2$.

The reason for the divergence of the function $g_{d/2}(z)$ at $z = 1$ is the singularity of the integrand (5) near $|k| = 0$. In spherical coordinates we see that at $z = 1$, in arbitrary dimension d, the contribution to N of the states near $|\mathbf{k}| = 0$ is

$$N \sim \frac{V}{(2\pi)^d} \int dk \, \frac{k^{d-1}}{\left(\dfrac{\hbar^2\beta}{2m}\right)k^2} \sim \frac{V}{(2\pi)^d}\left(\frac{2mk_\mathrm{B}T}{\hbar^2}\right) \int dk \, k^{d-3} \tag{25}$$

This integral is finite for $d > 2$, infinite for $d \leq 2$. The phase space factor k^{d-1} in low dimensions emphasizes the role of low-lying excitations and destroys the phase transition.

In the case of an interacting Bose gas, the low-lying excitations are, as we shall see, sound like (i.e., their energy is proportional to k rather than to k^2). The integral corresponding to (25) is then convergent, but there is no off-diagonal long-range order in the sense of (23) for $d \leq 2$. This situation has been encountered previously in connection with our discussion of the Kosterlitz–Thouless theory in Section 5.E.

7.B. SUPERFLUIDITY

The Bose fluid ^{4}He exhibits a phase transition to a superfluid phase at 2.17 K at atmospheric pressure. Some of the features of this transition are strongly suggestive of Bose condensation, although proper treatment of the ^{4}He liquid requires the inclusion, in the Hamiltonian, of interactions between the particles, in particular, of the short-range repulsive interactions. Calculation of the properties of the system, especially near the transition, constitutes a very difficult and, to date, not completely solved problem in statistical physics. Nevertheless, a great deal can be understood about the behavior of ^{4}He at very low temperatures. We discuss, in Section 7.B(a), a number of the most striking features of ^{4}He and attempt to show how they may be understood in terms of the qualitative features of a condensate and the spectrum of excitations in the normal fluid. In Section 7.B(b) we discuss the Bogoliubov theory of the quasiparticle spectrum. We return to this topic from a different point of view in Chapter 8, where we use the formalism of linear response theory to arrive at the same results.

(a) Qualitative Features of Superfluidity

The feature that has given superfluidity its name is the ability of the liquid to flow through pipes and capillaries without dissipation. We show first that the ideal Bose gas does *not* have this property. Suppose that our fluid is flowing with velocity **v** relative to the laboratory frame of reference, and that an excitation is created in the fluid through its interaction with the walls of the container. In the frame of reference of the fluid the particles are initially at rest. In an ideal Bose gas an excitation from the condensate is single particle-like, that is, a particle acquires energy $p^2/2m$ through interaction with the "moving" wall. In the laboratory frame of reference the energy of the moving fluid will, in the case of a single excitation, be

$$E_L = (N - 1)\frac{mv^2}{2} + \frac{(\mathbf{p} + m\mathbf{v})^2}{2m} \tag{1}$$

An excitation with momentum **p** (in the moving frame) will thus lower the total energy of the fluid in the laboratory frame if $E_L < Nmv^2/2$, which occurs if

$$\frac{p^2}{2m} + \mathbf{p}\cdot\mathbf{v} < 0 \tag{2}$$

If **p** is antiparallel to **v** the condition (2) becomes simply $p < 2mv$, which can easily be satisfied since the ideal gas allows states of arbitrarily low momentum. The energy of the moving system can thus be dissipated due to collisions with the container and the ideal Bose condensate is, therefore, not a superfluid.

The feature that makes ^{4}He a superfluid is the fact that the low-lying excitations of the system are *sound quanta* rather than single-particle excitations.

This can be understood intuitively by considering the effect of the strong repulsive interactions between a pair of atoms at short distances. It is obvious that a freely moving particle will undergo a large number of collisions with other particles and that such states cannot be eigenstates of the Hamiltonian. A cooperative displacement of all the particles, such as occurs in a sound wave, is a better candidate for an elementary excitation of the system. The elementary excitation spectrum was postulated by Landau (1941, 1947) and later justified by Feynman [1954]. (Feynman's results were actually obtained much earlier by Bijl [1940], whose suggestions unfortunately attracted little attention.) We will not discuss the Bijl–Feynman theory and, at this point, simply note that the excitation spectrum can be obtained experimentally by inelastic neutron scattering (Cowley and Woods, 1971). It is displayed schematically in Figure 7.6.

For low momenta the energy is linear in p, $\epsilon = cp$, with a sound velocity $c = 2.4 \times 10^4$ cm/s. There is, moreover, a local minimum in $\epsilon(\mathbf{p})$ at $p/h = 1.9 \, \overset{\circ}{A}{}^{-1} = p_0$. In the vicinity of the minimum the dispersion relation is given by

$$\epsilon(\mathbf{p}) = \Delta + \frac{(p - p_0)^2}{2m^*} \tag{3}$$

with $\Delta/k_B = 8.7$ K and $m^* = 0.16 m_{He}$. Excitations with p close to p_0 are called rotons (mainly due to historical accident).

It is now easy to show that a fluid with an excitation spectrum of the general shape shown in Figure 7.6 will display superfluidity. Suppose that the fluid is moving through a container and consider, in the rest frame of the fluid, the container to be a classical object with mass M and momentum with an initial value $\mathbf{P}$. The relative velocity of the container with respect to the fluid is thus $\mathbf{v} = \mathbf{P}/M$. It is possible to create excitations of energy $\epsilon(p)$ and momentum $\mathbf{p}$ in the fluid if

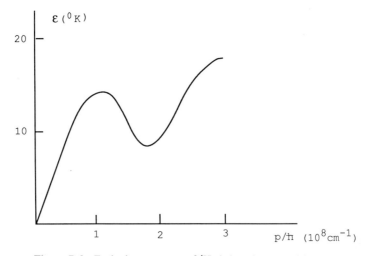

Figure 7.6 Excitation spectrum of ^{4}He below the λ transition.

$$\frac{P^2}{2M} - \frac{(\mathbf{P} - \mathbf{p})^2}{2M} = \epsilon(\mathbf{p}) \tag{4}$$

In the limit $M \to \infty$ we see that dissipation of energy is possible only if $v > \epsilon(p)/p$. From Figure 7.7 we see that there is a critical velocity v_{crit} below which it is not possible to dissipate energy through the creation of excitations by contact with the walls of the container through which the fluid is flowing.

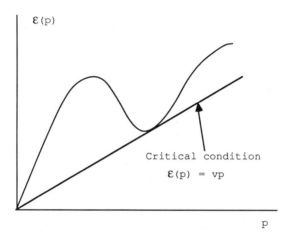

Figure 7.7   Critical velocity of superfluid flow from (4).

The critical velocity for superflow, given by the argument above and the shape of the measured excitation spectrum, is $v_{\text{crit}} = 60$ m/s. Experimentally, the critical velocity is much lower than this and strongly dependent on the size and shape of the container. The reason for this is that in most situations turbulent (vortex) excitations will occur before the phonon-like excitations considered above.

The argument above suggests that further properties of superfluid ^{4}He can be understood by analogy with Bose condensation. To be specific, we will, following Landau, assume that a superfluid can be considered to consist of two coexisting fluids. We write for the density

$$\rho = \rho_s + \rho_n \tag{5}$$

where ρ_s is the superfluid mass density and ρ_n is the density of the normal component. We incorporate the idea that there is a macroscopic condensation into a single quantum state by assuming that there exists an order parameter $[\Psi(\mathbf{r}) = (n_0 m)^{1/2}\psi$, where ψ is given by (7.A.23)] which is proportional to the wave function of this state. We express the order parameter in terms of an amplitude and a phase:

$$\Psi(\mathbf{r}) = a(\mathbf{r})e^{i\gamma(\mathbf{r})} \tag{6}$$

and identify the square of the amplitude with the superfluid density:

$$\rho_s = a^2 \tag{7}$$

With this interpretation we have for the mass current density of the superflow,

$$\mathbf{j}_s = -[\tfrac{1}{2}\Psi(-i\hbar\nabla)\Psi^* + \tfrac{1}{2}\Psi^*(i\hbar\nabla)\Psi] = \hbar a^2 \nabla\gamma(\mathbf{r}) \tag{8}$$

The associated velocity field is

$$\mathbf{u}_s = \frac{\hbar}{m}\nabla\gamma(\mathbf{r}) \tag{9}$$

A flow pattern in which the velocity field can be expressed as the gradient of a scalar function is referred to as *potential flow*. A characteristic feature of potential flow is that it is irrotational, that is,

$$\nabla \times \mathbf{u}_s = 0 \tag{10}$$

One might think that this condition on the flow would prohibit a superfluid from participating in any kind of rotation; for example, if a normal fluid rotates uniformly as a rigid body, we must have

$$\mathbf{v} = \boldsymbol{\omega} \times \mathbf{r}$$
$$\nabla \times \mathbf{v} = 2\boldsymbol{\omega} \tag{11}$$

where $\boldsymbol{\omega}$ is the angular velocity. The curl of the velocity field is commonly referred to as the vorticity and in the case of rigid-body rotation is given by the second equation in (11).

To check the foregoing assumptions, Osborne [1950] and others conducted a series of "rotating bucket" experiments.[2] In a normal fluid, rotating uniformly, the shape of the upper surface is determined by a balance between the centripetal and gravitational forces (see Figure 7.8), giving

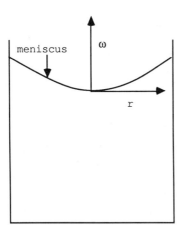

Figure 7.8 Shape of liquid surface in a rotating bucket.

[2] For an early review of the history of the hydrodynamics of rotating superfluid ^{4}He, see Andronikashvili and Mamaladze [1966].

$$z(r) = \frac{\omega^2 r^2}{2g} \tag{12}$$

It was expected that if a bucket of superfluid helium were rotated, only the normal component would participate in the rotation and that the shape of the meniscus would be given by

$$z(r) = \frac{\rho_n \omega^2 r^2}{2\rho g} \tag{13}$$

However, the observed shape of the surface conformed with (12) rather that (13). An observation made originally by Onsager [1949] makes it possible to reconcile the outcome of the rotating bucket experiments with our two-fluid picture of superfluidity.

The quantity $\gamma(\mathbf{r})$ differs from an ordinary velocity potential in that it is a *phase* and is only defined modulo 2π:

$$\gamma(\mathbf{r}) + 2\pi = \gamma(\mathbf{r})$$

This does not change the fact that since $\mathbf{u}_s \sim \nabla\gamma$, we must have $\nabla \times \mathbf{u}_s = 0$. It is possible, however, to have a nonzero circulation

$$\oint \mathbf{u}_s \cdot d\mathbf{l} \neq 0$$

without violating $\nabla \times \mathbf{u}_s = 0$, if in the region of phase singularity (γ is undefined at $\mathbf{r} = 0$) the superfluid density is zero (see Figure 7.9).

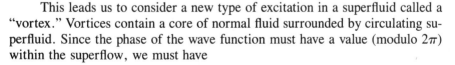

Figure 7.9   Picture of rotating ⁴He. The superfluid density is zero in the core region.

This leads us to consider a new type of excitation in a superfluid called a "vortex." Vortices contain a core of normal fluid surrounded by circulating superfluid. Since the phase of the wave function must have a value (modulo 2π) within the superflow, we must have

$$\oint \nabla\gamma \cdot d\mathbf{l} = 2\pi n \qquad n = \text{integer}$$

where the contour in the integration above is any closed path around the core. Equivalently,

$$\oint \mathbf{u}_s \cdot d\mathbf{l} = \frac{nh}{m} \tag{14}$$

that is, the circulation around a vortex is *quantized*. It is generally believed that the creation of quantized vortices is the mechanism by which superfluid flow is destroyed when the critical velocity is exceeded. In fact, there is a certain similarity between the breakdown of superfluidity through the formation of vortices and the onset of turbulence from laminar flow in ordinary fluids. It is also worth noting the similarity between the vortices discussed here and in Section 5.E. There is also, in this case, an analogy between (14) and Ampère's law in electromagnetic theory which can be used to estimate the physical properties of a simple vortex.

Another important property of superfluidity that can be understood by analogy with the two-fluid model of Bose condensation is that the superfluid component carries no entropy. Combining the Gibbs–Duhem equation

$$E = TS - PV + \mu N \tag{15}$$

with the thermodynamic relation

$$dE = T \, dS - P \, dV + \mu \, dN \tag{16}$$

we find

$$N \, d\mu = -S \, dT + V \, dP \tag{17}$$

Consider the experimental arrangement of Figure 7.10. Two vessels containing ^{4}He below the λ point are connected by a pipe that is clogged by some permeable obstacle to form a "superleak." The obstacle prevents the flow of normal liquid, but the superfluid component can pass back and forth. We now release a small amount of heat in the container on the left and consider how a new equilibrium state can be established. Since the superfluid can flow freely, the chemical potential must, in the end, be the same on the two sides. On the other hand, since the superfluid component carries no entropy, there is no heat flow and therefore no tendency for the temperature to equilibrate. Similarly, since the pipe is clogged as far as the normal component is concerned, there is nothing to prevent the establishment of a pressure differential. From (17) we then obtain

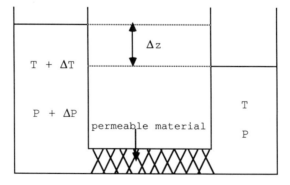

Figure 7.10   Superleak.

$$\frac{\Delta T}{\Delta P} = \frac{V}{S} \tag{18}$$

where ΔT is related to the amount of heat released and the heat capacity in the normal way. We conclude that in a superfluid a temperature differential is associated with a pressure difference. A dramatic manifestation of this is the *fountain effect* (Figure 7.11).

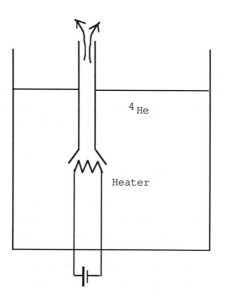

Figure 7.11   Fountain effect.

We next consider some aspects of the hydrodynamics of a two-component system. Consider first ordinary hydrodynamics. The conditions for mass and momentum balance in an isotropic fluid are

$$\frac{\partial \rho}{\partial t} + \mathbf{\nabla} \cdot \mathbf{j} = 0$$

$$\frac{\partial \mathbf{j}}{\partial t} + \nabla P = 0 \tag{19}$$

In (19), ρ is the mass density, $\mathbf{j}$ the mass current, and P the pressure. Taking the time derivative of the first equation, the divergence of the second, and subtracting, we find that

$$\frac{\partial^2 \rho}{\partial t^2} - \nabla^2 P = 0 \tag{20}$$

Usually, the pressure variations in a second wave are sufficiently fast that they can be taken to be *adiabatic,* so that we can relate pressure and density fluctuations through

$$dP = \left(\frac{\partial P}{\partial \rho}\right)_s d\rho \tag{21}$$

which, for small amplitudes, yields the wave equation

$$\frac{\partial^2 \rho}{\partial t^2} - c^2 \nabla^2 \rho = 0 \tag{22}$$

with

$$c = \sqrt{\left(\frac{\partial P}{\partial \rho}\right)_s} \tag{23}$$

In a superfluid the mass density has two components, ρ_n and ρ_s. It can be shown (see Landau and Lifshitz, 1980) that there are now two types of motion. In *first sound* the normal and superfluid components are in phase with each other just as in ordinary sound. In *second sound* the normal and superfluid components beat against each other in such a way that the mass density is constant. Since the normal component carries entropy while the superfluid does not, second sound is an entropy or thermal wave.

Another way of describing second sound is the following. If we think of helium as a degenerate Bose system, the normal component can be identified with collective excitations out of the ground state. These excitations represent first sound. Second sound, on the other hand, can be thought of as a sound wave in the gas of excitations. With this interpretation second sound should be observable in ordinary solids as well as in superfluids. Normally, the damping of sound waves is too large for second sound to be observable, but the phenomenon has indeed been observed in very pure crystals at very low temperatures [McNelly et al 1970] and in smectic liquid crystals (see de Gennes 1974).

It is interesting that the foregoing interpretation allowed Landau to make a definite prediction for the velocity of second sound in the low-temperature limit. If n_k is the number of excitations in the mode labeled by $\mathbf{k}$, the energy in this mode is $\hbar \omega n_k = \hbar c_1 k n_k$, where c_1 is the velocity of first sound. The number of longitudinal modes per unit volume is $d^3k/(2\pi)^3$, giving for the free energy of the gas of excitations,

$$A = \frac{V k_B T}{2\pi^2 c_1^3} \int_0^\infty d\omega \, \omega^2 \ln\left(1 - \exp\{-\beta \hbar \omega\}\right) \tag{24}$$

Comparing $P = -\partial A/\partial V$ and $E = \partial(\beta A)/\partial \beta$ (see Problem 2.8), we find the equation of state

$$PV = \frac{E}{3} \tag{25}$$

where E is the internal energy and P is the pressure. Writing

$$\rho = \frac{E}{c_1^2 V} \tag{26}$$

for the density of inertia ("mass density") of the gas, we find from (23)

$$c_2 = \frac{c_1}{\sqrt{3}} \tag{27}$$

The velocity of second sound found experimentally is sketched in Figure 7.12. The measurements are difficult at very low temperatures, but Landau's prediction about the second sound velocity appears to be essentially correct in the $T \to 0$ limit. On the other hand, over a fairly wide range of temperature, $1\text{ K} < T < 2\text{ K}$, the second sound velocity appears to be essentially constant with $c_2 \approx 20$ m/s compared to $c_1 \approx 240$ m/s for first sound.

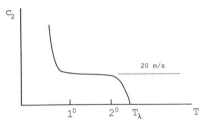

Figure 7.12 Second sound velocity as function of temperature.

(b) Bogoliubov Theory of the ^{4}He Excitation Spectrum

We conclude this section with an attempt to make plausible the sound wave–like excitation spectrum of superfluid helium in the long-wavelength limit. We do this by developing a theory due to Bogoliubov [1947] for a weakly interacting Bose system. For this purpose it is convenient to express the Hamiltonian in the second quantized form (see the Appendix). The grand canonical Hamiltonian in the plane-wave representation is

$$K = H - \mu N = \sum_{\mathbf{k}} \left(\frac{\hbar^2 k^2}{2m} - \mu \right) b_{\mathbf{k}}^+ b_{\mathbf{k}} + \frac{1}{2V} \sum_{\mathbf{k},\mathbf{k}',\mathbf{q}} v_{\mathbf{q}} b_{\mathbf{k}+\mathbf{q}}^+ b_{\mathbf{k}'-\mathbf{q}}^+ b_{\mathbf{k}'} b_{\mathbf{k}} \tag{28}$$

We will assume that the interaction between the particles, in the long-wavelength limit, may be approximated by a repulsive δ-function potential. In this case the Fourier transform of the potential, $v_{\mathbf{q}}$, is simply a constant, $v_{\mathbf{q}} = u_0$ for all $\mathbf{q}$. This assumption is not realistic in that the interatomic interaction has a hard core, and therefore, no Fourier transform. We are in effect making a "scattering length" approximation (Mahan, 1981). If the system were an ideal gas, the mean occupation number below the condensation temperature would be

$$\langle n_{\mathbf{k}} \rangle = N_0 \, \delta_{\mathbf{k},0} + \frac{1}{\exp\{\beta \epsilon(\mathbf{k})\} - 1} \tag{29}$$

where $N_0 \sim N$ is of the order of Avogadro's number. In this limit the creation and annihilation operators for the lowest single-particle state, $\mathbf{k} = 0$, can be treated approximately as scalars of magnitude $(N_0)^{1/2}$, that is,

$$b_0^+ |N_0\rangle = (N_0 + 1)^{1/2} |N_0 + 1\rangle \approx (N_0)^{1/2} |N_0\rangle$$
$$b_0 |N_0\rangle = (N_0)^{1/2} |N_0 - 1\rangle \approx (N_0)^{1/2} |N_0\rangle \tag{30}$$

It is then convenient to identify terms in the Hamiltonian (28) in which one or more of the operators act on the $\mathbf{k} = 0$ state. In the two-particle interaction term in (28) we find the following combinations of wave vectors which lead to the most important contributions:

$$\text{(a)} \ \mathbf{k} = \mathbf{k'} = \mathbf{q} = 0: \qquad \frac{1}{2V} u_0 N_0^2$$

$$\text{(b)} \ \mathbf{k} = \mathbf{k'} = 0, \ \mathbf{q} \neq 0: \qquad \frac{u_0}{2V} N_0 b_\mathbf{q}^+ b_{-\mathbf{q}}^+$$

$$\text{(c)} \ \mathbf{q} = -\mathbf{k}, \ \mathbf{k'} = 0 \qquad \frac{u_0}{2V} N_0 b_\mathbf{k}^+ b_\mathbf{k}$$

$$\text{(d)} \ \mathbf{q} = \mathbf{k'}, \ \mathbf{k} = 0: \qquad \frac{u_0}{2V} N_0 b_{\mathbf{k'}}^+ b_{\mathbf{k'}} \tag{31}$$

$$\text{(e)} \ \mathbf{k} = -\mathbf{k'} = -\mathbf{q}: \qquad \frac{u_0}{2V} N_0 b_\mathbf{k} b_{-\mathbf{k}}$$

$$\text{(f)} \ \mathbf{q} = \mathbf{k} = 0, \ \mathbf{k'} \neq 0: \qquad \frac{u_0}{2V} N_0 b_{\mathbf{k'}}^+ b_{\mathbf{k'}}$$

$$\text{(g)} \ \mathbf{q} = \mathbf{k'} = 0, \ \mathbf{k} \neq 0: \qquad \frac{u_0}{2V} N_0 b_\mathbf{k}^+ b_\mathbf{k}$$

All other terms are of lower order in N_0 and will be neglected. The grand canonical Hamiltonian, in this approximation becomes

$$K = -\mu N_0 + \frac{u_0 N_0^2}{2V} + \sum_\mathbf{k}{}' \left[\epsilon(\mathbf{k}) - \mu + \frac{u_0 N_0}{V} \right] b_\mathbf{k}^+ b_\mathbf{k}$$
$$+ \frac{u_0 N_0}{2V} \sum_\mathbf{k}{}' \left(b_\mathbf{k}^+ b_{-\mathbf{k}}^+ + b_{-\mathbf{k}} b_\mathbf{k} + b_\mathbf{k}^+ b_\mathbf{k} + b_{-\mathbf{k}}^+ b_{-\mathbf{k}} \right) \tag{32}$$

In (32) the notation Σ' indicates that the sum excludes terms with wave vector $\mathbf{k} = 0$. The chemical potential, μ, must be eliminated through the requirement that the overall density be independent of temperature. This is a nontrivial task and we refer the reader to the book by Fetter and Walecka [1971] for a derivation. The result is that for $T < < T_c$, $\mu = u_0 N_0/V$, which is obtained by minimizing the grand potential with respect to the variational parameter N_0. Assuming this result, we have

$$K = -\frac{u_0 N_0^2}{2V} + \frac{1}{2} \sum_\mathbf{k}{}' \left[\epsilon(\mathbf{k})\{b_\mathbf{k}^+ b_\mathbf{k} + b_{-\mathbf{k}}^+ b_{-\mathbf{k}}\} + \frac{u_0 N_0}{V} \{b_\mathbf{k}^+ b_{-\mathbf{k}}^+ + b_{-\mathbf{k}} b_\mathbf{k} \right.$$
$$\left. + b_\mathbf{k}^+ b_\mathbf{k} + b_{-\mathbf{k}}^+ b_{-\mathbf{k}}\} \right] \tag{33}$$

This "phonon" Hamiltonian can be diagonalized by means of a canonical transformation known as a *Bogoliubov transformation*. We write

$$b_{\mathbf{k}} = \eta_{\mathbf{k}} \cosh \theta_{\mathbf{k}} - \eta^{+}_{-\mathbf{k}} \sinh \theta_{\mathbf{k}}$$
$$b_{-\mathbf{k}} = \eta_{-\mathbf{k}} \cosh \theta_{\mathbf{k}} - \eta^{+}_{\mathbf{k}} \sinh \theta_{\mathbf{k}} \tag{34}$$

where $[\eta_k, \eta^+_k] = [\eta_{-k}, \eta^+_{-k}] = 1$. This transformation is canonical in that it preserves the commutation relations for any real θ_k:

$$[b_k, b^+_k] = \cosh^2\theta_{\mathbf{k}} - \sinh^2\theta_{\mathbf{k}} = 1$$

We determine $\theta_{\mathbf{k}}$ from the condition that the Hamiltonian be diagonal in the quasiparticle operators. Substituting (34) into (33), we obtain, for the off-diagonal piece

$$K_{\text{o.d.}} = - \sum_{\mathbf{k}}{}' \left\{ \epsilon(\mathbf{k}) + \frac{u_0 N_0}{V} \right\} \cosh \theta_{\mathbf{k}} \sinh \theta_{\mathbf{k}} (\eta^+_{\mathbf{k}} \eta^+_{-\mathbf{k}} + \eta_{-\mathbf{k}} \eta_{\mathbf{k}})$$

$$+ \frac{u_0 N_0}{V} \sum_{\mathbf{k}}{}' (\cosh^2 \theta_{\mathbf{k}} + \sinh^2 \theta_{\mathbf{k}})(\eta^+_{\mathbf{k}} \eta^+_{-\mathbf{k}} + \eta_{-\mathbf{k}} \eta_{\mathbf{k}}) \tag{35}$$

We see that $K_{\text{o.d.}} = 0$ if

$$\tanh 2\theta_{\mathbf{k}} = \frac{u_0 N_0/V}{\epsilon(\mathbf{k}) + u_0 N_0/V} \tag{36}$$

and note that this equation has a solution for all $\mathbf{k}$ only if $u_0 > 0$. Thus, the repulsive nature of the interparticle potential is crucial for condensation. Using (36), we obtain, for the diagonal part of (33),

$$K = - \frac{u_0 N_0^2}{2V} + \sum_{\mathbf{k}}{}' \sqrt{\epsilon^2(\mathbf{k}) + 2\frac{u_0 N_0}{V}\epsilon(\mathbf{k})}\ \eta^+_{\mathbf{k}} \eta_{\mathbf{k}}$$

$$+ \sum_{\mathbf{k}}{}' \left[\sqrt{\epsilon^2(\mathbf{k}) + 2\frac{u_0 N_0}{V}} - \epsilon(\mathbf{k}) - \frac{u_0 N_0}{V} \right] \tag{37}$$

Two features of (37) are worth noting. We see that the quasiparticle energy

$$E(\mathbf{k}) = \sqrt{\epsilon^2(\mathbf{k}) + 2\frac{u_0 N_0}{V}\epsilon(\mathbf{k})}$$

is, for small $|\mathbf{k}|$, dominated by the second term in the expression under the root and is linear in $|\mathbf{k}|$. We have therefore obtained the phonon dispersion relation for the excitations of a weakly interacting Bose gas. The third term in equation (37) is a shift of the ground-state energy, which, in fact, diverges. This divergence, which is due to the contribution from large wave vectors, is an artifact of our approximation $v_{\mathbf{q}} = u_0$ for all $\mathbf{q}$.

The excitation spectrum in this theory, while qualitatively correct at small wave vectors, is not quantitatively accurate. Moreover, it is difficult to improve the theory systematically. For a critique of such "pairing" approximations and a discussion of better theories for superfluid ^{4}He, the reader is referred to Mahan [1981].

7.C SUPERCONDUCTIVITY

The phenomenon of superconductivity is in some ways similar to superfluidity, and for this reason, we have chosen to discuss it in this chapter. The frictionless flow of superfluid ^{4}He has its analog in the persistent currents in a superconductor, and there are other similarities between the two systems. We begin this section by concentrating on the microscopic theory in order to show how a condensation similar to Bose condensation can occur in a system of interacting fermions. For a more detailed treatment, the reader is encouraged to consult the original papers (Cooper, 1956; Bardeen et al., 1957) or one of the monographs, such as that of Schrieffer [1964].

The crucial feature responsible for superconductivity is an effective attractive interaction between electrons due to "overscreening" by the ions of the metal. The bare Coulomb interactions between electrons is, of course, repulsive. The ions of the system respond to the motion of the electrons and, in certain circumstances, can produce an effective interaction between electrons which is attractive. This discovery was due to Fröhlich [1950]. Later, Cooper [1956] showed that this attractive interaction could produce a two-electron bound state in the presence of a Fermi sphere of Bloch electrons. These bound pairs have properties which are similar to those of bosons and, at sufficiently low temperatures, condense into the superconducting state. The outline of this section is as follows. We first present the solution of the Cooper problem. In section 7.C(b) we solve the BCS ground-state problem and, in Section 7.C(c) discuss the finite-temperature behavior of the system. Finally, in Section 7.C(d), we briefly describe the Landau–Ginzburg theory of superconductivity. For this part of the theory, Part 2 of the book by Landau and Lifshitz [1980] is an excellent reference.

(a) Cooper Problem

We consider the problem of two interacting electrons in the presence of a filled Fermi sea. The Hamiltonian of the two-electron system is

$$H = \sum_{\mathbf{k},\sigma} \epsilon(\mathbf{k}) c_{\mathbf{k},\sigma}^+ c_{\mathbf{k},\sigma} + \frac{1}{2} \sum_{\mathbf{k},\mathbf{k}',\mathbf{q},\sigma,\sigma'} v(\mathbf{q}) c_{\mathbf{k}+\mathbf{q},\sigma}^+ c_{\mathbf{k}'-\mathbf{q},\sigma'}^+ c_{\mathbf{k}',\sigma'} c_{\mathbf{k},\sigma} \tag{1}$$

where σ, σ' label the spin states of the particles and where the sum over wave vectors is restricted to $|\mathbf{k}| > k_\mathrm{F}$, the Fermi wave vector. In a metal, one can show (Fröhlich, 1950; see also Chapter 8) that the effective interaction $v(\mathbf{q})$ can be negative for a range of energy of extent of order $\hbar\omega_D$ near the Fermi energy, where ω_D is the Debye frequency. This "attractive" interaction is due to the overscreening alluded to in the introduction. Instead of using the full potential $v(\mathbf{q})$, we make the simplifying assumption (Cooper, 1956) that $v(\mathbf{q}) = -v$ for $\epsilon_F \leq \epsilon(q) \leq \epsilon_F + \hbar\omega_D$. We now attempt to find a two-electron bound state of the Hamiltonian (1). We take the zero of energy to be ϵ_F and consider a trial wave function of the form

$$|\psi\rangle = \sum_{\mathbf{k}} \alpha_{\mathbf{k}} c^+_{\mathbf{k},\frac{1}{2}} c^+_{-\mathbf{k},-\frac{1}{2}} |F\rangle$$

where

$$|F\rangle = \prod_{|\mathbf{k}| < k_F, \sigma} c^+_{\mathbf{k},\sigma} |0\rangle$$

is the wave function of the filled Fermi sphere. We demand that $(E - H)|\psi\rangle = 0$, which yields

$$0 = [E - 2\epsilon(\mathbf{k})]\alpha_{\mathbf{k}} - v\sum_{\mathbf{q}} \alpha_{\mathbf{k}+\mathbf{q}} \tag{2}$$

We define the constant Λ to be

$$\Lambda = \sum_{\mathbf{q}} \alpha_{\mathbf{k}+\mathbf{q}} = \int_0^{\hbar\omega_D} d\epsilon\, \rho(\epsilon)\alpha(\epsilon)$$

and thus obtain the equation

$$\Lambda = v\Lambda \int_0^{\hbar\omega_D} d\epsilon\, \frac{\rho(\epsilon)}{E - 2\epsilon} \tag{3}$$

or

$$1 \approx v\rho(\epsilon_F) \int_0^{\hbar\omega_D} d\epsilon\, \frac{1}{E - 2\epsilon} = v\rho(\epsilon_F)\ln\frac{E - 2\hbar\omega_D}{E} \tag{4}$$

where we have assumed that the density of states $\rho(\epsilon)$ does not vary appreciably in the range $(\epsilon_F, \epsilon_F + \hbar\omega_D)$. We may now solve (4) for the eigenvalue E to find

$$E = \frac{-2\hbar\omega_D}{\exp\{1/(v\rho(\epsilon_F))\} - 1} \approx -2\hbar\omega_D\exp\left\{\frac{-1}{v\rho(\epsilon_F)}\right\} \tag{5}$$

which demonstrates that no matter how weak the effective interaction, v, there always exists a two-particle bound state with energy less than the Fermi energy. This indicates that the Fermi sea is unstable against formation of paired states. We show below that these paired states form a condensate separated from the nearest single-particle states by a gap of width roughly E as given in

(5). It can also be shown (Schrieffer, 1964) that the paired state with zero center of mass momentum is the lowest-energy paired state. The BCS (Bardeen et al., 1957) theory of superconductivity focuses on such states.

(b) BCS Ground State

As we have seen in Section 7.C(a), the normal state of an electron gas is unstable with respect to formation of bound pairs near the Fermi surface. Bardeen et al. [1957] (BCS) considered an approximate Hamiltonian for the N-electron problem which contains the effective attraction due to overscreening and which contains, beside the Bloch energies, only interactions between paired electrons in singlet states. This "reduced" Hamiltonian is

$$H_{red} = \sum_{k, \sigma} \epsilon(k) c_{k, \sigma}^+ c_{k, \sigma} + \frac{1}{V_0} \sum_{k, k'} V_{kk'} b_k^+ b_{k'} \tag{6}$$

where V_0 is the volume, $c_{k, \sigma}^+, c_{k, \sigma}$ are creation and annihilation operators for electrons in Bloch states (k, σ), and $b_k^+ b_k$ are creation and annihilation for the pair $(k, +\frac{1}{2}; -k, -\frac{1}{2})$, that is, $b_k^+ = c_{k, 1/2}^+ c_{-k, -1/2}^+$ and $b_k = c_{-k, -1/2} c_{k, 1/2}$. As in Section 7.C(a) the interaction $V_{kk'}$ is assumed to be zero unless both $\epsilon(k)$ and $\epsilon(k')$ are in a shell of width $\hbar \omega_D$ centered on the Fermi energy. If we restrict our set of basis states to paired states, that is, the single-particle states $(k, +\frac{1}{2})$ and $(-k, -\frac{1}{2})$ are either both occupied or both empty, the Hamiltonian (6) can be expressed as

$$H = 2 \sum_k \epsilon(k) b_k^+ b_k + \frac{1}{V_0} \sum_{k, k'} V_{kk'} b_k^+ b_{k'} \tag{7}$$

The normal state $|F\rangle$ falls into this category as

$$|F\rangle = \prod_{|k| < k_{F,\sigma}} c_{k, \sigma}^+ |0\rangle = \prod_{|k| < k_F} b_k^+ |0\rangle$$

The commutation relations of the pair creation and annihilation operators are

$$\left[b_k, b_{k'} \right] = \left[b_k^+, b_{k'}^+ \right] = 0$$

$$\left[b_k, b_{k'}^+ \right] = \delta_{kk'} \left\{ 1 - n_{k, 1/2} - n_{-k, -1/2} \right\} \tag{8}$$

If the term $n_{k, 1/2} + n_{-k, -1/2}$ in the last commutator were absent, these commutation relations would be those of boson operators. In fact, the commutation relations (8) are those of a set of spin-$\frac{1}{2}$ operators with the formal correspondence

$$b_k^+ = S_{k, x} + i S_{k, y} = S_k^+$$

$$b_k = S_{k, x} - i S_{k, y} = S_k^- \tag{9}$$

$$1 - n_{\mathbf{k}, 1/2} - n_{-\mathbf{k}, -1/2} = 2S_{\mathbf{k}, z}$$

This identification of equivalent spin operators holds only for our restricted set of basis states in which the operator $n_{\mathbf{k}, 1/2} + n_{-\mathbf{k}, -1/2}$ has only the eigenvalues 0 and 2. In this spin language the Hamiltonian (7) takes the form

$$H = 2 \sum_{\mathbf{k}} \epsilon(\mathbf{k}) \left(S_{\mathbf{k}, z} + \frac{1}{2} \right) + \frac{1}{V_0} \sum_{\mathbf{k}, \mathbf{k}'} V_{\mathbf{k}\mathbf{k}'} (S_{\mathbf{k}, x} S_{\mathbf{k}', x} + S_{\mathbf{k}, y} S_{\mathbf{k}', y}) \tag{10}$$

which is an XY model with a "Zeeman" energy due to a "magnetic field" in the z direction. The two-dimensional rotational symmetry of the Hamiltonian is obvious in this notation. In magnetic language, if $V_{\mathbf{k}\mathbf{k}'}$ is large enough, the ground state will have some nonzero magnetization $\mathbf{M} = \Sigma \mathbf{S}_{\mathbf{k}}$, and this vector can be rotated about the z axis without cost of energy. The order parameter is a two-component order parameter and the transition from the normal to the superconducting state will be in the universality class of XY model. We also see, from this analogy, that conventional long-range order in two-dimensional systems (thin metallic films) will be excluded (see also Section 5.E).

Following BCS we now construct a trial wave function for the ground state and minimize the resulting expression for the energy subject to the constraint that the mean number of electrons is N. The trial wave function

$$|\psi\rangle = \prod_{\mathbf{k}} \frac{1 + g_k b_{\mathbf{k}}^+}{\sqrt{1 + g_k^2}} |0\rangle \tag{11}$$

contains paired states with an indefinite number of particles and is normalized. The function g_k is our variational parameter. The expectation value of the grand canonical Hamiltonian

$$K = H - \mu N$$

in the state (11) is

$$K_0 = \langle \psi | K | \psi \rangle$$

$$= 2 \sum_{\mathbf{k}} [\epsilon(\mathbf{k}) - \mu] \frac{g_k^2}{1 + g_k^2} + \frac{1}{V_0} \sum_{\mathbf{k}, \mathbf{k}'} V_{\mathbf{k}\mathbf{k}'} \frac{g_k g_{k'}}{(1 + g_k^2)(1 + g_{k'}^2)} \tag{12}$$

We note that the normal state corresponds to the value $g_k = \infty$ for $|\mathbf{k}| < k_F$ and $g_k = 0$ for $|\mathbf{k}| > k_F$. We now define the functions u_k and v_k through

$$u_k = \frac{1}{\sqrt{1 + g_k^2}}$$

$$v_k = \frac{g_k}{\sqrt{1 + g_k^2}} \tag{13}$$

with $u^2 + v^2 = 1$. In the normal state $u_k = 0$, $v_k = 1$ for $|\mathbf{k}| < k_F$ and $u_k = 1$, $v_k = 0$ for $|\mathbf{k}| > k_F$. The variation of K_0 with respect to u_k and v_k yields

$$\delta K_0 = 4 \sum_k [\epsilon(\mathbf{k}) - \mu] v_k \delta v_k + \frac{2}{V_0} \sum_{\mathbf{k},\mathbf{k}'} V_{\mathbf{k}\mathbf{k}'}(u_{k'} v_{k'})\{u_k \delta v_k + v_k \delta u_k\}$$

$$= 4 \sum_k [\epsilon(\mathbf{k}) - \mu] v_k \delta v_k + \frac{2}{V_0} \sum_{\mathbf{k},\mathbf{k}'} V_{\mathbf{k}\mathbf{k}'}(u_{k'} v_{k'}) \frac{u_k^2 - v_k^2}{u_k} \delta v_k \tag{14}$$

where we have used the condition $u_k \delta u_k + v_k \delta v_k = 0$. Requiring the coefficient of δv_k to be zero for all k, we obtain

$$2[\epsilon(\mathbf{k}) - \mu] v_k + \frac{1}{V_0} \sum_{\mathbf{k}'} V_{\mathbf{k}\mathbf{k}'}(u_{k'} v_{k'}) \frac{u_k^2 - v_k^2}{u_k} = 0 \tag{15}$$

This equation may be simplified by defining new variables. We let

$$u_k v_k = \frac{\Delta_k}{2E_k} \tag{16}$$

where $E_k = [\{\epsilon(\mathbf{k}) - \mu\}^2 + \Delta_k^2]^{1/2}$. Using the condition $u_k^2 + v_k^2 = 1$, we may reexpress u_k and v_k in terms of the single unknown Δ_k:

$$u_k^2 = \frac{1}{2}\left[1 + \frac{\epsilon(\mathbf{k}) - \mu}{E_k}\right]$$
$$v_k^2 = \frac{1}{2}\left[1 - \frac{\epsilon(\mathbf{k}) - \mu}{E_k}\right] \tag{17}$$

The choice of sign inside the brackets of (17) is determined by the condition $u_k = 0$ for $|k| < k_F$ in the normal state ($\Delta_k = 0$, $\mu = \epsilon_F$). Substituting into (15), we obtain an equation for the zero-temperature "gap," Δ_k:

$$\Delta_k = -\frac{1}{V_0} \sum_{\mathbf{k}'} V_{\mathbf{k}\mathbf{k}'} \frac{\Delta_{k'}}{2E_{k'}} \tag{18}$$

At this point we simplify equation (18) by taking $V_{\mathbf{k}\mathbf{k}'}$ to be a constant, $-V$, for $|\epsilon(k) - \epsilon_F| < \hbar\omega_D/2$, $|\epsilon(k') - \epsilon_F| < \hbar\omega_D/2$, and $\Delta_k = \Delta$, independent of $\mathbf{k}$ in the same region of k-space in which V is nonzero. Converting the sum in (18) to an integral over the density of states, $\rho(\epsilon)$, we finally obtain

$$1 = \frac{V}{2} \int_{\epsilon_F - \hbar\omega_D/2}^{\epsilon_F + \hbar\omega_D/2} d\epsilon \, \rho(\epsilon) \frac{1}{\sqrt{(\epsilon - \mu)^2 + \Delta^2}} \tag{19}$$

It can be shown that the shift in chemical potential is small. Thus, replacing μ by ϵ_F and assuming that the density of states is essentially constant over the small range $\hbar\omega_D$ near the Fermi surface, we find, from (19),

$$1 = V\rho(\epsilon_F) \sinh^{-1} \frac{\hbar\omega_D}{2\Delta} \tag{20}$$

or

$$\Delta = \frac{\hbar\omega_D}{2 \sinh[1/(V\rho(\epsilon_F))]} \approx \hbar\omega_D \exp\left\{\frac{-1}{V\rho(\epsilon_F)}\right\} \tag{21}$$

in the weak-coupling limit $V\rho(\epsilon_F) << 1$. The reader will note that (21) is a singular function of the interaction strength V. Thus any attempt to arrive at the superconducting ground state by perturbative methods would be doomed to failure.

Having obtained an expression for the variational parameter Δ, we must still show that the ground-state energy for nonzero Δ is lower than the normal ($\Delta = 0$) energy. We again take $\mu = \epsilon_F$ and take the zero of energy to be ϵ_F. The difference in energy between the variational ground state and the normal state is given by

$$E_0(\Delta) - E_N = 2 \sum_{\mathbf{k}} [\epsilon(\mathbf{k}) - \epsilon_F] v_k^2 + \frac{1}{V_0} \sum_{\mathbf{k},\mathbf{k}'} V_{\mathbf{k}\mathbf{k}'}(u_k v_k)(u_{k'} v_{k'})$$

$$- 2 \sum_{|\mathbf{k}|<k_F} [\epsilon(\mathbf{k}) - \epsilon_F] \qquad (22)$$

Using the expression (17) to substitute for v_k^2, equations (16) and (18) to eliminate the sum over k' in the second term on the right-hand side, and converting the remaining sums to integrals over the density of states, we find that

$$E_0(\Delta) - E_N = -\frac{V_0}{2}\rho(\epsilon_F)\Delta^2 \qquad (23)$$

indicating that the variational ground state is of lower energy than the normal state.

In Section 7.C(c) we shall see that the quantities E_k are approximate two-particle eigenvalues of the Hamiltonian. At the Fermi surface they are separated from the nearest excited states (the normal states with no pairing) by a gap in energy 2Δ. This gap is responsible for the absence of dissipation in current flow.

Finally, we show that the BCS ground state is a state of *broken symmetry*. To see this we simply calculate the expectation value of the operators $b_\mathbf{k}, b_\mathbf{k}^+$:

$$\langle \psi_0 | b_\mathbf{k} | \psi_0 \rangle = u_k v_k = \langle \psi_0 | b_\mathbf{k}^+ | \psi_0 \rangle \qquad (24)$$

In terms of the "pseudo-spin" picture (9) this corresponds to the system being "magnetized" in the x direction. The BCS Hamiltonian (7) is invariant under the coherent rotation of all the spins

$$\text{(i.e., } b_\mathbf{k} \to b_\mathbf{k} \exp\{i\phi\}, \ b_\mathbf{k}^+ \to b_\mathbf{k}^+ \exp\{-i\phi\}\text{).}$$

This symmetry is absent in the ground state.

(c) Finite-Temperature BCS Theory

We found in Section 7.C(b) that the ground state of a superconductor is characterized by broken symmetry and by a nonzero expectation value of the pair creation and annihilation operators b_k, b_k^+: $\langle b_k \rangle = \langle b_k^+ \rangle = \Delta_k/(2E_k)$. We will now use these results to construct an approximate (but very successful) mean field theory of superconductivity for $T \neq 0$. We will confine ourselves to a derivation of the temperature-dependent gap equation. A much more detailed discussion of superconductivity can be found in Abrikosov et al. [1963], Schrieffer [1964], or Mahan [1981].

We begin, again, with the reduced grand canonical Hamiltonian

$$K = \sum_{\mathbf{k}, \sigma} \epsilon(\mathbf{k}) n_{\mathbf{k}, \sigma} + \frac{1}{V_0} \sum_{\mathbf{k}, \mathbf{k}'} V_{\mathbf{k}\mathbf{k}'} b_{\mathbf{k}}^+ b_{\mathbf{k}'} \tag{25}$$

where we have absorbed the chemical potential, μ, into the single-particle energies $\epsilon(\mathbf{k})$. In the spirit of mean field theory (Chapter 3), we now approximate the interaction term in equation (25) by the decoupled form

$$\sum_{\mathbf{k}, \mathbf{k}'} V_{\mathbf{k}\mathbf{k}'} b_{\mathbf{k}}^+ b_{\mathbf{k}'} \approx \sum_{\mathbf{k}, \mathbf{k}'} V_{\mathbf{k}\mathbf{k}'} (b_{\mathbf{k}}^+ \langle b_{\mathbf{k}'} \rangle + \langle b_{\mathbf{k}}^+ \rangle b_{\mathbf{k}'} - \langle b_{\mathbf{k}}^+ \rangle \langle b_{\mathbf{k}'} \rangle) \tag{26}$$

where the last term in (26) has been included to compensate for double counting. The expectation values $\langle b_{\mathbf{k}}^+ \rangle$, $\langle b_{\mathbf{k}} \rangle$ will be evaluated with respect to the grand canonical probability density and will, of course, be temperature dependent. We assume that, as in the ground state, $\langle b_{\mathbf{k}}^+ \rangle$ is real and hence equal to $\langle b_{\mathbf{k}} \rangle$. This assumption is justified in BCS theory but needs to be modified if the BCS Hamiltonian is replaced by the full electron–phonon interaction. We note, in passing, that the nonzero expectation value of a particle nonconserving operator such as $b_{\mathbf{k}}^+ = c_{\mathbf{k}, 1/2}^+ c_{-\mathbf{k}, -1/2}^+$ is referred to as off-diagonal long-range order (Yang, 1962) and is an essential property of superfluids and superconductors (see also Section 7.A and the Bogoliubov theory of the weakly interacting Bose gas).

In analogy with the ground-state calculations, we define the quantity $\Delta_{\mathbf{k}}(T)$ through

$$\Delta_{\mathbf{k}}(T) = -\frac{1}{V_0} \sum_{\mathbf{k}'} V_{\mathbf{k}\mathbf{k}'} \langle b_{\mathbf{k}'} \rangle \tag{27}$$

Using (26) and (27), we find, for the reduced Hamiltonian,

$$K = \sum_{\mathbf{k}} [\epsilon(\mathbf{k})(c_{\mathbf{k}, 1/2}^+ c_{\mathbf{k}, 1/2} + c_{-\mathbf{k}, -1/2}^+ c_{-\mathbf{k}, -1/2})$$
$$- \Delta_{\mathbf{k}}(c_{\mathbf{k}, 1/2}^+ c_{-\mathbf{k}, 1/2}^+ + c_{-\mathbf{k}, -1/2} c_{-\mathbf{k}, -1/2} - \langle b_{\mathbf{k}}^+ \rangle)] \tag{28}$$

An operator such as (28) which is bilinear in fermion operators can always be diagonalized by a Bogoliubov transformation (Section 7.B). Our object is to bring (28) into the form

$$K = \sum_{\mathbf{k}} E(\mathbf{k}) \xi_{\mathbf{k}}^+ \xi_{\mathbf{k}} + \text{constant} \tag{29}$$

where the quasiparticle energies $E(\mathbf{k})$ and the new fermion operators are to be determined. Once we have achieved this form, the partition function is simply that of a set of noninteracting fermions. We assume a transformation of the form

$$c_{k, 1/2} = \alpha \xi_k + \beta \xi^+_{-k}$$
$$c^+_{k, 1/2} = \alpha \xi^+_k + \beta \xi_{-k}$$
$$c_{-k, -1/2} = \gamma \xi_{-k} + \delta \xi^+_k$$
$$c^+_{-k, 1/2} = \gamma \xi^+_{-k} + \delta \xi_k$$

$$(30)$$

where $[\xi_k, \xi^+_{k'}]_+ = \delta_{kk'}$, $[\xi_k, \xi_{k'}]_+ = [\xi^+_k, \xi^+_{k'}]_+ = 0$.

This condition—that the c and ξ operators both obey fermion commutation relations, [i.e., that the transformation (30) is canonical]—produces three equations:

$$\alpha^2 + \beta^2 = 1$$
$$\gamma^2 + \delta^2 = 1 \qquad (31)$$
$$\alpha\delta + \beta\gamma = 0$$

The fourth equation for the unknown coefficients α . . . δ comes from the requirement that the Hamiltonian should be of the form (29). Substituting (30) into (28), we obtain, for the nondiagonal piece of K,

$$K_{\text{o.d.}} = \sum_k [\epsilon(k)(\alpha\beta - \gamma\delta) - \Delta_k (\alpha\gamma + \beta\delta)](\xi^+_k \xi^+_{-k} + \xi_{-k} \xi_k) \qquad (32)$$

If we require this expression to be identically zero, we find, using equation (31), that $\gamma = \alpha$, $\delta = -\beta$, and

$$\alpha^2 = \frac{1}{2}\left[1 + \frac{\epsilon(k)}{E(k)}\right]$$

$$\beta^2 = \frac{1}{2}\left[1 - \frac{\epsilon(k)}{E(k)}\right]$$

$$(33)$$

with $E(k) = [\epsilon^2(k) + \Delta^2_k]^{1/2}$. The sign convention in (33) is such that the quasiparticle energy becomes the usual single-particle energy if $\Delta = 0$ and $k > k_F$. For $k < k_F$ the quasiparticle energy is that of a hole (missing electron). We complete the specification of the coefficients by taking α, $\beta > 0$, $\delta < 0$.

With these coefficients substituted into (28), the remaining diagonal piece of K becomes

$$K = \sum_k [E(k)(\xi^+_k \xi_k + \xi^+_{-k} \xi_{-k}) + \epsilon(k) - E(k) - \Delta_k \langle b^+_k \rangle] \qquad (34)$$

The lowest-energy state is always the state in which no quasiparticles are present and we see at once that if $\Delta_k = 0$, the ground-state energy of the normal system is obtained.

We are now in a position to calculate Δ_k self-consistently. The expectation value $\langle b_k \rangle$ is given by

$$\langle b_k \rangle = \frac{\text{Tr } b_k \exp\{-\beta K\}}{\text{Tr } \exp\{-\beta K\}} = \alpha\delta\frac{\text{Tr } (\xi_k^+ \xi_k - \xi_{-k} \xi_{-k}^+) \exp\{-\beta K\}}{\text{Tr } \exp\{-\beta K\}} \tag{35}$$

$$= \alpha\delta[\langle n_k \rangle + \langle n_{-k} \rangle - 1] \tag{36}$$

where we have used the fact that expectation values of operators such as $\xi_k \xi_{-k}$ vanish. Since (34) is the Hamiltonian of a set of noninteracting "particles," we have

$$\langle n_k \rangle = \frac{1}{\exp\{\beta E(k)\} + 1}$$

and

$$\langle b_k \rangle = \frac{\Delta_k}{2E(k)} \tanh \frac{\beta E(k)}{2} \tag{37}$$

and finally,

$$\Delta_k(T) = -\frac{1}{V_0} \sum_{k'} \frac{V_{kk'} \Delta_{k'}}{2E(k')} \tanh \frac{\beta E(k')}{2} \tag{38}$$

If we now assume, as in Section 7.C(b), that $V_{kk'} = -V$, independent of k in the shell of width $\hbar\omega_D$ centered on the Fermi energy, then Δ_k is also independent of k and given by

$$1 = \frac{V}{2V_0} \sum_k \frac{\tanh\left[\frac{1}{2}\beta\sqrt{\epsilon^2(k) + \Delta^2}\right]}{\sqrt{\epsilon^2(k) + \Delta^2}} \tag{39}$$

which is the finite-temperature generalization of formula (19). From (39) we may obtain the critical temperature of the system. As $T \to T_c$, $\Delta \to 0$ and we have

$$1 = \frac{V\rho(\epsilon_F)}{2} \int_{-\hbar\omega_D/2}^{\hbar\omega_D/2} d\epsilon \frac{\tanh(\beta_c \epsilon/2)}{\epsilon} \tag{40}$$

The integeral on the right-hand side can be evaluated exactly in the weak-coupling limit (Fetter and Walecka, 1971), with the result

$$k_B T_c = 0.567\hbar\omega_D \exp\left\{\frac{-1}{\rho(\epsilon_F)V}\right\} \tag{41}$$

Since $\Delta(T = 0) = \hbar\omega_D \exp\{-1/[\rho(\epsilon_F)V]\}$, we have

$$\frac{\Delta(T = 0)}{k_B T_c} = 1.764 \tag{42}$$

which is an important quantitative prediction of the theory. The parameters ω_D, V which appear in the BCS theory are not well known; indeed, the use of the Debye frequency to limit the region in energy over which pairs form is certainly not precise. Nevertheless, experimental data, for a number of weak-coupling superconductors, (mostly elemental), is consistent with (42), but significant deviations are sometimes found. These discrepancies are accounted for in the "strong-coupling" theories of superconductivity, which are also based on BCS ideas but incorporate a more realistic form of the electron–phonon interaction and of the Fermi surface. For an introduction to these theories, see Mahan [1981].

The theory, as we have presented it above, is quite obviously a form of mean field theory, and it should not be surprising that the order parameter of the system [$\Delta(T)$] has a typical mean field temperature dependence near the critical point:

$$\frac{\Delta(T)}{\Delta(0)} \approx 1.74\left(1 - \frac{T}{T_c}\right)^{1/2} \tag{43}$$

As well, the specific heat has a discontinuity at the critical point. The demonstration of this is left as an exercise.

(d) Landau–Ginzburg Theory of Superconductivity

The usefulness of the Landau–Ginzburg approach to superconductivity lies in the fact that it provides a direct route to the handling of problems associated with fluctuations and inhomogeneities. The approach also permits a description of the electrodynamics of superconductors and their response to electromagnetic fields is, of course, what makes these materials interesting from a practical point of view.

The BCS theory, which we have outlined above, is a mean field theory that exhibits a second-order phase transition. As such, the theory can be described in the language of Chapter 3. In particular, the gap equation (39) is analogous to the self-consistent equations encountered in the Weiss molecular field treatment of magnets or in the mean field theory of the Maier–Saupe model. As shown explicitly in Section 7.C(b), the symmetry of the ordered phase is that of the XY model (i.e., the ordered phase is characterized both by a phase and an amplitude). A description in terms of a one-component order parameter such as the energy gap Δ (or equivalently, by a condensate mass density ρ_s in analogy with the two-fluid model of Bose condensation) therefore cannot account for all the physics.

The free-energy density must not depend on the phase of the order parameter and we expect that near T_c the free-energy density, $g = G/V_0$, of a bulk superconductor will be of the form

$$g = g_n + a(T)|\Psi|^2 + \tfrac{1}{2}b(T)|\Psi|^4 + \cdots \tag{44}$$

where $a(T) = \alpha(T - T_c)$ and where α and $b(T)$ are only weakly temperature dependent. Minimizing this free-energy density, we find that g is a minimum for $T < T_c$ for

$$|\Psi|^2 = \frac{\alpha(T_c - T)}{b} \tag{45}$$

The difference in free-energy density between the superconducting and normal states is then

$$g_S - g_N = -\frac{\alpha^2}{2b}(T_c - T)^2 \qquad T < T_c \tag{46}$$

and the specific heat has a discontinuity at the transition given by

$$C_S - C_N = \frac{\alpha^2 T_c}{b} \tag{47}$$

The normalization of the order parameter is, in principle, arbitrary. In accordance with our intuitive picture of the transition as a condensation of Cooper pairs, we write, in analogy with (7.B.6) and (7.B.7),

$$\Psi = (\tfrac{1}{2}n_s)^{1/2} e^{i\gamma} \tag{48}$$

where n_s is the superconducting electron density. This density n_s can be determined by subtracting, from the conduction electron density, the inertia of the gas of quasiparticle excitations in much the same way as outlined in Problem 7.3 for a superfluid. The parameters α and b can then be determined by comparing (47) and the transition temperature T_c with the predictions of the BCS theory. The results are $\alpha = b\pi^2 T_c/[7\zeta(3)\epsilon_F]$, where ζ is the Riemann zeta function, and $b = \alpha T_c/n$, where $n = k_F^3/(3\pi^2)$ is the conduction electron density.

We are now in a position to discuss inhomogeneities and to derive an expression for the *coherence length*. Recalling our intuitive picture of a condensate of Cooper pairs, each with charge $-2e$ and mass $2m$ we write, in analogy with (7.B.8), for the superconducting current associated with a spatially varying phase:

$$\mathbf{j}_s = \frac{ie\hbar}{2m}(\Psi^* \nabla\Psi - \Psi \nabla\Psi^*) \tag{49}$$

We next identify the gradient term in the Landau–Ginzburg free-energy density (3.J.1) with the kinetic energy density of the moving charges and thus obtain

$$G = G_N + \int d^3r\left(\frac{\hbar^2}{4m}|\nabla\Psi|^2 + a|\Psi|^2 + \frac{1}{2}b|\Psi|^4 + \cdots\right) \tag{50}$$

Remembering that $|\Psi|^2 = \Psi^*\Psi$ and requiring that the functional derivative of G with respect to Ψ^* be zero, we obtain

$$-\frac{\hbar^2}{4m}\nabla^2\Psi + a\Psi + b|\Psi|^2\Psi = 0 \tag{51}$$

We have encountered equations of the form (51) earlier in this text, in Section 3.J and in the discussion of the liquid vapor interface (Section 4.E). We therefore simply note that (51) predicts a coherence length (referred to in other contexts as a correlation length)

$$\xi = \frac{\hbar}{[8m|a|]^{1/2}} \tag{52}$$

for $T < T_c$. This is the natural length scale for spatial variations of the order parameter.

We conclude our discussion of Landau–Ginzburg theory by noting that there is another length scale which is relevant to the properties of superconductors, namely the *London penetration depth*. Let us consider a superconductor in the presence of an applied magnetic field $\mathbf{B} = \nabla \times \mathbf{A}$. The expression for the current density now becomes

$$\mathbf{j} = \frac{ie\hbar}{2m}\left[\Psi^*\left(\nabla + \frac{2ie}{\hbar c}\mathbf{A}\right)\Psi - \Psi\left(\nabla - \frac{2ie}{\hbar c}\mathbf{A}\right)\Psi^*\right] \tag{53}$$

If the length scale over which $\mathbf{A}$ varies is short compared to the coherence length ξ, the magnitude $|\Psi|$ of the order parameter is approximately constant and the expression (53) simplifies to

$$\mathbf{j} = -\left(\frac{2e^2}{mc}\mathbf{A} + \frac{e\hbar}{m}\nabla\gamma\right)|\Psi|^2 \tag{54}$$

Taking the curl of both sides of (54) then leads to the London equation

$$\nabla \times \mathbf{j} = -\frac{2e^2}{mc}|\Psi|^2(\nabla \times \mathbf{A}) \tag{55}$$

If we combine (55) with Ampère's law,

$$\nabla \times \mathbf{B} = \frac{4\pi}{c}\mathbf{j} \tag{56}$$

we find

$$\nabla^2\mathbf{B} = \frac{8\pi|\Psi|^2e^2}{mc^2}\mathbf{B}$$

$$\nabla^2\mathbf{j} = \frac{8\pi|\Psi|^2e^2}{mc^2}\mathbf{j} \tag{57}$$

Equation (57) typically has solutions with exponentially decaying currents and fields (Meissner effect). The appropriate length scale is then the London penetration depth, δ, which appears in (57):

$$\delta = \left(\frac{mc^2}{8\pi|\Psi|^2 e^2}\right)^{1/2} = \left[\frac{mc^2 b}{8\pi e^2 \alpha (T_c - T)}\right]^{1/2} \tag{58}$$

The analysis above assumes that $\xi >> \delta$. Materials for which this assumption holds are called *type I superconductors*. A type I superconductor will expel all currents and external fields H except for a thin layer of thickness δ near the surface. The diamagnetic energy cost per unit volume associated with this expulsion is $H_c^2/8\pi$ and we see immediately from (46) that the critical field, above which superconductivity is suppressed, is $H_c = (4\pi\alpha^2/b)^{1/2}(T_c - T)$.

If the coherence length is not long in comparison with the penetration depth, the problem becomes more complicated. Equation (51) must then be replaced by

$$-\frac{\hbar^2}{4\pi}\left(\nabla + \frac{2ie}{\hbar c}\mathbf{A}\right)^2 \Psi + a\Psi + b|\Psi|^2\Psi = 0 \tag{59}$$

Equation (59) together with Ampère's law (56) constitute the full Landau–Ginzburg equations. The solution of these equations is beyond the scope of this book and we refer to Landau and Lifshitz [1980] for further reading. We only mention that if $\kappa = \delta/\xi > 2^{1/2}$, the Landau–Ginzburg equations can have vortex solutions in analogy with superfluids. A superconductor that allows penetration of magnetic fields by forming vortex lines is called a *type II superconductor*.

In Chapter 3 we showed, by means of the Ginzburg criterion, that mean field theories are self-consistent only for spatial dimensionalities $d \geq 4$. Nevertheless, the BCS theory describes real three-dimensional superconductors with very high accuracy, even close to T_c. The reason for this can also be understood in terms of the Ginzburg criterion. By using it to estimate the reduced temperature $(T_c - T)/T_c$ at which the BCS theory should break down in three dimensions, we find (Kadanoff et al., 1967) that this will occur at $(T_c - T)/T_c \approx 10^{-14}$, a temperature deviation well beyond current experimental resolution.

We have, in this section, barely scratched the surface of the theory of superconductivity but have demonstrated another phenomenon associated with Bose condensation. Another fermion system in which such condensation occurs is liquid ^{3}He. An introductory discussion of this system can be found in Mahan [1981].

PROBLEMS

7.1. *Critical Properties of the Ideal Bose Gas.* Consider an ideal Bose system at fixed density. The entropy $S(T, V, \mu)$ is given by (7.A.17) and the specific heat at fixed volume and fixed number of particles, $C_{V,N}$, is

$$C_{V,N} = T\left(\frac{\partial S}{\partial T}\right)_{V,N} = T\left(\frac{\partial S}{\partial T}\right)_{V,\mu} + T\left(\frac{\partial S}{\partial \mu}\right)_{V,T}\left(\frac{\partial \mu}{\partial T}\right)_{V,N}$$

$$= C_{V,\mu} - T\frac{\left(\frac{\partial S}{\partial \mu}\right)_T\left(\frac{\partial N}{\partial T}\right)_\mu}{\left(\frac{\partial N}{\partial \mu}\right)_T}$$

For $T < T_c$, $(\partial\mu/\partial T)_{V,N} = 0$ and $C_{V,N} = C_{V,\mu}$.

(a) Show that as $T \to T_c^+$, the second term in the expression for $C_{V,N}$ vanishes and that the specific heat is therefore continuous at the transition.

(b) Show further that $dC_{V,N}/dT$ is discontinuous at T_c. *Hint:* It may be useful to derive the limiting form of the function $g_\nu(z)$ for $z \to 1$ and for $\nu \le 1$ as it diverges at $z = 1$ for these values of the index ν. One way of approaching this is to assume that $g_\nu(z) \sim z(1 - z)^{-\gamma(\nu)}$ and to compare the Taylor expansion of this function with the exact form (7.A.6) to find the exponent $\gamma(\nu)$ (see also Section 5.B).

(c) Analyze, by the methods of parts (a) and (b), the behavior of the pressure near T_c of an ideal Bose gas at fixed density.

7.2. *Bose Condensation of Spin-Aligned Atomic Hydrogen.* Under normal conditions, atomic hydrogen cannot be maintained at high density since the atoms would rapidly combine to form molecular H_2. It is possible, however, to inhibit the recombination process by means of a magnetic field strong enough to spin polarize the electronic states. This has the effect of suppressing the attractive tail of the H–H interaction. For this reason many researchers have considered spin-polarized hydrogen to be a suitable system for the observation of Bose condensation. Let us for the moment ignore the many technical considerations which complicate this task (for recent reviews of the properties of spin polarized H, see Greytak and Kleppner [1984] and Silvera and Walraven [1986]) and consider the following hypothetical experiment.

Atomic hydrogen is not miscible in ^{4}He and forms bubbles when injected into liquid helium. Assuming that the H atoms constitute an ideal Bose gas, one can calculate the size of a bubble containing N particles by equating the expansion force due to internal pressure with the contraction force due to the surface tension. The latter can be calculated by assuming that the energy of a bubble is σA, where $\sigma = 0.37$ erg/cm^2 and A is the surface area of the bubble.

(a) Show that if N is large enough, a quasi-equilibrium state can exist with the bubble containing a normal nondegenerate atomic hydrogen gas. At any finite temperature this gas will be unstable since the hydrogen atoms will eventually recombine to form molecules.

(b) Assume that any molecules which form and the excess heat is instantaneously removed by the surrounding fluid so that the process can be taken to be isothermal. The bubble then gradually contracts and the pressure increases. Eventually, the critical pressure for Bose condensation will be reached and at this point the bubble will collapse. Find the size and number of atoms in the bubble at this critical pressure at 0.1 K.

7.3. *λ Point of Helium.* The theory of superfluidity presented in Section 7.B was based on an analogy with Bose condensation. This theory does not adequately describe the nature of the transition to the superfluid state. Nevertheless, it is possible to estimate the critical temperature to remarkable accuracy by the following crude argument (Landau and Lifshitz, Part 2, 1980).

Imagine a quasiparticle gas with a Bose distribution function $n(\epsilon(\mathbf{p}) - \mathbf{p} \cdot \mathbf{v})$ (i.e., which is moving with velocity $\mathbf{v}$ with respect to the superfluid). The momentum of this gas is in the low-velocity limit given by

$$\mathbf{P} = -\sum_{\mathbf{p}} \mathbf{p}n(\epsilon - \mathbf{p} \cdot \mathbf{v}) \approx \sum_{\mathbf{p}} \mathbf{p}(\mathbf{p} \cdot \mathbf{v})\frac{dn(\epsilon)}{d\epsilon} = \frac{1}{3}\sum_{\mathbf{p}} p^2\left[-\frac{dn(\epsilon)}{d\epsilon}\right]\mathbf{v}$$

The last step follows if we split $\mathbf{p}$ into components parallel and perpendicular to $\mathbf{v}$ and note that the perpendicular component averages to zero, while the spherical average of $\cos^2 \theta = \frac{1}{3}$. The ratio between the momentum and velocity is the inertia or mass of the quasiparticle gas. The inertia of this gas increases with temperature and when the inertia of the gas is equal to the mass of the helium, there is no superfluid component left. This occurs when

$$M_{\text{He}} = \frac{1}{3}\sum_{\mathbf{p}} p^2\left[-\frac{dn(\epsilon)}{d\epsilon}\right]$$

(a) Show that for a phonon gas with velocity of sound c in volume V

$$M_{\text{ph}} = \frac{2\pi^2 VT^4 k_B^4}{45\hbar^3 c^5}$$

(b) Near the transition temperature, the thermodynamic properties are dominated not by the phonons but rather by the rotons with energy

$$\epsilon(p) = \Delta + \frac{(p - p_0)^2}{2m^*}$$

Assume that the rotons have a Boltzmann distribution and that m^* is small enough that we may take $p = p_0$ everywhere in thermal averages except in the Boltzmann factor. Show that

$$M_r = \frac{2(m^*)^{1/2} p_0^4 \exp\{-\beta\Delta\}}{3(2\pi)^{3/2}(k_B T)^{1/2}\hbar^3}V$$

(c) Estimate the critical temperature assuming that the density of He is 0.145 gm/cm^3, $m^* = 0.16\, m_{\text{He}}$, $c = 2.4 \times 10^4$ cm/s, $p_0/h = 1.9$ Å^{-1}, and $\Delta/k_B = 8.7$ K.

(d) Can the argument above be used to calculate the condensation temperature of the ideal Bose gas?

8

Linear Response Theory

In this chapter we consider the response of a system to an external perturbation. At this point we are concerned only with weak perturbations in which case a linear approximation to the effect of the perturbation is adequate. In Section 8.A we define the dynamic structure factor, introduce the concept of generalized susceptibility, and derive an important result, the *fluctuation-dissipation* theorem, which relates equilibrium fluctuations to dissipation in the linear (ohmic) regime. We next show how the thermodynamic properties of a quantum many-body system in which the particles interact through two-body potentials can be derived from the response function. Our discussion here is analogous to our treatment of classical liquids in Section 4.B. We illustrate the formalism in Section 8.B by means of a number of simple examples. We show that a simple mean field theory yields results which are equivalent to those derived in Chapter 7 by means of the Bogoliubov transformation for the weakly interacting Bose gas at low temperatures. We also discuss the dielectric response of an interacting electron gas. We then turn to a discussion of the electron–phonon interaction in metals. The purpose of this discussion is to derive a result originally due to Fröhlich [1950] that electrons can interact by exchanging phonons to produce an effective interaction that can be attractive in certain circumstances. This result was used in Chapter 7 in our treatment of the BCS theory of superconductivity.

In Section 8.C we return to the development of the formalism by considering *steady-state* situations in which there is a current flowing in response to a field. We derive the Kubo relations between the appropriate conductivities and equilibrium current–current correlation functions. On the basis of an assumption of microscopic reversibility, we next demonstrate the Onsager reciprocity theorem. Finally, in Section 8.D, we briefly discuss the Boltzmann equation approach to transport.

8.A EXACT RESULTS

In this section we develop the formalism of linear response theory quite generally and exhibit a number of interesting and useful properties. Within the framework of the linear approximation, the results of this section are exact.

(a) Generalized Susceptibility, Dynamic Structure Factor, and the Fluctuation-Dissipation Theorem

We consider a system of particles subject to an external perturbation which may be time dependent. The Hamiltonian is

$$H_{tot} = H + H_{ext} = H_0 + H_1 + H_{ext} \tag{1}$$

where H_0 contains the kinetic energy and the single-particle potential and H_1 the interparticle potential energy. H_0 will usually represent an ideal Bose or Fermi gas. The pair interaction H_1 can be treated only approximately, and we will generally assume that a mean field type of theory will be adequate. We assume that the external perturbation couples linearly to the density:

$$H_{ext} = \int d^3x \, n(\mathbf{x})\phi_{ext}(\mathbf{x}, t) \tag{2}$$

where ϕ_{ext} is a scalar function of position and time.

The system will respond to the perturbation (2) through an induced change in the particle density

$$\langle \delta n(\mathbf{x}, t) \rangle = \langle n(\mathbf{x}, t) \rangle_{H_{tot}} - \langle n(\mathbf{x}, t) \rangle_H \tag{3}$$

The expectation value $\langle n(\mathbf{x}, t) \rangle_{H_{tot}}$ in (3) is the density at $\mathbf{x}$, at time t, when the system has evolved from an equilibrium state of H (at $t = -\infty$) with the external perturbation switched on for all finite times. This expectation value thus involves a trace over an equilibrium density matrix (at $t = -\infty$) and a modification of the states of the system due to H_{ext}. We assume that H_{ext} is small enough that we can treat it in first-order perturbation theory. This assumption is frequently harmless, for example, if H_{ext} represents an infinitesimal "test field" introduced formally to probe the equilibrium dynamic or static response of the system.

We suppose that we have determined the eigenstates of H in the Heisenberg picture and denote them by $|\psi_H\rangle$. With the perturbation switched on at

$t = -\infty$, the time-dependent Schrödinger equation for an arbitrary state $|\psi\rangle$ is

$$i\hbar \frac{\partial}{\partial t} |\psi\rangle = H_{\text{tot}} |\psi(t)\rangle \equiv H_{\text{tot}} \exp\left\{-\frac{i}{\hbar} Ht\right\} |\psi_H(t)\rangle \qquad (4)$$

The time dependence in $|\psi_H(t)\rangle$ is due to the external perturbation, and the state $|\psi_H(t)\rangle$ therefore obeys the differential equation

$$i\hbar \frac{\partial}{\partial t} |\psi_H(t)\rangle = H_{\text{ext}}(t) |\psi_H(t)\rangle \qquad (5)$$

where

$$\begin{aligned} H_{\text{ext}}(t) &= e^{iHt/\hbar} H_{\text{ext}} e^{-iHt/\hbar} \\ &= e^{iKt/\hbar} H_{\text{ext}} e^{-iKt/\hbar} \end{aligned} \qquad (6)$$

where K is the "grand Hamiltonian"

$$K = H - \mu N \qquad (7)$$

The last equality in (6) follows if the external perturbation conserves particle number (i.e., N commutes with H_{ext}). Solving (5) to lowest order in H_{ext}, we find

$$|\psi_H(t)\rangle = |\psi_H\rangle - \frac{i}{\hbar} \int_{-\infty}^{t} dt' \, H_{\text{ext}}(t') |\psi_H\rangle \qquad (8)$$

Thus

$$\begin{aligned} \langle \psi(t) | n(\mathbf{x}) | \psi(t) \rangle &= \langle \psi_H | e^{iHt/\hbar} n(\mathbf{x}) e^{-iHt/\hbar} | \psi_H \rangle \\ &\quad - \frac{i}{\hbar} \int_{-\infty}^{t} dt' \, \langle \psi_H | e^{iHt/\hbar} n(\mathbf{x}) e^{-iHt/\hbar} H_{\text{ext}}(t') \\ &\qquad - H_{\text{ext}}(t') e^{iHt/\hbar} n(\mathbf{x}) e^{-iHt/\hbar} | \psi_H \rangle \\ &= \langle \psi_H | n(\mathbf{x}, t) | \psi_H \rangle \\ &\quad - \frac{i}{\hbar} \int_{-\infty}^{t} dt' \int d^3x' \, \phi_{\text{ext}}(\mathbf{x}', t') \langle \psi_H | [n(\mathbf{x}, t), n(\mathbf{x}', t')] | \psi_H \rangle \end{aligned} \qquad (9)$$

where we use the notation that operators such as $n(\mathbf{x}, t)$ with an explicit time dependence are Heisenberg operators:

$$n(\mathbf{x}, t) = e^{iKt/\hbar} n(\mathbf{x}) e^{-iKt/\hbar}$$

while $n(\mathbf{x})$ is in the Schrödinger picture and $[A, B] = AB - BA$.

As mentioned above, the system was in equilibrium at $t = -\infty$. Hence we average equation (9) over the grand canonical probability density to obtain finally,

$$\langle \delta n(\mathbf{x}, t) \rangle = \frac{i}{\hbar} \int_{-\infty}^{t} dt' \int d^3x' \, \phi_{\text{ext}}(\mathbf{x}', t') \langle [n(\mathbf{x}', t'), n(\mathbf{x}, t)] \rangle_H \qquad (10)$$

It is convenient to express the response in terms of the retarded density–density correlation function, D^R, defined through

$$D^R(\mathbf{x}, t; \mathbf{x}', t') \equiv -\frac{i}{\hbar} \theta(t - t')\langle[n(\mathbf{x}, t), n(\mathbf{x}', t')]\rangle_H \qquad (11)$$

where $\theta(t)$ is the Heaviside step function. This correlation function (also called a propagator or Green's function) is independent of the perturbing potential and thus can be used to describe the response of the system to any external perturbation which couples linearly to the density. Substituting in (10), we obtain

$$\langle\delta n(\mathbf{x}, t)\rangle = \int d^3x' \int_{-\infty}^{\infty} dt' \, D^R(\mathbf{x}, t; \mathbf{x}', t')\phi_{\text{ext}}(\mathbf{x}', t') \qquad (12)$$

If the unperturbed system is homogeneous in space and time, $D^R(\mathbf{x}, t; \mathbf{x}', t')$ can only be a function of $\mathbf{x} - \mathbf{x}'$ and $t - t'$ and the integral in (12) is simply a four-dimensional convolution. Thus if we define Fourier transforms according to

$$\langle\delta n(\mathbf{x}, t)\rangle = \frac{1}{2\pi V} \sum_{\mathbf{q}} \int_{-\infty}^{\infty} d\omega \, \delta\rho(\mathbf{q}, \omega) \, e^{i\mathbf{q}\cdot\mathbf{x}}e^{-i\omega t} \qquad (13)$$

$$D^R(\mathbf{x}, t; \mathbf{x}', t') = \frac{1}{2\pi V} \sum_{\mathbf{q}} \int_{-\infty}^{\infty} d\omega \, \chi^R(\mathbf{q}, \omega)e^{i\mathbf{q}\cdot(\mathbf{x}-\mathbf{x}')}e^{-i\omega(t-t')} \qquad (14)$$

and

$$\phi_{\text{ext}}(\mathbf{x}', t') = \frac{1}{2\pi V} \sum_{\mathbf{q}} \int_{-\infty}^{\infty} d\omega \, \phi_{\text{ext}}(\mathbf{q}, \omega)e^{i\mathbf{q}\cdot\mathbf{x}'}e^{-i\omega t'} \qquad (15)$$

we have

$$\delta\rho(\mathbf{q}, \omega) = \chi^R(\mathbf{q}, \omega)\phi_{\text{ext}}(\mathbf{q}, \omega) \qquad (16)$$

The function χ^R is called the *generalized susceptibility*.

Before discussing the susceptibility and its relation to the dynamic structure factor, we pause to express the Fourier transformed quantities in second quantized form. The density operator $n(\mathbf{x})$ may be written in the form

$$n(\mathbf{x}) = \psi^+(\mathbf{x})\psi(\mathbf{x}) = \frac{1}{V} \sum_{\mathbf{k},\mathbf{q},\sigma} c^+_{\mathbf{k}-\mathbf{q},\sigma}c_{\mathbf{k},\sigma}e^{i\mathbf{q}\cdot\mathbf{x}} = \frac{1}{V} \sum_{\mathbf{q}} \rho(\mathbf{q})e^{i\mathbf{q}\cdot\mathbf{x}} \qquad (17)$$

where

$$c^+_{\mathbf{k},\sigma}, c_{\mathbf{k},\sigma} = \begin{cases} a^+_{\mathbf{k},\sigma}, a_{\mathbf{k},\sigma} & \text{fermions} \\ b^+_{\mathbf{k},\sigma}, b_{\mathbf{k},\sigma} & \text{bosons} \end{cases}$$

are the creation and annihilation operators introduced in the Appendix. In the case of spinless bosons, the sum over σ in (17) should, of course, be omitted.

The time-dependent density operator can then be written in the form

$$
n(\mathbf{x}, t) = \frac{1}{V} \sum_{\mathbf{q}} e^{iKt/\hbar} \rho(\mathbf{q}) e^{-iKt/\hbar} e^{i\mathbf{q} \cdot \mathbf{x}} = \frac{1}{V} \sum_{\mathbf{q}} \rho(\mathbf{q}, t) e^{i\mathbf{q} \cdot \mathbf{x}}
$$

$$
= \frac{1}{2\pi V} \sum_{\mathbf{q}} \int_{-\infty}^{\infty} d\omega \, \rho(\mathbf{q}, \omega) e^{i\mathbf{q} \cdot \mathbf{x}} e^{-i\omega t} \tag{18}
$$

and

$$
\rho(\mathbf{q}, \omega) = \int_{-\infty}^{\infty} dt \, \rho(\mathbf{q}, t) e^{i\omega t} = \int d^3x \int_{-\infty}^{\infty} dt \, n(\mathbf{x}, t) e^{-i\mathbf{q} \cdot \mathbf{x}} e^{i\omega t} \tag{19}
$$

The expression (16) for the frequency-dependent response is consistent with the conventional definition of the magnetic susceptibility, which is the ratio between the magnetization and the magnetic field, H (rather than the magnetic induction B). The magnetic field is usually interpreted as being due to an external source. On the other hand, the conventional definition of the electric susceptibility is the ratio of the polarization to the electric field E, not to the source field D.

We now define the *dynamic structure factor*, $S(\mathbf{q}, \omega)$:

$$
S(\mathbf{q}, \omega) = \int_{-\infty}^{\infty} dt \, e^{i\omega t} \langle \rho(\mathbf{q}, t) \rho(-\mathbf{q}, 0) \rangle \tag{20}
$$

The generalized susceptibility $\chi^R(\mathbf{q}, \omega)$ (14) can be expressed in terms of this function as we now demonstrate. We first note that the expectation value

$$
\langle n(\mathbf{x}, t) n(\mathbf{x}', t') \rangle = \frac{1}{V^2} \sum_{k_1, k_2} \langle \rho(\mathbf{k}_1, t) \rho(\mathbf{k}_2, t') \rangle e^{i\mathbf{k}_1 \cdot \mathbf{x}} e^{i\mathbf{k}_2 \cdot \mathbf{x}'}
$$

is a function only of $|\mathbf{x} - \mathbf{x}'|$. Letting $\mathbf{x} = \mathbf{x}' + \mathbf{y}$ and integrating over $\mathbf{x}'$, we obtain

$$
\frac{1}{V} \int d^3x' \langle n(\mathbf{x}' + \mathbf{y}, t) n(\mathbf{x}', t') \rangle = \frac{1}{V^2} \sum_{\mathbf{k}} \langle \rho(\mathbf{k}, t) \rho(-\mathbf{k}, t') \rangle e^{i\mathbf{k} \cdot \mathbf{y}}
$$

$$
= \frac{1}{V^2} \sum_{\mathbf{k}} \langle \rho(\mathbf{k}, t - t') \rho(-\mathbf{k}, 0) \rangle e^{i\mathbf{k} \cdot \mathbf{y}} \tag{21}
$$

Hence

$$
\chi^R(\mathbf{q}, \omega) = -\frac{i}{\hbar V^2} \sum_{\mathbf{k}} \int d^3y \int_{-\infty}^{\infty} d\tau \, e^{-i(\mathbf{k}+\mathbf{q}) \cdot \mathbf{y} + i\omega\tau} \theta(\tau) \langle \rho(\mathbf{k}, \tau) \rho(-\mathbf{k}, 0)
$$

$$
- \rho(-\mathbf{k}, 0) \rho(\mathbf{k}, \tau) \rangle
$$

$$
= -\frac{i}{\hbar V} \int_{-\infty}^{\infty} d\tau \, e^{i\omega\tau} \theta(\tau) \langle \rho(-\mathbf{q}, \tau) \rho(\mathbf{q}, 0) - \rho(\mathbf{q}, 0) \rho(-\mathbf{q}, \tau) \rangle \tag{22}
$$

Using the integral representation for the Heaviside function,

$$\theta(\tau) = \frac{1}{2\pi i} \int_{-\infty}^{\infty} d\omega' \frac{e^{i\omega'\tau}}{\omega' - i\eta} \tag{23}$$

where η is a positive infinitesimal, we find

$$\chi^R(\mathbf{q}, \omega) = \frac{1}{2\pi \hbar V} \int_{-\infty}^{\infty} d\tau \int_{-\infty}^{\infty} d\omega' \, e^{i(\omega' + \omega)\tau}$$
$$\frac{\langle \rho(-\mathbf{q}, \tau)\rho(\mathbf{q}, 0) - \rho(\mathbf{q}, 0)\rho(-\mathbf{q}, \tau)\rangle}{\omega' - i\eta} \tag{24}$$

Substituting $\omega' = -\omega + \omega''$ and noting, from the definition of the expectation values, that for any pair of operators A, B,

$$\langle A(0)B(t)\rangle = \langle A(-t)B(0)\rangle \tag{25}$$

and, moreover, that

$$\langle \rho(-\mathbf{k}, \tau))\rho(\mathbf{k}, 0)\rangle = \langle \rho(\mathbf{k}, \tau)\rho(-\mathbf{k}, 0)\rangle$$

we have

$$\chi^R(\mathbf{q}, \omega) = \frac{1}{\hbar V} \int_{-\infty}^{\infty} \frac{d\omega'}{2\pi} \frac{S(\mathbf{q}, \omega') - S(\mathbf{q}, -\omega')}{\omega - \omega' + i\eta} \tag{26}$$

We now examine the dynamic structure factor in more detail. Let $|n\rangle$ be a complete set of eigenstates of the grand Hamiltonian K:

$$K|n\rangle = K_n |n\rangle \tag{27}$$

Using these basis states we find, from the definition (20),

$$S(\mathbf{q}, \omega) = \int_{-\infty}^{\infty} dt \, e^{i\omega t} \sum_n \langle n| e^{-\beta K} e^{iKt/\hbar} \rho(\mathbf{q}, 0) e^{-iKt/\hbar} \rho(-\mathbf{q}, 0) |n\rangle / Z_G$$
$$= \sum_{n,m} \frac{e^{-\beta K_n}}{Z_G} \int_{-\infty}^{\infty} dt \, e^{i\{\omega + (K_n - K_m)/\hbar\}t} \langle n|\rho(\mathbf{q}, 0)|m\rangle\langle m|\rho(-\mathbf{q}, 0)|n\rangle$$
$$= \frac{2\pi}{\hbar} \sum_{n,m} \frac{e^{-\beta K_n}}{Z_G} |\langle n|\rho(\mathbf{q}, 0)|m\rangle|^2 \, \delta(\hbar\omega + K_n - K_m) \tag{28}$$

From this equation we see that $S(\mathbf{q}, \omega)$ is real and positive semidefinite:

$$S(\mathbf{q}, \omega) \geq 0 \tag{29}$$

and that

$$S(-\mathbf{q}, \omega) = S(\mathbf{q}, \omega) = e^{\beta\hbar\omega}S(\mathbf{q}, -\omega) \tag{30}$$

Equation (30) is an expression of the *principle of detailed balance*. We interpret (30) in the following way. The quantum-mechanical transition rate between two states is independent of the direction of the transition; there is no distinction between emission and absorption. The dynamic structure factor, on the other hand, is a thermal average of transition rates, as can be seen from (28), and hence is the quantum transition rate weighted by the average occupation number of the initial state. This accounts for (30).

The function $S(\mathbf{q}, \omega)$ represents the frequency spectrum of density fluctuations. On the other hand, the imaginary part of the response function χ^R has a physical interpretation in terms of energy dissipation. Combining (26) and (30) and using

$$\frac{1}{\omega + i\eta} = P\left(\frac{1}{\omega}\right) - i\pi\delta(\omega) \tag{31}$$

where P indicates principal value, we obtain the *fluctuation-dissipation* theorem:

$$(1 - e^{-\beta\hbar\omega})S(\mathbf{q}, \omega) = -2\hbar V \operatorname{Im} \chi^R(\mathbf{q}, \omega) \tag{32}$$

Before proceeding to some examples, we generalize the formalism above by considering the effect of an external field that couples to an arbitrary dynamical variable A rather than specifically to the density. Thus

$$H_{\text{ext}} = \phi_{\text{ext}}(t)A \tag{33}$$

The response of another observable B to this perturbation can be derived, in strict analogy with (10), to be

$$\langle \delta B(t) \rangle = \frac{i}{\hbar} \int_{-\infty}^{t} dt' \langle [A(t'), B(t)] \rangle_H \phi_{\text{ext}}(t') \tag{34}$$

and the appropriate Green's function, or propagator, is given by

$$D_{BA}^R(t, t') = -\frac{i}{\hbar} \theta(t - t') \langle [B(t), A(t')] \rangle_H \tag{35}$$

which is the generalization of (11). Proceeding as before, we define

$$\langle \delta B(t) \rangle = \int_{-\infty}^{\infty} \frac{d\omega}{2\pi} \langle \delta B(\omega) \rangle e^{-i\omega t}$$

$$D_{BA}^R(t, t') = \int_{-\infty}^{\infty} \frac{d\omega}{2\pi} \chi_{BA}^R(\omega) e^{-i\omega(t-t')} \tag{36}$$

$$\phi_{\text{ext}}(t') = \int_{-\infty}^{\infty} \frac{d\omega}{2\pi} \phi_{\text{ext}}(\omega) e^{-i\omega t'}$$

and obtain

$$\langle \delta B(\omega) \rangle = \chi_{BA}^R(\omega) \phi_{\text{ext}}(\omega) \tag{37}$$

We may, as above, express the response function in terms of an equilibrium correlation function. We let

$$S_{BA}(\omega) = \int_{-\infty}^{\infty} dt \langle B(t)A(0)\rangle_H e^{i\omega t}$$

$$S_{AB}(\omega) = \int_{-\infty}^{\infty} dt \langle A(t)B(0)\rangle_H e^{i\omega t}$$

(38)

and after some manipulations similar to (20)–(26) obtain the result

$$\chi_{BA}^R(\omega) = \frac{1}{2\pi\hbar} \int_{-\infty}^{\infty} d\omega' \frac{S_{BA}(\omega') - S_{AB}(-\omega')}{\omega - \omega' + i\eta}$$

(39)

Using arguments similar to those leading to (30), we may easily relate $S_{AB}(-\omega)$ to $S_{BA}(\omega)$. The result after some algebra is

$$S_{AB}(-\omega) = e^{-\beta\hbar\omega} S_{BA}(\omega)$$

(40)

To obtain the fluctuation dissipation theorem (32), we need to restrict the formalism somewhat. For the case $B = A^+$ we see that S_{AB} is real and positive semidefinite and, using (31), we find that

$$(1 - e^{-\beta\hbar\omega})S_{BA}(\omega) = -2\hbar \operatorname{Im} \chi_{BA}^R(\omega)$$

(41)

We shall be using this more general formalism in Section 8.B(c) where we derive the magnon spectrum of the Heisenberg ferromagnet.

(b) Thermodynamic Properties and the Dynamic Structure Factor

In Section 4.B we expressed the equation of state, the internal energy, and the compressibility of a system of classical particles, interacting through two-body forces, in terms of the pair correlation function, or equivalently, in terms of the static structure factor. We now wish to derive similar relations for a quantum many-body system. For simplicity we limit ourselves to an isotropic system and assume that the Hamiltonian can be written in the form

$$H = H_0 + H_1 = \sum_{i=1}^{N} \frac{p_i^2}{2m} + \sum_{i<j} v(\mathbf{r}_i - \mathbf{r}_j)$$

(42)

The equipartition theorem allowed us, in Section 4.B, to consider the two terms $\langle H_0 \rangle$ and $\langle H_1 \rangle$ separately. For a quantum system this is no longer possible and we must take a slightly more roundabout route. Let us first consider the ground-state energy and rewrite the Hamiltonian (42) in the form

$$H_\lambda = H_0 + \lambda H_1$$

(43)

For the physical system $\lambda = 1$, but we can imagine intermediate values. In the latter case we assume that the ground state $|\lambda\rangle$ has energy E_λ and is normalized so that $\langle \lambda | \lambda \rangle = 1$. Thus

$$E_\lambda = \langle \lambda | H_\lambda | \lambda \rangle \tag{44}$$

Differentiation with respect to λ yields

$$\frac{\partial E_\lambda}{\partial \lambda} = \left(\frac{\partial}{\partial \lambda} \langle \lambda | \right) H_\lambda | \lambda \rangle + \langle \lambda | \frac{\partial H_\lambda}{\partial \lambda} | \lambda \rangle + \langle \lambda | H_\lambda \frac{\partial}{\partial \lambda} | \lambda \rangle$$

The first and third terms combine to give

$$E_\lambda \frac{\partial}{\partial \lambda} \langle \lambda | \lambda \rangle = 0$$

since the state $| \lambda \rangle$ is normalized. We thus have

$$\frac{\partial E_\lambda}{\partial \lambda} = \frac{E_{int}(\lambda)}{\lambda}$$

where E_{int} is the ground-state expectation value of the two-body potential. Integrating, we finally obtain

$$E = E_0 + \int_0^1 \frac{d\lambda}{\lambda} E_{int}(\lambda) \tag{45}$$

This argument can easily be generalized to finite temperatures [see also Section 3.D, equations (3.D.11)–(3.D.15)]. We have, for example, for the Helmholtz free energy,

$$A(N, V, T) = A_0(N, V, T) + \int_0^1 \frac{d\lambda}{\lambda} \langle \lambda H_1 \rangle_{\lambda,c} \tag{46}$$

where the subscripts indicate that the expectation value is to be evaluated in the canonical ensemble at temperature T, volume V, for an N-particle system. In (46), $A_0(N, V, T)$ is the corresponding free energy of the noninteracting system. The interaction energy can be related to the static (or geometric) structure factor of Section 4.B through

$$E_{int}(\lambda) = \frac{N\lambda}{2V} \sum_{\mathbf{q} \neq 0} v_{-\mathbf{q}} [S_\lambda(\mathbf{q}) - 1] + \frac{N^2 \lambda}{2V} v_0 \tag{47}$$

where $v_{\mathbf{q}}$ is the Fourier transform of the pair potential. The geometric structure factor can, in turn, be related to the dynamic structure factor through

$$S(\mathbf{q}) = \frac{V}{N} \int_{-\infty}^{\infty} \frac{d\omega}{2\pi} S(\mathbf{q}, \omega) \tag{48}$$

It is now a straightforward matter to relate the ground-state energy, or the free energy, to the structure factor, and from the free energy one can obtain all other thermodynamic properties. A fairly serious complication is the fact that one needs to know the structure factor at all intermediate coupling strengths $\lambda < 1$. We will return to this point in the mean field approximation of the next section and discuss a case in which the integration over λ is quite straightforward.

(c) Sum Rules and Inequalities

We next derive some exact relationships that must be satisfied by the response functions. Some of these "sum rules" are useful in checking the validity of approximations, while others offer valuable insights into basic principles.

Let us first consider the consequences of *causality*. Since the response of a stable system may not precede the disturbance, we required (11) that $D^R(\mathbf{x}, t; \mathbf{x}', t') = 0$ for $t < t'$. The theory of analytic functions of a complex variable tells us that $\chi^R(\mathbf{q}, \omega)$ then is an analytic function of the complex variable ω in the upper halfplane and, in particular, can have no poles there. Poles on the real axis, on the other hand, would correspond to nondissipative resonances and can be ruled out if we require that a finite source field gives rise to a finite response. These analytic properties, together with the assumption that the response function falls off sufficiently rapidly as $\omega \to \infty$, are sufficient to show that χ must satisfy Kramers–Kronig relations. We simply carry out the contour integral

$$0 = \frac{1}{2\pi i} \oint_C dz\, \frac{\chi^R(\mathbf{q}, z)}{z - \omega}$$

where the contour C is shown in Figure 8.1. Taking the radius R of the large semicircle to infinity, and that of the small one, ρ, to zero we obtain the Kramers–Kronig relations:

$$\text{Re } \chi^R(\mathbf{q}, \omega) = P \int_{-\infty}^{\infty} \frac{d\omega'}{\pi} \frac{\text{Im } \chi^R(\mathbf{q}, \omega')}{\omega - \omega'}$$

$$\text{Im } \chi^R(\mathbf{q}, \omega) = -P \int_{-\infty}^{\infty} \frac{d\omega'}{\pi} \frac{\text{Re } \chi^R(\mathbf{q}, \omega')}{\omega - \omega'} \tag{49}$$

where P indicates the principal value of the integrals.

Another important relationship, the f-sum rule, holds for systems in which the interparticle potential is independent of velocity. The density operator (17) and its Fourier transform commute with H_1 in (42) but not with

$$H_0 = \sum_{\mathbf{k}, \sigma} \frac{\hbar^2 k^2}{2m} c_{\mathbf{k},\sigma}^+ c_{\mathbf{k},\sigma} \tag{50}$$

It is easy to verify that

$$\langle [[H, \rho(\mathbf{q})], \rho(\mathbf{q})] \rangle = \langle [[H_0, \rho(\mathbf{q})], \rho(\mathbf{q})] \rangle = \frac{\hbar^2 q^2}{m} \langle N \rangle$$

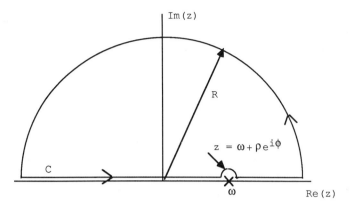

Figure 8.1   Contour used to derive Kramers–Kronig relations.

Moreover, we also have

$$\langle [[H, \rho(\mathbf{q})], \rho(\mathbf{q})]\rangle = \sum_{n,m} |\langle n|\rho(\mathbf{q})|m\rangle|^2 (E_n - E_m)(e^{-\beta E_n} - e^{-\beta E_m})$$

Using (28), (32), and (48), we find, after some algebra,

$$\int_{-\infty}^{\infty} \frac{d\omega}{\pi} \omega \, \mathrm{Im} \, \chi^R(\mathbf{q}, \omega) = - \frac{q^2 \langle N\rangle}{mV} \tag{51}$$

which is the f-sum rule.

Another self-consistency condition can be derived from the relationship between the pair distribution function and the response function. Since the pair distribution function has the interpretation of a probability, it must be nonnegative. This imposes restrictions on the response function which can be difficult to satisfy in approximate theories. We refer the reader to the texts by Mahan [1981] and by Pines and Nozières [1966] and references therein for discussion of this question and for further sum rules that can be derived.

8.B MEAN FIELD RESPONSE

To this point our treatment is exact within the linear response approximation. The dynamic structure factor and the response functions can usually not be evaluated exactly since the statistical treatment of a system of interacting particles is in general an unsolved problem. However, the approach of Section 8.A forms a useful starting point for some of the most successful approximation schemes for many-particle systems. The most common approach is that of mean field theory or generalizations thereof. In its most straightforward and simple form, mean field theory is based on the assumption that a system responds as a system of free particles to an effective potential which is determined self-consistently in terms of the externally applied potential. This method has been applied in many different situations and, consequently, has many different names, such as the random phase approximation, time-dependent Hartree approximation, and Lindhard and self-consistent field approximation. The Vlasov equation of plasma physics is another example of an approximation made in the same spirit. We illustrate this approach through a few simple examples.

(a) Dielectric Function of Charged Particles in an Isotropic Medium

Assume an externally controlled electrostatic potential energy $\phi_{ext}(\mathbf{q}, \omega)$, (8.A.15). The response of the system is formally given by

$$\langle \delta\rho(\mathbf{q}, \omega) \rangle = \chi^R(\mathbf{q}, \omega)\phi_{ext}(\mathbf{q}, \omega) \tag{1}$$

where χ^R, in (1), is the exact response function. Since a density fluctuation $\langle \delta\rho(\mathbf{q}, \omega) \rangle$ produces a Coulomb potential given by

$$\phi_{ind}(\mathbf{q}, \omega) = \int d^3r \int d^3r' \frac{e^2}{|\mathbf{r} - \mathbf{r}'|} \langle \delta n(\mathbf{r}', \omega) \rangle e^{i\mathbf{q}\cdot\mathbf{r}}$$

$$= \frac{4\pi e^2}{q^2} \langle \delta\rho(\mathbf{q}, \omega) \rangle \tag{2}$$

we have an effective potential

$$\phi_{eff}(\mathbf{q}, \omega) = \phi_{ext}(\mathbf{q}, \omega) + \phi_{ind}(\mathbf{q}, \omega) \tag{3}$$

In (2) we have assumed that each particle has charge e. We define the dielectric function $\epsilon(\mathbf{q}, \omega)$ through

$$\phi_{eff}(\mathbf{q}, \omega) = \frac{\phi_{ext}(\mathbf{q}, \omega)}{\epsilon(\mathbf{q}, \omega)} \tag{4}$$

and obtain

$$\frac{1}{\epsilon(\mathbf{q}, \omega)} = 1 + \frac{4\pi e^2}{q^2}\chi^R(\mathbf{q}, \omega) \tag{5}$$

which relates the dielectric constant to the exact response function. We note, in passing, that in our formulation it is $\epsilon^{-1}(\mathbf{q}, \omega) - 1$, not $\epsilon(\mathbf{q}, \omega)$, which is the response function. For this reason $(\epsilon^{-1}(\mathbf{q}, \omega) - 1)$ must satisfy the Kramers–Kronig dispersion relations (8.A.49). We can imagine a situation in which our polarizable material is exposed to a fixed potential rather than to an external test charge (Figure 8.2). In this case the polarization response is proportional to $(\epsilon(\mathbf{q}, \omega) - 1)$ and, as a consequence, the dielectric function must satisfy Kramers–Kronig relations. However, we can only attach the capacitor plates externally with a macroscopic separation. Causality therefore imposes the Kramers–Kronig relations only in the long-wavelength limit. For a more detailed discussion of this point, see Kirzhnits [1976] and Dolgov et al. [1981].

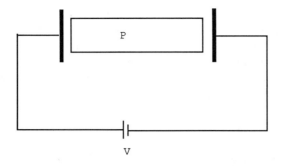

Figure 8.2   Polarization of a sample by a fixed external potential.

We can easily construct a mean field approximation for $\chi^R(\mathbf{q}, \omega)$ by writing

$$\langle \delta\rho(\mathbf{q}, \omega) \rangle = \chi_0^R(\mathbf{q}, \omega)\phi_{\text{eff}}(\mathbf{q}, \omega) \tag{6}$$

where χ_0 is the response function of a noninteracting system. Using (3) and (4), we obtain

$$\langle \delta\rho(\mathbf{q}, \omega) \rangle = \chi_0^R(\mathbf{q}, \omega)\phi_{\text{ext}}(\mathbf{q}, \omega) + \chi_0^R(\mathbf{q}, \omega)\frac{4\pi e^2}{q^2}\langle \delta\rho(\mathbf{q}, \omega) \rangle$$

or   (7)

$$\langle \delta\rho(\mathbf{q}, \omega) \rangle = \frac{\chi_0^R(\mathbf{q}, \omega)}{1 - (4\pi e^2/q^2)\chi_0^R(\mathbf{q}, \omega)}\phi_{\text{ext}}(\mathbf{q}, \omega)$$

and the mean field theory response function is given by

$$\chi_{\text{MF}}(\mathbf{q}, \omega) = \frac{\chi_0^R(\mathbf{q}, \omega)}{1 - (4\pi e^2/q^2)\chi_0^R(\mathbf{q}, \omega)} \tag{8}$$

Similarly, the mean field version of the dielectric constant is given by

$$\epsilon_{\text{MF}}(\mathbf{q}, \omega) = 1 - \frac{4\pi e^2}{q^2}\chi_0^R(\mathbf{q}, \omega) \tag{9}$$

The mean field approach is clearly not exact. The local potential seen by a particle will not be equal to the average potential $\phi_{\text{eff}}(\mathbf{q}, \omega)$ but will be modified by exchange and correlation effects. It is customary to write

$$\langle \delta\rho(\mathbf{q}, \omega) \rangle = \chi_0^R(\mathbf{q}, \omega)\psi_{\text{eff}}(\mathbf{q}, \omega) \tag{10}$$

where the effective potential, as seen by one of the particles, is given by

$$\psi_{\text{eff}}(\mathbf{q}, \omega) = \phi_{\text{ext}}(\mathbf{q}, \omega) + \frac{4\pi e^2}{q^2}f(\mathbf{q}, \omega)\langle \delta\rho(\mathbf{q}, \omega) \rangle \tag{11}$$

The *local field* correction[1] $f(\mathbf{q}, \omega)$ (often taken to be independent of frequency) has been estimated in many different ways, the most popular being that of Singwi and co-workers (Vashista and Singwi [1972] and references therein; see also Mahan [1981]). In this approximation the particles respond to a weak perturbation in the same way as independent particles would respond to an effective potential

$$\psi_{eff}(\mathbf{q}, \omega) = \frac{\phi_{ext}(\mathbf{q}, \omega)}{1 - (4\pi e^2/q^2) f(\mathbf{q}, \omega) \chi_0^R(\mathbf{q}, \omega)} \tag{12}$$

It is worth noting that the ratio between ϕ_{ext} and ψ_{eff} is not the dielectric function. This function is still given by (4). After some straightforward algebra we obtain

$$\epsilon(\mathbf{q}, \omega) = 1 - \frac{4\pi e^2}{q^2} \frac{\chi_0^R(\mathbf{q}, \omega)}{1 - (4\pi e^2/q^2)(1 - f(\mathbf{q}, \omega))\chi_0^R(\mathbf{q}, \omega)} \tag{13}$$

In the absence of local field corrections $f(\mathbf{q}, \omega) = 1$ and (13) reduces to the simple mean field result (9).

(b) Weakly Interacting Bose Gas

Our discussion, here, of the weakly interacting Bose gas is complementary to the treatment of Section 7.B. The approximations made are in the same spirit, but the formalism is that of linear response theory. We begin by calculating the dynamic structure factor $S_0(\mathbf{q}, \omega)$ of the noninteracting (ideal) Bose gas.

At $T = 0$, the only state that contributes to the ensemble average (8.A.20) is the ground state $|N\rangle$, which has all the particles condensed into the zero-momentum state. In the occupation number representation (see the Appendix) we have

$$\rho(\mathbf{q}) = \sum_{\mathbf{k}} b_{\mathbf{k}-\mathbf{q}}^+ b_{\mathbf{k}} \tag{14}$$

where $b_{\mathbf{k}}^+$, $b_{\mathbf{k}}$ are the creation and annihilation operators for particles with momentum $\hbar\mathbf{k}$. Since $b_{\mathbf{k}}|N\rangle = 0$ unless $\mathbf{k} = 0$ we see that the state $\rho(\mathbf{q})|N\rangle$ is one in which a single particle has been excited into a state of energy

$$\epsilon_0(\mathbf{q}) = \frac{\hbar^2 q^2}{2m} \tag{15}$$

---

[1] The term "local field correction" is also used to describe effects due to the lack of complete translation symmetry of a real metal. Equation (10) is then no longer correct and one must include "umklapp" terms, where the wave vector q on the left-hand side differs from that on the right-hand side of the equation by a reciprocal lattice vector. For a discussion, see Adler [1962] and Wiser [1963].

Using

$$b^+_{-\mathbf{q}}b_0 |N\rangle = \sqrt{N}b^+_{-\mathbf{q}} |N - 1\rangle$$

and (8.A.20), we find for $\mathbf{q} \neq 0$,

$$S_0(\mathbf{q}, \omega) = 2\pi\hbar N\delta(\hbar\omega - \epsilon_0(\mathbf{q})) \tag{16}$$

Similarly,

$$S_0(\mathbf{q}, -\omega) = 2\pi\hbar N \delta(\hbar\omega + \epsilon_0(\mathbf{q}))$$

Thus, using (8.A.26), we obtain

$$\chi^R_0(\mathbf{q}, \omega) = \frac{N}{V} \int_{-\infty}^{\infty} d\omega' \, \frac{\delta(\hbar\omega' - \epsilon_0(\mathbf{q})) - \delta(\hbar\omega' + \epsilon_0(\mathbf{q}))}{\omega - \omega' + i\eta}$$

$$= \frac{N}{V} \frac{q^2/m}{(\omega + i\eta)^2 - \hbar^2 q^4/4m^2} \tag{17}$$

or

$$\chi^R_0(\mathbf{q}, \omega - i\eta) = \frac{N}{V} \frac{2\epsilon_0(\mathbf{q})}{(\hbar\omega)^2 - \epsilon^2_0(\mathbf{q})} \tag{18}$$

Assume next that the particles interact through a pair potential which has a spatial Fourier transform $v_\mathbf{q}$. As in Section 8.B(a), we obtain, with $4\pi e^2/q^2$ replaced by $v_\mathbf{q}$,

$$\chi_{\mathrm{MF}}(\mathbf{q}, \omega - i\eta) = \frac{\chi^R_0(\mathbf{q}, \omega - i\eta)}{1 - v_\mathbf{q}\chi^R_0(\mathbf{q}, \omega - i\eta)}$$

$$= \frac{N}{V} \frac{\epsilon_0(\mathbf{q})}{\epsilon(\mathbf{q})} \left[\frac{1}{\hbar\omega - \epsilon(\mathbf{q})} - \frac{1}{\hbar\omega + \epsilon(\mathbf{q})} \right] \tag{19}$$

where

$$\epsilon(\mathbf{q}) = \sqrt{\epsilon^2_0(\mathbf{q}) + \frac{2N}{V} v_\mathbf{q}\epsilon_0(\mathbf{q})}$$

and, by comparison with (16),

$$S(\mathbf{q}, \omega) = 2\pi N \hbar \frac{\epsilon_0(\mathbf{q})}{\epsilon(\mathbf{q})} \delta(\hbar\omega - \epsilon(\mathbf{q})) \tag{20}$$

is the dynamic structure factor for $\mathbf{q} \neq 0$, in mean field theory, of weakly interacting bosons at $T = 0$. Thus the response of the system to an external perturbation which couples to the density is to create sound waves with energy

$$\epsilon(\mathbf{q}) \sim \sqrt{\frac{2N}{V} v_0 \frac{\hbar^2}{m}} q$$

for small wave vectors, where we have assumed (see also Section 7.B) that the limit as $\mathbf{q} \to 0$ of v_q exists and is finite. The energy spectrum is identical to that found in Section 7.B by means of the Bogoliubov transformation.

(c) Excitation Spectrum of the Heisenberg Ferromagnet: Magnons

We consider the anisotropic Heisenberg model at low temperature. The Hamiltonian is

$$H = - \sum_{\langle ij \rangle} [J_z S_{iz} S_{jz} + J_{xy}(S_{ix} S_{jx} + S_{iy} S_{jy})] - mB \sum_i S_{iz} \qquad (21)$$

where the sum in the first term is over nearest-neighbor pairs on a lattice with coordination number ν and where $mS\hbar$ is the magnetic moment per atom. The isotropic Heisenberg ferromagnet corresponds to $J_z = J_{xy} > 0$. Here we simply assume that $J_z \geq J_{xy} > 0$ and we have taken the magnetic field to be in the z direction, which, with our choice of coupling constants, is the direction of the ground-state magnetization. The spin operators obey the usual angular momentum commutation relations:

$$[S_i^+, S_{jz}] = -\hbar S_i^+ \delta_{ij}$$

$$[S_i^-, S_{jz}] = \hbar S_i^- \delta_{ij} \qquad (22)$$

$$[S_i^+, S_j^-] = 2\hbar S_{iz} \delta_{ij}$$

where $S^+ = S_x + iS_y$, $S^- = S_x - iS_y$ are the raising and lowering operators.

We note that the operator $M_z = \sum_i S_{iz}$ commutes with H. The eigenstates of H can therefore be partially indexed by the z component of the magnetization M_z. In particular, the ground state $|0\rangle$ is the state with all spins fully aligned in the positive z direction which we have taken to be the direction of the external field:

$$S_{iz} |0\rangle = S\hbar |0\rangle \qquad (23)$$

for all i.

We now suppose an external perturbation of the form

$$H_{\text{ext}}(t) = - \sum_j h_j(t)(S_j^+ + S_j^-) \qquad (24)$$

and calculate the response of the system. A physical realization of such a perturbation could be a beam of neutrons polarized in the x direction. We apply the formalism (8.A.33)–(8.A.40) to this situation. In particular, we shall calculate the correlation functions

$$S_{+-}(\mathbf{q}, \omega) = \int_{-\infty}^{\infty} dt \, \langle S^+(\mathbf{q}, t)S^-(\mathbf{q}, 0)\rangle$$

$$\qquad (25)$$

$$S_{-+}(\mathbf{q}, \omega) = \int_{-\infty}^{\infty} dt \, \langle S^-(\mathbf{q}, t)S^+(\mathbf{q}, 0)\rangle$$

where

$$S^+(\mathbf{q}, t) = \frac{1}{\sqrt{N}} \sum_j S_j^+(t) e^{i\mathbf{q} \cdot \mathbf{r}_j}$$

$$S^-(\mathbf{q}, t) = \frac{1}{\sqrt{N}} \sum_j S_j^-(t) e^{-i\mathbf{q} \cdot \mathbf{r}_j} = [S^+(\mathbf{q}, t)]^+ \tag{26}$$

We begin by considering the zero-temperature case and calculate, for $B = 0$,

$$HS^-(\mathbf{q}, 0) \, |0\rangle = -\frac{1}{\sqrt{N}} \sum_m e^{-i\mathbf{q} \cdot \mathbf{r}_m} \sum_{\langle ij \rangle} \left[J_z S_{iz} S_{jz} + \frac{J_{xy}}{2} (S_i^+ S_j^- + S_j^+ S_i^-) \right] S_m^- \, |0\rangle \tag{27}$$

We examine (27) term by term:

$$J_z S_{iz} S_{jz} S_m^- \, |0\rangle = J_z S^2 \hbar^2 S_m^- \, |0\rangle \qquad \text{if } i \ne m \text{ and } j \ne m$$

$$= J_z S(S - 1) \hbar^2 S_m^- \, |0\rangle \qquad \text{if } i = m \text{ or } j = m$$

Therefore,

$$-\sum_{\langle ij \rangle} J_z S_{iz} S_{jz} S_m^- \, |0\rangle = E_0 S_m^- \, |0\rangle + \nu J_z \hbar^2 S S_m^- \, |0\rangle \tag{28}$$

where $E_0 = -\nu N J_z \hbar^2 S^2 / 2$ is the zero-field ground-state energy. Also, since i, j are nearest neighbors,

$$J_{xy} S_i^+ S_j^- S_m^- \, |0\rangle = 0 \qquad \text{if } i \ne m$$

$$= 2 J_{xy} \hbar^2 S S_j^- \, |0\rangle \qquad \text{if } i = m$$

Therefore,

$$\sum_{\langle ij \rangle} \frac{J_{xy}}{2} (S_i^+ S_j^- + S_j^+ S_i^-) S_m^- \, |0\rangle = J_{xy} \hbar^2 S \sum_j{}' S_j^- \, |0\rangle \tag{29}$$

where Σ' indicates that the summation extends only over nearest neighbors of site m. Substituting (28) and (29) into (27), we obtain

$$HS^-(\mathbf{q}, 0) \, |0\rangle = E_0 S^-(\mathbf{q}, 0) \, |0\rangle + \frac{\hbar S}{\sqrt{N}} \sum_{m, \delta} (J_z - J_{xy} e^{i\mathbf{q} \cdot \delta}) S_m^- e^{-i\mathbf{q} \cdot \mathbf{r}_m} \, |0\rangle \tag{30}$$

where δ is a nearest-neighbor vector. Therefore,

$$HS^-(\mathbf{q}, 0) \, |0\rangle = [E_0 + \epsilon_0(\mathbf{q})] S^-(\mathbf{q}, 0) \, |0\rangle \tag{31}$$

with

$$\epsilon_0(\mathbf{q}) = \hbar^2 S \sum_\delta (J_z - J_{xy} e^{i\mathbf{q} \cdot \delta}) \tag{32}$$

which is the energy of the spin wave or magnon with wave vector $\mathbf{q}$. For the simple cubic lattice, with nearest-neighbor spacing a, the dispersion relation (32) takes the form

$$\epsilon_0(\mathbf{q}) = 6\hbar^2 S(J_z - J_{xy})$$
$$+ 6\hbar^2 SJ_{xy}[1 - \tfrac{1}{3}(\cos q_x a + \cos q_y a + \cos q_z a)] \tag{33}$$

The anisotropy in the coupling constants leads to a gap $\Delta = 6\hbar^2 S(J_z - J_{xy})$ in the spin wave spectrum. In the case of a finite magnetic field in the z direction this gap would be increased by $mB\hbar$, as is easily shown. The gap leads to an exponential variation of the order parameter at low temperatures:

$$\Delta M_z = M_z(0) - M_z(T) \sim e^{-\Delta/k_B T} \tag{34}$$

In the case of the isotropic Heisenberg model, $J = J_z = J_{xy}$ the spectrum is free–particle like for small wave vectors q:

$$\epsilon_0(\mathbf{q}) \approx J\hbar Sq^2 a^2 \tag{35}$$

If we assume that the spin waves or magnons are noninteracting bosons, we can then show that at low temperatures and zero applied magnetic field, the temperature dependence of the magnetization is given by (see Problem 8.5)

$$\Delta M_z(T) \sim T^{3/2} \tag{36}$$

The assumption that magnons are noninteracting bosons is only approximately valid but becomes more and more exact as the temperature is lowered, and equation (36) can be shown to hold in this limit (see Mattis [1981] or the original paper by Dyson [1958] for a complete discussion).

We note, in passing, that in contrast to the case of the Ising model, where the statistical mechanics of the antiferromagnet (at least on lattices such as the simple cubic which consist of interpenetrating sublattices, see 3.C.) is identical to that of the ferromagnet, the Heisenberg antiferromagnet poses a much more difficult problem than the ferromagnet. Indeed, not even the ground state is known in three or two dimensions. Mattis [1981] provides a thorough discussion of this problem. We mention only that antiparallel ordering on two sublattices produces a phonon like (linear in q) excitation spectrum rather than the free-particle spectrum (35).

Returning to the correlation functions (25), we immediately find

$$S_{+-}(\mathbf{q}, \omega) = 4\pi\hbar S\delta\left(\omega - \frac{\epsilon_0(\mathbf{q})}{\hbar}\right) \tag{37}$$

and $S_{-+}(\mathbf{q}, \omega) = 0$ for the ground state aligned in the $+z$ direction. The response function or transverse susceptibility is therefore given by (8.A.39)

$$\chi_{+-}^R(\mathbf{q}, \omega) = 2\hbar S \int_{-\infty}^{\infty} d\omega' \, \frac{\delta(\omega' - \epsilon_0(\mathbf{q})/\hbar)}{\omega - \omega' + i\eta} = \frac{2\hbar S}{\omega - \epsilon_0(\mathbf{q})/\hbar + i\eta} \tag{38}$$

Specializing to a static field, independent of position [i.e., $h_j(t) = h$ in (24)], we find, for the static transverse susceptibility,

$$\chi^S_{+-} = - \lim_{\omega,\mathbf{q}\to 0} \frac{2\hbar S}{\omega - \epsilon_0(\mathbf{q})/\hbar + i\eta} = \frac{2\hbar S}{6\hbar S(J_z - J_{xy}) + mB} \tag{39}$$

which diverges in the isotropic case as the field in the z direction approaches zero. This is simply an indication of the infinite degeneracy of the ground state—there is no energy cost associated with the coherent rotation of all N spins.

(d) Screening and Plasma Resonances in an Electron Gas at T = 0

As in the case of the weakly interacting Bose gas [Section 8.B(b)], we must first calculate the dynamic structure factor of the noninteracting system. Again, at $T = 0$, the only state that contributes to the ensemble average (8.A.20) is the ground state, which has plane wave states occupied up to a spherical Fermi surface of radius k_F. Consider the states $|n\rangle$ which can be populated through the action of the density operator

$$\rho_q = \sum_{k,\sigma} a^+_{\mathbf{k}+\mathbf{q},\sigma} a_{\mathbf{k},\sigma} \tag{40}$$

on the ground state $|0\rangle$. The state $|n\rangle = a^+_{\mathbf{k}+\mathbf{q},\sigma} a_{\mathbf{k},\sigma} |0\rangle$ is one in which a particle with wave vector $\mathbf{k}$ *inside* the Fermi surface is excited to a state of wave vector $\mathbf{k} + \mathbf{q}$ outside the Fermi surface as illustrated in Figure 8.3. The energy of this state is

$$E_n = E_0 + \frac{\hbar^2(\mathbf{k} + \mathbf{q})^2}{2m} - \frac{\hbar^2 k^2}{2m} \tag{41}$$

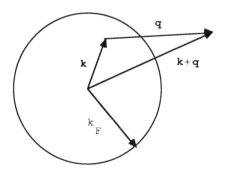

Figure 8.3 Single-particle excitation from a filled Fermi sphere.

where E_0 is the energy of the filled Fermi sea. Using (41) we find, for the dynamic structure factor of the noninteracting system (note a factor of 2 for spin),

$$S_0(\mathbf{q}, \omega) = \frac{4\pi\hbar}{V} \sum_{\mathbf{k},\mathbf{q}} \theta(k_F - k)\theta(|\mathbf{k} + \mathbf{q}| - k_F)$$

$$\times \delta\left\{\hbar\omega - \frac{\hbar^2}{2m}[(\mathbf{k} + \mathbf{q})^2 - k^2]\right\} \quad (42)$$

where $\theta(x) = 1$ for $x > 0$, 0 for $x < 0$.

The free-particle response function is given by (8.A.26), which, with (42), becomes

$$\chi_0^R(\mathbf{q}, \omega) = \frac{2}{V} \sum_{k<k_F, |\mathbf{k}+\mathbf{q}|>k_F} \left\{ \frac{1}{\hbar\omega - (\hbar^2/2m)[(\mathbf{k} + \mathbf{q})^2 - k^2] + i\eta} \right.$$

$$\left. - \frac{1}{\hbar\omega - (\hbar^2/2m)[(\mathbf{k} + \mathbf{q})^2 - k^2] + i\eta} \right\} \quad (43)$$

If we make the substitution $\mathbf{k} = -(\mathbf{p} + \mathbf{q})$ in the second term and then relabel $\mathbf{p} \to \mathbf{k}$ the denominators in (43) become identical. Moreover, we use the identity

$$\theta(k_F - k)\theta(|\mathbf{k} + \mathbf{q}| - k_F) - \theta(k - k_F)\theta(k_F - |\mathbf{k} + \mathbf{q}|)$$

$$= \theta(k_F - k) - \theta(k_F - |\mathbf{k} + \mathbf{q}|) \quad (44)$$

to simplify (43) and obtain

$$\chi_0^R(\mathbf{q}, \omega) = \frac{2}{V} \sum_{\mathbf{k}} \frac{\theta(k_F - k) - \theta(k_F - |\mathbf{k} + \mathbf{q}|)}{\hbar\omega - (\hbar^2/2m)[(\mathbf{k} + \mathbf{q})^2 - k^2] + i\eta} \quad (45)$$

The function $\chi_0^R(\mathbf{q}, \omega)$ is commonly referred to as the Lindhard function. If the sum over $\mathbf{k}$ is transformed into an integral, analytical expressions can be found for both the real and imaginary parts. The resulting expressions are rather complicated and we will here exhibit only two special cases. Explicit formulas for the general case can be found in Mahan [1981], Fetter and Walecka [1971], or Pines and Nozières [1966].

(i) $\omega \to 0$ $\mathbf{q} \neq 0$. This limit is important for the theory of static screening. It is also the limit in which one usually studies the electron phonon interaction, since lattice vibrational energies are generally small compared with typical electronic excitation energies. Substituting

$$\frac{1}{\hbar\omega + E + i\eta} = P \frac{1}{\hbar\omega + E} - i\pi\delta(\hbar\omega + E) \quad (46)$$

where P indicates principal value, we find for $\omega = 0$,

$$\text{Im } \chi_0^R(\mathbf{q}, 0) = -\frac{2\pi}{V} \sum_{\mathbf{k}} (\theta(k_F - k) - \theta(k_F - |\mathbf{k} + \mathbf{q}|))$$

$$\times \delta\left\{\frac{\hbar^2}{2m}[k^2 - (\mathbf{k} + \mathbf{q})^2]\right\} = 0 \qquad (47)$$

since the two step functions cancel when $|\mathbf{k}| = |\mathbf{k} + \mathbf{q}|$. Thus the static susceptibility is real for all $\mathbf{q}$ and

$$\chi_0^R(\mathbf{q}, 0) = \frac{m}{2\pi^3\hbar^2}\left[P\int_{k<k_F} \frac{d^3k}{k^2 - (\mathbf{k} + \mathbf{q})^2} - P\int_{|\mathbf{k}+\mathbf{q}|<k_F} \frac{d^3k}{k^2 - (\mathbf{k} + \mathbf{q})^2}\right]$$

$$(48)$$

If we make the substitution $\mathbf{k} + \mathbf{q} = -\mathbf{p}$ in one of the integrals, we see that the two terms give identical contributions and have

$$\chi_0^R(\mathbf{q}, 0) = -\frac{m}{\pi^3\hbar^2}P\int_{k<k_F} \frac{d^3k}{q^2 + 2\mathbf{k} \cdot \mathbf{q}} \qquad (49)$$

In spherical coordinates, with $\mu = \cos\theta$, (49) becomes

$$\chi_0^R(\mathbf{q}, 0) = -\frac{2m}{\pi^2\hbar^2}\int_0^{k_F} dk\, k^2\, P\int_{-1}^{1} d\mu \frac{1}{q^2 + 2kq\mu}$$

$$= -\frac{m}{q\pi^2\hbar^2}P\int_0^{k_F} dk\, k\ln\left|\frac{q + 2k}{q - 2k}\right| \qquad (50)$$

$$= -\frac{m}{q\pi^2\hbar^2}P\int_{-k_F}^{k_F} dk\, k\ln|q + 2k|$$

The last integral can easily be evaluated through integration by parts to yield, finally,

$$\chi_0^R(\mathbf{q}, 0) = -\frac{mk_F}{\pi^2\hbar^2}\left(\frac{1}{2} + \frac{4k_F^2 - q^2}{8qk_F}\ln\left|\frac{q + 2k_F}{q - 2k_F}\right|\right) \qquad (51)$$

In the mean field approximation the static dielectric function is therefore given by (9)

$$\epsilon(\mathbf{q}, 0) = 1 - \frac{4\pi e^2}{q^2}\chi_0^R(\mathbf{q}, 0) \qquad (52)$$

Defining the function

$$u_\mathbf{q} = \left(\frac{1}{2} + \frac{4k_F^2 - q^2}{8qk_F}\ln\left|\frac{q + 2k_F}{q - 2k_F}\right|\right) \qquad (53)$$

we thus have

$$\epsilon(\mathbf{q}, 0) = 1 + \frac{k_{TF}^2}{q^2} u_{\mathbf{q}} \tag{54}$$

where k_{TF} is the *Thomas–Fermi wave vector*

$$k_{TF}^2 = \frac{4me^2}{\pi \hbar^2} k_F \tag{55}$$

The function u_q is sketched in Figure 8.4. We note that for small q, $u_q \approx 1$ while $u_q \to 0$ as $q \to \infty$. In the vicinity of $2k_F$, u_q varies rapidly and there is a logarithmic singularity in the derivative with respect to q at $2k_F$.

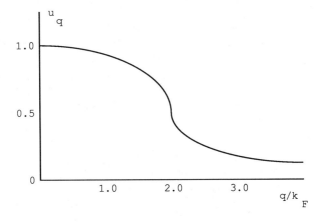

Figure 8.4 Function u_q of equation (53).

Consider next the effective potential due to a point charge e in the mean field linear response approximation:

$$\phi_{\text{eff}}(\mathbf{q}) = \frac{4\pi e^2}{q^2 + k_{TF}^2 u_q} \tag{56}$$

Normally, the spatial dependence of a function at large r will be determined primarily by the behavior of the Fourier transformed function at small q. Since $u_q \to 1$ as $q \to 0$, one might expect that for large r the effective potential will behave like the Fourier transform of the $q = 0$ limit of (56), which is (see Problem 8.1)

$$\phi_{TF}(\mathbf{r}) = \frac{e^2}{r} e^{-k_{TF}r} \tag{57}$$

The foregoing argument is, however, not correct because of the singularity in u_q at $2k_F$. It can be shown (see, e.g., Fetter and Walecka, 1971) that for large r

$$\phi_{\text{eff}}(\mathbf{r}) \sim \frac{1}{r^3} \cos 2k_F r \tag{58}$$

The oscillations in $\phi_{\text{eff}}(r)$ are commonly referred to as *Friedel oscillations*. The result (58) is gratifying from the point of view of understanding the cohesion of metals. If the effective interaction of ions in an electron gas had been purely repulsive, as in the Thomas–Fermi approximation (57), it would be difficult to see what keeps a metal together. The potential (58), however, leads to an ion–ion interaction which is attractive at certain distances, with an energy minimum at the equilibrium ionic positions of the resulting crystal. Unfortunately, this model is too crude to be useful quantitatively for the description of real metals. We next consider another interesting limiting case.

   (ii) $|\mathbf{q}|$ **small,** $|\omega| >> \hbar q^2|2m$.   In this limit the denominator in (45) cannot vanish and we rewrite this equation in the form

$$\chi_0^R(\mathbf{q}, \omega) = \frac{2}{V} \sum_k \theta(k_F - k)\left\{\frac{1}{\hbar\omega - (\hbar^2/2m)[(\mathbf{k} + \mathbf{q})^2 - k^2]}\right.$$

$$\left. - \frac{1}{\hbar\omega + (\hbar^2/2m)[(\mathbf{k} - \mathbf{q})^2 - k^2]}\right\} \quad (59)$$

Using $(\mathbf{k} \pm \mathbf{q})^2 - k^2 = q^2 \pm 2\mathbf{k} \cdot \mathbf{q}$ and combining the two terms, we obtain

$$\chi_0^R(\mathbf{q}, \omega) = \frac{2}{V} \sum_{k<k_F} \frac{q^2/m}{[\omega - (\hbar/2m)(q^2 + 2\mathbf{k} \cdot \mathbf{q})][\omega - (\hbar/2m)(q^2 - 2\mathbf{k} \cdot \mathbf{q})]}$$

$$\approx \frac{q^2 N}{mV} \frac{1}{\omega^2} \quad (60)$$

Using this result, we find the dielectric function in the mean field approximation:

$$\epsilon(\mathbf{q}, \omega) = 1 - \frac{4\pi e^2}{q^2}\chi_0^R(\mathbf{q}, \omega) \approx 1 - \frac{4\pi e^2 N}{mV} = 1 - \frac{\Omega_p^2}{\omega^2} \quad (61)$$

where $\Omega_p = (4\pi e^2 N/mV)^{1/2}$ is the *plasma frequency*. We note that Planck's constant does not appear in the plasma frequency. Indeed, the plasma frequency obtained in this approximation is identical to that of a classical system of charged particles in the Drude model (see e.g. Aschcroft and Mermin 1976). This is easily seen from the following elementary argument. Suppose that a classical free electron system is subject to a time-dependent electric field of the form

$$\mathbf{E}(t) = \mathbf{E}_0 e^{-i\omega t} \quad (62)$$

The resulting force on a particle is $\mathbf{f} = -e\mathbf{E}(t)$ and the mean displacement is $\langle \mathbf{x}(t) \rangle = e\mathbf{E}/\omega^2 m$. Therefore, the induced polarization is given by

$$\mathbf{P} = -\frac{Ne}{V} \langle \mathbf{x} \rangle = -\frac{ne^2}{m\omega^2}\mathbf{E} \quad (63)$$

The dielectric function, in turn, is given by

$$\mathbf{D} = \epsilon \mathbf{E} = \mathbf{E} + 4\pi \mathbf{P} \qquad (64)$$

Combining (63) and (64), we recover the previous result (61).

From $\phi_{\text{eff}}(\mathbf{q}, \omega) = \phi_{\text{ext}}(\mathbf{q}, \omega)/\epsilon(\mathbf{q}, \omega)$, we see that a zero in the dielectric function corresponds to a resonant response. The resulting excitation is a longitudinal charge density wave which for nonzero $\mathbf{q}$ propagates with a frequency $\Omega_{\text{pl}}(\mathbf{q})$ which reduces to the plasma frequency Ω_p (61) in the long-wavelength limit. It can be seen, by expanding (60) to higher order in q, that for small q, $\Omega_{\text{pl}}(\mathbf{q})$ differs from Ω_p by an amount proportional to q^2. The corresponding quantum of excitation or quasiparticle is called a *plasmon* and has energy $\hbar\Omega_{\text{pl}}(\mathbf{q})$.

(e) Excitation Spectrum and Exchange and Correlation Energy of the Electron Gas at $T = 0$

We now return to the case of arbitrary values of $\mathbf{q}$ and ω. From the expression (42) for the dynamic structure factor $S_0(\mathbf{q}, \omega)$ and the inequalities, for $k < k_F$,

$$q^2 + 2k_F q \geq q^2 + 2kq \geq (\mathbf{k} + \mathbf{q})^2 - k^2 \geq q^2 - 2kq \geq q^2 - 2k_F q$$

we see that $S_0(\mathbf{q}, \omega)$ is nonzero in the shaded region of Figure 8.5.

If $\chi_0(\mathbf{q}, \omega)$ is continued analytically to complex values of ω,

$$\chi_0^R(\mathbf{q}, \omega) = 2 \int \frac{d^3k}{(2\pi)^3} \frac{\theta(k_F - k) - \theta(k_F - |\mathbf{k} + \mathbf{q}|)}{\hbar\omega + \epsilon_\mathbf{k} - \epsilon_{\mathbf{k}+\mathbf{q}}} \qquad (65)$$

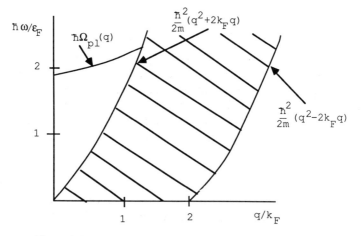

Figure 8.5 Excitation spectrum of an interacting electron gas.

and we see that for ω on the real axis,

$$\text{Im } \chi_0^R(\mathbf{q}, \omega) = \tfrac{1}{2}[\chi_0^R(\mathbf{q}, \omega + i\eta) - \chi_0^R(\mathbf{q}, \omega - i\eta)] \tag{66}$$

that is, Im $\chi_0(\mathbf{q}, \omega)$ is $\tfrac{1}{2}$ times the discontinuity across a branch cut with extension

$$-\frac{\hbar^2}{2m}(q^2 + 2k_F q) < \hbar\omega < \frac{\hbar^2}{2m}(q^2 + 2k_F q) \qquad q < 2k_F$$

$$\frac{\hbar^2}{2m}(q^2 - 2k_F q) < |\hbar\omega| < \frac{\hbar^2}{2m}(q^2 + 2k_F q) \qquad q > 2k_F$$

The expression

$$\text{Im } \frac{\chi_0^R(\mathbf{q}, \omega)}{1 - v_q \chi_0^R(\mathbf{q}, \omega)} = \text{Im } \frac{1}{\epsilon(q, \omega)}$$

will have a branch cut for the same range of ω as well as a pole at the plasmon frequency. Here $v_q = 4\pi e^2/q^2$. The residue at this pole is largest for $q = 0$ and gradually goes to zero as the pole merges with particle–hole continuum (shaded region in Figure 8.5).

We next turn to the problem of evaluating the ground-state energy of the electron gas. By combining (8.A.47), the $T = 0$ limit of (8.A.41) and (5), we find

$$E_{\text{int}}(\lambda) = -\sum_q \left\{ \int_0^\infty \frac{\hbar d\omega}{2\pi} \text{Im } \frac{1}{\epsilon_\lambda(\mathbf{q}, \omega)} + \frac{N v_q}{2V} \right\} \tag{67}$$

From (66) we see that the integration over frequency in (67) is equivalent to following the contour C in Figure 8.6. We can add a semicircle to this contour since the contribution to the integral will vanish in the limit of large radius. The analytical continuation (65) will not have any singularities except on the real axis. We can therefore deform the contour to lie along the imaginary axis (C') and obtain, with $\omega = iu$, in the mean field approximation:

$$E = \frac{3}{5} \frac{\hbar^2 k_F^2}{2m} N - \sum_q \left\{ \frac{N v_q}{2V} + \int_{-\infty}^\infty \frac{\hbar du}{4\pi} \int_0^1 d\lambda \frac{v_q \chi_0^R(\mathbf{q}, iu)}{1 - \lambda v_q \chi_0^R(\mathbf{q}, iu)} \right\}$$

$\chi_0^R(\mathbf{q}, iu)$ turns out to be real and, after performing the integration over λ, we finally obtain

$$E = \frac{3}{5} \frac{\hbar^2 k_F^2}{2m} N - \sum_q \left\{ \frac{N v_q}{2V} - \int_{-\infty}^\infty \frac{\hbar du}{4\pi} \ln\left[1 - v_q \chi_0^R(\mathbf{q}, iu)\right] \right\} \tag{68}$$

This approximate expression for the ground-state energy was first obtained by Gell-Mann and Brueckner [1957]. It can easily be evaluated numerically and turns out to give rise to a significant contribution to the cohesive energy of metals. From the expression for the energy, other quantities, such as

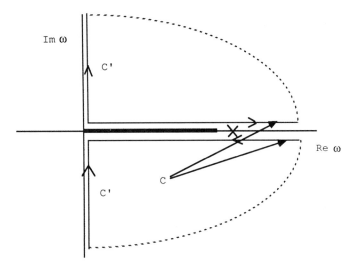

Figure 8.6 Contour used in the integration (67). The branch cut is shown as a bold line.

the compressibility, can be obtained using thermodynamic identities. We refer the interested reader to the texts by Mahan [1981] and Pines and Nozières [1966] for further discussion of the electron liquid.

(f) Phonons in Metals

We now consider an idealized model of a metal in which we treat the system as a two-component plasma of N_i ions and $N_e = ZN_i$ electrons. We will calculate the *charge* response

$$\langle -e\delta\rho^e(\mathbf{q}, \omega) + Ze\delta\rho^i(\mathbf{q}, \omega)\rangle$$

to an external potential (voltage) $v_{ext}(\mathbf{q}, \omega)$ and let $\chi_0^i(\mathbf{q}, \omega)$ and $\chi_0^e(\mathbf{q}, \omega)$ be, respectively, the free ion and free electron response functions as defined earlier. We find, for the mean field response, suppressing explicit reference to the dependence on $\mathbf{q}$ and ω,

$$\langle \delta\rho^i\rangle = \chi_0^i(Zev_{ext} + \frac{4\pi Ze}{q^2}\langle Ze\,\delta\rho^i - e\,\delta\rho^e\rangle) \qquad (69)$$

with a similar formula for the response of the electrons. After some straight-forward algebra, we obtain

$$\langle -\delta\rho^e + Z\,\delta\rho^i\rangle = \frac{-e\chi_0^e + Ze\chi_0^i}{1 - (4\pi e^2/q^2)(Z^2\chi_0^i + \chi_0^e)}v_{ext} \qquad (70a)$$

$$\langle \delta\rho^i\rangle = \frac{Ze\chi_0^i v_{ext}}{1 - (4\pi e^2/q^2)(Z^2\chi_0^i + \chi_0^e)} \qquad (70b)$$

$$\langle \delta \rho^e \rangle = \frac{(-e)\chi_0^e v_{\text{ext}}}{1 - (4\pi e^2/q^2)(Z^2 \chi_0^i + \chi_0^e)} \tag{70c}$$

Because of their large mass the ions respond much more slowly than do the electrons. We will therefore assume that we are dealing with frequencies at which the high-frequency approximation (60) is valid for the ion response function:

$$\chi_0^i(\mathbf{q}, \omega) \approx \frac{q^2 N}{MV} \frac{1}{\omega^2} \tag{71}$$

where M is the mass of an ion. We next assume that the electronic response is sufficiently fast that the static approximation (51), (53):

$$\chi_0^e(\mathbf{q}, 0) = -\frac{mk_F}{\pi^2 \hbar^2} u_{\mathbf{q}} \tag{72}$$

is valid. With these approximations, the ionic response becomes

$$\langle \delta \rho^i \rangle = \frac{1}{\omega^2 - 4\pi e^2 Z^2 N_i q^2/(q^2 + k_{\text{TF}}^2 u_{\mathbf{q}})V} \frac{ZeN}{MV} \frac{q^2 v_{\text{ext}}(\mathbf{q}, \omega)}{q^2 + k_{\text{TF}}^2 u_{\mathbf{q}}} \tag{73}$$

If it were not for the screening of the ions by the electrons, the lattice vibrational frequency would be the ion plasma frequency $\Omega_i = (4\pi N_i Z^2 e^2/M)^{1/2}$. With screening, we find instead an acoustic mode, the quantum of which is the *phonon*, with frequency given by

$$\omega^2(\mathbf{q}) = \frac{4\pi e^2 Z^2 N_i}{(q^2 + k_{\text{TF}}^2 u_{\mathbf{q}})V} q^2 \tag{74}$$

In the limit of long wavelength, $u_{\mathbf{q}} \to 1$ and (74) reduces to $\omega = cq$, where

$$c \approx \left(\frac{4\pi Z^2 e^2 N_i}{k_{\text{TF}}^2 V} \right)^{1/2} \tag{75}$$

is the Bohm–Staver sound velocity.

The model that we have used to derive this result is extremely crude. Nevertheless, the resulting velocity of sound is of the right order of magnitude for most metals. We also note that for any nonzero frequency $\chi_0^e(\mathbf{q}, \omega)$ will have a small but nonzero imaginary part. The theory thus predicts attenuation of acoustic waves.

From (70a) we find for the dielectric function in the mean field approximation:

$$\epsilon_{\text{MF}}(\mathbf{q}, \omega) = 1 - \frac{4\pi e^2}{q^2}[Z^2 \chi_0^i(\mathbf{q}, \omega) + \chi_0^e(\mathbf{q}, \omega)] \tag{76}$$

Once again, using the approximations (71) and (72), we obtain

$$\epsilon_{\text{MF}}(\mathbf{q}, \omega) \approx \left(1 + \frac{k_{\text{TF}}^2 u_{\mathbf{q}}}{q^2} \right)\left[1 + \frac{\omega^2(\mathbf{q})}{\omega^2 - \omega^2(\mathbf{q})} \right] \tag{77}$$

and we see that for frequencies less than $\omega(\mathbf{q})$ the dielectric function is negative. This has the consequence that the quasielastic interaction between electrons in states near the Fermi surface is attractive. As discussed in Section 7.C, such an attractive interaction in the BCS theory produces a Fermi surface instability, which is the mechanism generally believed to be responsible for the phenomenon of superconductivity.

8.C ENTROPY PRODUCTION, THE KUBO FORMULA, AND THE ONSAGER RELATIONS FOR TRANSPORT COEFFICIENTS

In this book we are concerned primarily with equilibrium properties of matter. For systems sufficiently close to equilibrium that linear response theory is adequate, it is possible to relate *transport properties* to equilibrium correlation functions. We begin the discussion of transport by deriving the Kubo formula for the special case of electrical conductivity. In Section 8.C(b) we generalize the concepts of currents and fields and also introduce the notions of microscopic reversibility and local equilibrium. In Section 8.C(c) we use these ideas to derive the Onsager relations between transport coefficients.

(a) Kubo Formula

Consider a system of particles, each with charge e and with positions specified by $\mathbf{r}_m$. The electrical current density at point $\mathbf{x}$ is given by

$$\mathbf{j}(\mathbf{x}) = \sum_m e\mathbf{v}_m\delta(\mathbf{x} - \mathbf{r}_m) \tag{1}$$

where $\mathbf{v}_m$ is the velocity of particle m. We assume an external electric field, $\mathbf{E}(t)$, which does not vary with position. In this case the current density will also be time but not position dependent and the perturbing Hamiltonian is

$$H_{\text{ext}}(t) = -e\sum_m \mathbf{r}_m \cdot \mathbf{E}(t) \tag{2}$$

Using (8.A.34) with $B = j_\alpha$, the αth component of the current density, we have

$$j_\alpha(t) = -\frac{i}{\hbar V}\int_{-\infty}^{t} dt' \sum_{m,n,\gamma} \langle[r_{m\gamma}(t'), \dot{r}_{n\alpha}(t)]\rangle_H e^2 E_\gamma(t') \tag{3}$$

where V is the volume of the system. Substituting

$$\langle\phi\rangle = \text{Tr}\,\rho\phi$$

where ρ is the density matrix of the unperturbed system, and using the fact that the trace of a product of operators is invariant under cyclic permutation of the operators, we may write (3) in the form

$$j_\alpha(t) = -\frac{i}{\hbar} \int_{-\infty}^{t} dt' \, \text{Tr} \, j_\alpha(t - t') \sum_{m\gamma} [r_{m\gamma}, \rho] e E_\gamma(t') \tag{4}$$

We simplify this expression by rewriting the commutator. For an arbitrary operator A we have

$$Z_G[A, \rho] = [A, e^{-\beta K}] \equiv e^{-\beta K} \phi(\beta) \tag{5}$$

where

$$\phi(\lambda) = e^{\lambda K} A e^{-\lambda K} - A \tag{6}$$

Differentiating (6) with respect to λ and integrating from 0 to β, we obtain

$$\phi(\beta) = \int_0^\beta d\lambda \, e^{\lambda K}[K, A]e^{-\lambda K}$$

or, in (4),

$$[r_{m\gamma}, \rho] = \rho \int_0^\beta d\lambda \, e^{\lambda K}[K, r_{m\gamma}]e^{-\lambda K} = \rho \int_0^\beta d\lambda \, \frac{\hbar}{i} \dot{r}_{m\gamma}(-i\hbar\lambda) \tag{7}$$

where we have used the Heisenberg equation of motion

$$[K, A] = \frac{\hbar}{i} e^{-iKt/\hbar} \frac{dA}{dt} e^{iKt/\hbar}$$

Finally,

$$\begin{aligned}
j_\alpha(t) &= -V \sum_\gamma \int_{-\infty}^t dt' \, \text{Tr} \, j_\alpha(t - t')\rho \int_0^\beta d\lambda \, j_\gamma(-i\hbar\lambda) E_\gamma(t') \\
&= V \sum_\gamma \int_{-\infty}^t d\tau' \int_0^\beta d\lambda \, \langle j_\gamma(-i\hbar\lambda) j_\alpha(t - t') \rangle E_\gamma(t')
\end{aligned} \tag{8}$$

If we specialize to a time-independent field $\mathbf{E}(t') = \mathbf{E}$ and let $t' = t - \tau$, we have

$$j_\alpha(t) = j_\alpha = -V \sum_\gamma \int_0^\infty d\tau \int_0^\beta d\lambda \, \langle j_\gamma(-i\hbar\lambda) j_\alpha(\tau) \rangle = \sum_\gamma \sigma_{\alpha\gamma} E_\gamma \tag{9}$$

or

$$\sigma_{\alpha\gamma} = -V \int_0^\infty d\tau \int_0^\beta d\lambda \, \langle j_\gamma(-i\hbar\lambda) j_\alpha(\tau) \rangle \tag{10}$$

which expresses the conductivity tensor, σ, in terms of the *equilibrium* current–current correlation function. In the classical limit ($\hbar \to 0$, which is equivalent to the high-temperature limit $\beta \to 0$) we may approximate (10) by the formula

$$\sigma_{\alpha\gamma} = -\frac{V}{k_\mathrm{B} T} \int_0^\infty d\tau \, \langle j_\gamma(0) j_\alpha(\tau) \rangle \tag{11}$$

Equations (10) and (11) are known as Kubo formulas and it is clear that similar expressions can be derived for other transport coefficients. The Kubo formulation of transport theory is complementary to the Boltzmann equation approach. We refer the interested reader to the original work of Kubo [1957] and the book of Mahan [1981].

(b) Entropy Production and Generalized Currents and Forces

We now wish to extend our formalism to a wider class of transport phenomena. As before, we limit ourselves to nonequilibrium phenomena which are sufficiently close to equilibrium that the macroscopic intensive variables such as pressure, temperature, chemical potential, and electrostatic or vector potential are well defined locally. We will further assume that these variables are related to the densities associated with the conjugate extensive variables through the equilibrium equations of state. In equilibrium the temperature, pressure, and chemical potential must be constant throughout the system. Here we consider instead the situation in which a gradient in these variables is maintained, resulting in currents of the corresponding density variables. To develop a systematic approach, we must first select appropriate variables. In equilibrium we have, from (1.C.5), the differential for the entropy density

$$ds = \frac{1}{T} de - \frac{1}{T} \sum_i X_i \, d\xi_i - \frac{1}{T} \sum_j \mu_j \, dn_j = \sum_k \phi_k \, d\rho_k \tag{12}$$

In (12), e is the energy density, X_i a generalized force, ξ_i the density coupled to that force, and n_j the particle density of species j. For notational convenience we have combined these terms into a set of generalized potential variables ϕ_k and density variables ρ_k.

We now consider the effect of spatially varying potentials ϕ_k, which in turn generate current densities j_k and assume that there is a conservation law associated with each density ρ_k so that

$$\frac{\partial \rho_k}{\partial t} + \boldsymbol{\nabla} \cdot \mathbf{j}_k = 0 \tag{13}$$

We may also associate an entropy current density j_s with the currents j_k, that is,

$$\mathbf{j}_s = \sum_k \phi_k \mathbf{j}_k \tag{14}$$

The rate of change in the entropy density is then given by

$$\frac{ds}{dt} = \frac{\partial s}{\partial t} + \nabla \cdot \mathbf{j}_s \tag{15}$$

where

$$\frac{\partial s}{\partial t} = \sum_k \phi_k \frac{\partial \rho_k}{\partial t} \tag{16}$$

It is a straightforward consequence of (13)–(16) that

$$\frac{ds}{dt} = \sum_k \nabla \phi_k \cdot \mathbf{j}_k \tag{17}$$

Equation (17) specifies the appropriate conjugate generalized *force fields* $\nabla \phi_k$ and *current densities* $\mathbf{j}_k$.

We now specialize to systems in which there is a *steady-state* current in response to a static field. The reader will be familiar with a number of phenomenological laws which are used to describe this type of situation. We have already come across Ohm's law for the electrical conductivity. Other examples are Fick's law for diffusion and the Seebeck and Peltier relations for the thermoelectric effects discussed in the next section. In general, we will, in linear response theory, assume relationships of the form

$$\mathbf{j}_i = \sum_k L_{ik} \nabla \phi_k + O(\nabla \phi_k)^2) \tag{18}$$

where the coefficients $L_{ik}(\rho_k)$ are called *kinetic coefficients* and where we shall ignore the second- and higher-order terms. In the next Section we shall discuss Onsager's analysis of these coefficients, which led him to the conclusion that under the very reasonable assumption of *microscopic reversibility*, these coefficients must satisfy the Onsager reciprocity relations:[2]

$$L_{ik} = L_{ki} \tag{19}$$

(c) Microscopic Reversibility:
The Onsager Relations

In the following discussion we give a derivation very similar to that in Onsager's [1931] original articles. Consider first the specific situation of heat flow in an insulator. We denote the heat currents in the three Cartesian directions by j_1, j_2, and j_3 and write

$$j_i = \sum_j L_{ij} (\nabla T^{-1})_j \tag{20}$$

---

[2] We have assumed that the generalized force field is the gradient of a scalar potential (i.e., a polar vector). The magnetic field $\mathbf{B} = \nabla \times \mathbf{A}$, on the other hand, is an axial vector and odd under time reversal. For this reason, the Onsager relations in the presence of a magnetic field are $L_{ij}(\mathbf{B}) = L_{ji}(-\mathbf{B})$.

where the coefficients L_{ij} are transport coefficients (the thermal conductivity tensor) which are independent of the temperature gradients in the linear approximation. Note that (12) implies that it is T^{-1}, not T, which is the potential field conjugate to the energy density. The coefficients L_{ij} are not all independent. If the crystal structure of our insulator is hexagonal with $i = 3$ the c axis, for example, the most general form of equation (20) is

$$j_1 = L_{11}\left(\frac{\partial T^{-1}}{\partial x}\right) + L_{12}\left(\frac{\partial T^{-1}}{\partial y}\right)$$

$$j_2 = -L_{12}\left(\frac{\partial T^{-1}}{\partial x}\right) + L_{11}\left(\frac{\partial T^{-1}}{\partial y}\right) \qquad (21)$$

$$j_3 = L_{33}\left(\frac{\partial T^{-1}}{\partial z}\right)$$

where the relations $L_{21} = -L_{12}$ and $L_{22} = L_{11}$ are a consequence of the point group symmetry (Problem 8.7) of the lattice. The Onsager relations, which we shall derive for this particular example, are $L_{ij} = L_{ji}$, which has the further consequence that $L_{12} = 0$.

Following Onsager [1931], we consider a fluctuation in the energy density. This fluctuation will, in general, result in nonzero values for the components of the first moments

$$\alpha_i = \int d^3r \, r_i \epsilon(\mathbf{r}) \qquad (22)$$

where $\epsilon(\mathbf{r})$ is the energy density and where all coordinates are measured from the center of the crystal. One expects the thermal averages of these moments to have the properties

$$\langle \alpha_i \rangle = 0 \qquad \langle \alpha_i \alpha_j \rangle = \delta_{ij}\langle \alpha^2 \rangle \qquad (23)$$

for a spherical crystal. The second of these equations is important for what follows and basically contains the statement that the equilibrium fluctuations are spatially isotropic, whereas the thermal conductivity, which depends on the connectivity of the lattice, may not be isotropic. Later in this section we formulate (23) more generally.

Suppose next that at a given time t variable α_1 takes on a specific value $\alpha_1(t)$. The correlation functions that characterize the decay of this fluctuation are the expectation values of the functions

$$\alpha_1(t + \Delta t)\alpha_1(t) \qquad \alpha_2(t + \Delta t)\alpha_1(t) \qquad \alpha_3(t + \Delta t)\alpha_1(t)$$

Consider, for example, the correlation function

$$\langle \alpha_2(t + \Delta t)\alpha_1(t) \rangle = \langle \alpha_2(t)\alpha_1(t) \rangle + \Delta t \langle \dot{\alpha}_2(t)\alpha_1(t) \rangle$$

$$= \Delta t \langle \dot{\alpha}_2(t)\alpha_1(t) \rangle \qquad (24)$$

where we have used (23) to eliminate the first term. The time derivative $d\alpha_2/dt$ is proportional to the heat current j_2 and we may now use (20) to obtain

$$\langle\alpha_2(t + \Delta t)\alpha_1(t)\rangle = \Delta t\, L_{21}\left\langle\left(\frac{\partial T^{-1}}{\partial x}\right)\alpha_1(t)\right\rangle \tag{25}$$

Moreover, $\partial T^{-1}/\partial x = -C\alpha_1/T^2$ and we have

$$\langle\alpha_2(t + \Delta t)\alpha_1(t)\rangle = -\frac{\Delta t\, L_{21}C\langle\alpha^2\rangle}{T^2} \tag{26}$$

Similarly,

$$\langle\alpha_1(t + \Delta t)\alpha_2(t)\rangle = \frac{\Delta t\, L_{21}C\langle\alpha^2\rangle}{T^2} \tag{27}$$

where we have used the hexagonal symmetry (21).

We now define a joint probability $P(\alpha_2', t + \Delta t|\alpha_1', t)$, which is the probability that variable α_2 takes on the specific value α_2' at time $t + \Delta t$ and that α_1 has value α_1' at time t. In terms of this function, the expectation value (26) is given by

$$\langle\alpha_2(t + \Delta t)\alpha_1(t)\rangle = \int d\alpha_1 \int d\alpha_2\, \alpha_1\, \alpha_2\, P(\alpha_2, t + \Delta t|\alpha_1, t) \tag{28}$$

The principle of microscopic reversibility states that

$$P(\alpha_1, t + \Delta t|\alpha_2, t) = P(\alpha_2, t + \Delta t|\alpha_1, t) \tag{29}$$

This equation is very plausible. If the velocities of all particles in a given configuration with a specific value of α_2, resulting from a configuration with value α_1 at time $t - \Delta t$ are reversed, then at time $t + \Delta t$ we will once again have the configuration of time $t - \Delta t$. Equation (29) states that these velocity-reversed configurations are equally probable. Clearly, then,

$$\langle\alpha_1(t + \Delta t)\alpha_2(t)\rangle = \langle\alpha_2(t + \Delta t)\alpha_1(t)\rangle \tag{30}$$

or

$$L_{12} = 0 \tag{31}$$

We now generalize the discussion by considering an arbitrary fluctuation in an isolated system. From the discussion of the preceding section we see that an appropriate choice of these fluctuating variables are the densities $\rho_k(r)$ and the entropy is then a functional of these variables. To avoid the complications of functional differentiation we assume that the variables $\rho_k(r)$ have been expressed in terms of a set of discrete variables α_i, $i = 1, 2, \ldots$, which are zero in the equilibrium state of maximum entropy. The number of such variables is, in principle, equal to the total number of degrees of freedom of the system less the number of extensive bulk thermodynamic variables. The variables α_i could, for example, be coefficients of a Fourier expansion or of an expansion in a set of orthogonal polynomials. The probability that the system is in a state with specific values of these variables is

$$P(\alpha_1, \alpha_2, \ldots, \alpha_n) = \frac{\exp\{S(\alpha_1, \alpha_2, \ldots, \alpha_n)/k_B\}}{\int d\alpha_1\, d\alpha_2 \cdots d\alpha_n \exp\{S(\alpha_1, \alpha_2, \ldots, \alpha_n)/k_B\}} \tag{32}$$

Since $(\partial S/\partial \alpha_i)|_{\{\alpha\}=0} = 0$, we have, quite generally,

$$P(\alpha_1, \alpha_2, \ldots, \alpha_n) = C \exp\left\{-\frac{1}{2}\sum_{j,m} \frac{g_{jm}\alpha_j\alpha_m}{k_B}\right\} \tag{33}$$

where C is a normalizing constant and where the coefficients g_{jm} are symmetric: $g_{jm} = g_{mj}$. This expression immediately implies that $\langle \alpha_j \rangle = 0$. Thermodynamic stability also implies that the determinant of the matrix g is greater than zero and that all its eigenvalues are positive.

We now obtain a simple expression for the correlation functions. Noting that

$$\frac{\partial P(\{\alpha\})}{\partial \alpha_i} = \frac{1}{k_B}\frac{\partial S}{\partial \alpha_i}P$$

we find

$$\left\langle \alpha_i \frac{\partial S}{\partial \alpha_i} \right\rangle = \int d\alpha_1\, d\alpha_2 \cdots d\alpha_n\, \alpha_i \frac{\partial S}{\partial \alpha_i} P(\{\alpha\})$$

$$= k_B \int d\alpha_1\, d\alpha_2 \cdots d\alpha_n\, \alpha_i \frac{\partial P}{\partial \alpha_i} \tag{34}$$

and integrating by parts,

$$\left\langle \alpha_i \frac{\partial S}{\partial \alpha_i} \right\rangle = -k_B \tag{35}$$

and for $i \neq m$,

$$\left\langle \alpha_i \frac{\partial S}{\partial \alpha_m} \right\rangle = 0 \tag{36}$$

Thus

$$\sum_m g_{im}\langle \alpha_m \alpha_j \rangle = k_B \delta_{ij} \tag{37}$$

We now return to the transport equations. The time derivatives of the variables α_i are proportional to observable currents such as the heat currents considered in the first part of this subsection. We assume again that a linearized equation

$$j_i = \frac{d\alpha_i}{dt} = \sum_j L_{ij}X_j \tag{38}$$

will describe the response of the system to generalized forces X_j. These generalized forces can be expressed in terms of the entropy through $X_j = \partial S / \partial \alpha_j$. Thus

$$\frac{d\alpha_i}{dt} = \sum_j L_{ij}\left(\frac{\partial S}{\partial \alpha_j}\right) \tag{39}$$

Consider, once again, the expectation value

$$\langle \alpha_i(t + \Delta t)\alpha_j(t)\rangle = \langle \alpha_i(t)\alpha_j(t)\rangle + \Delta t \langle \dot{\alpha}_i(t)\alpha_j(t)\rangle$$

$$= \langle \alpha_i(t)\alpha_j(t)\rangle + \Delta t \sum_m L_{im}\left\langle \frac{\partial S}{\partial \alpha_m}\alpha_j\right\rangle \tag{40}$$

$$= \langle \alpha_i(t)\alpha_j(t)\rangle - \Delta t\, L_{ij}k_B$$

Similarly,

$$\langle \alpha_j(t + \Delta t)\alpha_i(t)\rangle = \langle \alpha_j(t)\alpha_i(t)\rangle - \Delta t\, L_{ji}k_B \tag{41}$$

which implies, by the principle of microscopic reversibility, that

$$L_{ij} = L_{ji} \tag{42}$$

Thus, quite generally, we have arrived at the symmetry relations of the linear transport coefficients.

8.D THE BOLTZMANN EQUATION

In this section we discuss transport theory from the point of view of the Boltzmann equation. This approach lacks the general validity of the Kubo formalism. On the other hand, the different terms in the Boltzmann equation have a straightforward physical interpretation and the approach leads to explicit results. Our discussion will be rather brief and we will limit ourselves to situations in which linear response theory is applicable. We refer the reader to Ziman [1964] and Callaway [1974] for a more extensive discussion. We develop the formalism in Section 8.D(a), discuss dc conductivity in Section 8.D(b), and thermoelectric effects in Section 8.D(c).

(a) Fields, Drift, and Collisions

We assume, in the spirit of mean field theory, that the system of interest can be adequately described in terms of the single-particle distribution, $f_p(\mathbf{r})$, which is the density of particles with momentum $\mathbf{p}$ at position $\mathbf{r}$. The distribution is normalized (in accordance with our phase space definitions of Section 2.A), so that

$$\frac{1}{h^3}\int d^3p \int d^3r\, f_p(\mathbf{r}) = N \tag{1}$$

where N is the number of particles.

In the case of a quantum system, such as electrons in a crystalline solid, a particle with momentum **p** is represented as a Bloch function with wave vector $\mathbf{k} = \mathbf{p}/\hbar$, energy $\epsilon_\mathbf{k}$, and velocity $\mathbf{v}_\mathbf{p} = \partial\epsilon_\mathbf{k}/\partial\hbar\mathbf{k}$. Because of the uncertainty principle, we must assume that $f_\mathbf{p}(\mathbf{r})$ is coarse grained over a sufficiently large volume that **r** can be considered to be a *macroscopic* variable. This is consistent with the fundamental assumption, made elsewhere in this chapter, that densities and fields are sufficiently slowly varying that they are well defined locally. We will take a *semiclassical* approach to quantum systems, that is, we assume that in the presence of macroscopic electric and magnetic fields the equation of motion for particles of charge e is

$$\dot{\mathbf{p}} = \hbar\dot{\mathbf{k}} = e\left(\mathbf{E} + \frac{1}{c}\mathbf{v}_\mathbf{k} \times \mathbf{B}\right) \tag{2}$$

If the acceleration of the particles due to the fields were the only effect causing changes in the distribution function, we would have

$$f_\mathbf{p}(t + \delta t) = f_{\mathbf{p}-\delta\mathbf{p}}(t)$$

or

$$\left.\frac{\partial f_\mathbf{p}(\mathbf{r})}{\partial t}\right|_{\text{field}} = -e\left(\mathbf{E} + \frac{1}{c}\mathbf{v} \times \mathbf{B}\right)\frac{\partial f_\mathbf{p}}{\partial \mathbf{p}} \tag{3}$$

If we take the fields to be slowly varying, we can visualize the states as wave packets which are accelerated by the fields according to the classical equations of motion. This assumption is difficult to justify rigorously and we will not attempt to do so. In some cases, such as in inhomogeneous semiconductors, near surfaces, or in insulators subjected to intense fields, the electric fields are strong enough to cause *tunneling*. The semiclassical approach is then not appropriate.

Particles from time to time undergo *collisions* with obstacles in their path. Electrons in solids are scattered by impurities, vacancies, dislocations, and phonons. Because of screening, the interactions responsible for scattering are generally short ranged and this implies that collisions should be treated quantum mechanically. Let $W(\mathbf{k}, \mathbf{k}')$ be the transition rate from state **k** to state **k**'. This transition rate is typically calculated approximately by means of the golden rule (Problem 8.8). The distribution function then changes in time due to transitions *into* and *out of* state k and we may write

$$\left(\frac{\partial f_\mathbf{k}}{\partial t}\right)_{\text{coll}} = \sum_{\mathbf{k}'} \left[f_{\mathbf{k}'}(1 - f_\mathbf{k})W(\mathbf{k}', \mathbf{k}) - f_\mathbf{k}(1 - f_{\mathbf{k}'})W(\mathbf{k}, \mathbf{k}')\right]$$

From the principle of detailed balance, we have, for elastic processes, $W(\mathbf{k}, \mathbf{k}') = W(\mathbf{k}', \mathbf{k})$ and

$$\left(\frac{\partial f_\mathbf{k}}{\partial t}\right)_{\text{coll}} = \sum_{\mathbf{k}'} \{f_{\mathbf{k}'} - f_\mathbf{k}\}W(\mathbf{k}', \mathbf{k}) \tag{4}$$

In the rest of this section we consider only a system of fermions and will label states by wave vectors $\mathbf{k}$ rather than momenta $\mathbf{p}$, as we have done in (4). In what follows we also limit ourselves to the simplest treatment of collisions and work within the *relaxation time* approximation.

We assume that the external fields produce only a small change in the distribution function and write

$$f_{\mathbf{k}} = f_{\mathbf{k}}^0 + g_{\mathbf{k}} \tag{5}$$

where $f_{\mathbf{k}}^0$ is the equilibrium (zero-field) distribution given, as appropriate by the Boltzmann, Bose–Einstein, or, in our case, the Fermi–Dirac distributions. In the relaxation-time approximation one assumes that if the external fields were switched off, the nonequilibrium part of the distribution function would decay exponentially with time:

$$g_{\mathbf{k}}(t) = g_{\mathbf{k}}(0)e^{-t/\tau}$$

where τ is the relaxation time. We thus obtain

$$\frac{\partial f_{\mathbf{k}}^0}{\partial t}\bigg|_{\text{coll}} = 0 \qquad \frac{\partial g_{\mathbf{k}}}{\partial t} = -\frac{g_{\mathbf{k}}}{\tau} \tag{6}$$

If the distribution is inhomogeneous, it will change in time due to *drift*. If the particles are not subject to any forces,

$$f_{\mathbf{k}}(\mathbf{r}, t + \delta t) = f_{\mathbf{k}}(\mathbf{r} - \mathbf{v}_{\mathbf{k}}\,\delta t, t) \tag{7}$$

and hence

$$\frac{\partial f_{\mathbf{k}}}{dt}\bigg|_{\text{drift}} = -\mathbf{v}_{\mathbf{k}} \cdot \frac{\partial f_{\mathbf{k}}}{\partial \mathbf{r}} \tag{8}$$

Combining the various terms, we obtain the *Boltzmann equation* for the distribution function:

$$\frac{df_{\mathbf{k}}}{dt} = \frac{\partial f_{\mathbf{k}}}{\partial t}\bigg|_{\text{field}} + \frac{\partial f_{\mathbf{k}}}{\partial t}\bigg|_{\text{coll}} + \frac{\partial f_{\mathbf{k}}}{\partial t}\bigg|_{\text{drift}} \tag{9}$$

In a steady-state situation we require $df_{\mathbf{k}}/dt = 0$.

(b) DC Conductivity of a Metal

We next assume that for a weak electric field the nonequilibrium part, $g_{\mathbf{k}}$, is proportional to the field and linearize the Boltzmann equation. This yields

$$-\frac{e\mathbf{E}}{\hbar} \cdot \frac{\partial f_{\mathbf{k}}^0}{\partial \mathbf{k}} - \frac{g_{\mathbf{k}}}{\tau} = 0$$

or

$$g_{\mathbf{k}} = -\frac{e\mathbf{E}\tau}{\hbar} \frac{\partial f_{\mathbf{k}}^0}{\partial \epsilon_{\mathbf{k}}} \cdot \frac{\partial \epsilon_{\mathbf{k}}}{\partial \mathbf{k}} = -\tau e\mathbf{E} \cdot \mathbf{v}_{\mathbf{k}} \frac{\partial f_{\mathbf{k}}^0}{\partial \epsilon_{\mathbf{k}}} \tag{10}$$

Note that we have taken the charge of an electron to be e. For a metal at not too high a temperature,

$$\frac{\partial f^0}{\partial \epsilon} \approx -\delta(\epsilon - \epsilon_F) \tag{11}$$

The electrical current density is given by

$$\mathbf{j} = \frac{2}{V} \sum_{\mathbf{k}} e \mathbf{v_k} g_{\mathbf{k}} \tag{12}$$

Let S_ϵ be a constant energy surface, that is, the surface in k-space for which $\epsilon_{\mathbf{k}} = \epsilon$ and let $\hat{n}$ be the unit vector normal to S_ϵ. We then have

$$\sum_{\mathbf{k}} = \frac{V}{(2\pi)^3} \int d^3k = \frac{V}{(2\pi)^3} \int d\epsilon \int dS_\epsilon \left(\hat{n} \cdot \frac{\partial \mathbf{k}}{\partial \epsilon_{\mathbf{k}}} \right) \tag{13}$$

and obtain, for the current density,

$$\mathbf{j} = \frac{e^2 \tau}{4\pi^3 \hbar} \int \frac{dS_F}{|v_{\mathbf{k}}|} \mathbf{v_k}(\mathbf{v_k} \cdot \mathbf{E}) \tag{14}$$

where S_F is the Fermi surface. This in turn yields the conductivity (in dyadic form):

$$\sigma = \frac{e^2 \tau}{4\pi^3 \hbar} \int dS_F \frac{\mathbf{v_k} : \mathbf{v_k}}{|\mathbf{v_k}|} \tag{15}$$

This formula illustrates that the conductivity is, in general, a tensor. In component form

$$j_i = \sum_m \sigma_{im} E_m \tag{16}$$

and we have, for example,

$$\sigma_{xx} = \frac{e^2 \tau}{4\pi^3 \hbar} \int dS_F \frac{(v_{\mathbf{k}})_x^2}{|\mathbf{v_k}|} \tag{17}$$

In discussions of systems that lack spherical symmetry, one finds that some authors use anisotropic relaxation times $\tau(\mathbf{k})$. The relaxation time must then be kept inside the integral. In the references listed at the beginning of this section it is pointed out that there are serious difficulties with such an approach.

In an isotropic system the conductivity tensor is diagonal and

$$\sigma = \frac{e^2 \tau}{12\pi^3 \hbar} \int dS_F \, v_F \tag{18}$$

We write $v_F = \hbar k_F/m^*$ for the Fermi velocity, where m^* is the *effective mass* and, using $n = N/V = k_F^3/3\pi^2$, obtain

$$\sigma = \frac{ne^2 \tau}{m^*} \tag{19}$$

which is of the same form as the conductivity of the simple Drude model (see, e.g., Ashcroft and Mermin, 1976).

It is useful to interpret the result

$$f_k = f_k^0 + g_k = f_k^0 - \tau e (\mathbf{E} \cdot \mathbf{v_k}) \frac{\partial f_k^0}{\partial \epsilon_k} \tag{20}$$

in a different way. To first order in E, we can rewrite (20) in the form

$$f_k = f_k^0 \{\epsilon_k - \tau e (\mathbf{E} \cdot \mathbf{v_k})\} \tag{21}$$

The right-hand side of this equation is simply the equilibrium distribution of the system with all energies shifted by an amount

$$\delta \epsilon_k = \tau e (\mathbf{E} \cdot \mathbf{v_k}) \tag{22}$$

which is precisely the amount expected classically for particles moving with constant velocity $\mathbf{v}$ for a time τ in a force field $e\mathbf{E}$. The extra energy gained in this way can be interpreted in terms of a drift velocity $\delta \mathbf{v_k}$ in the direction of the field so that

$$\delta \mathbf{v_k} \cdot \frac{\partial \epsilon_k}{\partial \mathbf{v_k}} = e\tau (\mathbf{v_k} \cdot \mathbf{E}) \tag{23}-$$

If

$$\frac{\partial \epsilon_k}{\partial \mathbf{v_k}} = \mathbf{p_k} = m^* \mathbf{v_k} \tag{24}$$

we obtain

$$\delta \mathbf{v_k} = \frac{e\tau}{m^*} \mathbf{E} \tag{25}$$

For n particles per unit volume, we have for the current,

$$\mathbf{j} = ne \, \delta \mathbf{v} \tag{26}$$

and we recover (19) for the conductivity.

In the case of a metal the drift velocities are typically very small compared to the Fermi velocity v_F, mainly because the electric fields inside a metal tend to be small. In a semiconductor one sometimes deals with fields which are large enough that nonohmic effects are important. It is then not adequate to linearize the Boltzmann equation and one must consider the nonlinear problem. In such situations collisions often occur so frequently that one cannot describe them as independent events and the whole Boltzmann approach becomes suspect.

(c) Thermal Conductivity and Thermoelectric Effects

To this point we have only considered a situation in which the distribution function was spatially uniform. To give an example of the use of the Boltzmann equation when the drift term comes into play, we discuss the case of a time-independent temperature gradient maintained across a metallic sample. From Fourier's law we expect that, in analogy with Ohm's law, there will be a heat current whenever there is a temperature gradient:

$$\mathbf{j}_Q = L_{QQ} \nabla \left(\frac{1}{T} \right) = -\kappa \nabla T \tag{27}$$

where κ is the thermal conductivity and L_{QQ} the kinetic coefficient defined in Section 8.C. We also allow for an electric field $\mathbf{E}$, with corresponding scalar potential $\phi(\mathbf{r})$. From thermodynamics we have

$$T \, ds = du - \mu' dn \tag{28}$$

where $\mu' = \mu + e\phi(\mathbf{r})$ is the electrochemical potential at point $\mathbf{r}$ and where s, u, n are the entropy, energy, and particle densities. Hence

$$\mathbf{j}_Q = \mathbf{j}_\epsilon - \mu' \mathbf{j}_N \tag{29}$$

We assume that the heat current is due entirely to the motion of electrons and neglect the lattice thermal conductivity. In the presence of the electrostatic potential $\phi(\mathbf{r})$ the electronic energies will be "locally" shifted by an amount $e\phi(\mathbf{r})$ so that the *energy current* density is then given by

$$\mathbf{j}_\epsilon(\mathbf{r}) = \frac{2}{(2\pi)^3} \int d^3k \, (\epsilon_\mathbf{k} + e\phi(\mathbf{r})) \mathbf{v}_\mathbf{k} f_\mathbf{k}(\mathbf{r}) \tag{30}$$

and the particle current density is

$$\mathbf{j}_N(\mathbf{r}) = \frac{2}{(2\pi)^3} \int d^3k \, \mathbf{v}_\mathbf{k} f_\mathbf{k}(\mathbf{r}) \tag{31}$$

which, of course, also implies an electrical current density $\mathbf{j}_c = e\mathbf{j}_N$. Thus the heat current (29) is given by

$$\mathbf{j}_Q(\mathbf{r}) = \frac{2}{(2\pi)^3} \int d^3k [\epsilon_\mathbf{k} - \mu] \mathbf{v}_\mathbf{k} f_\mathbf{k}(\mathbf{r}) \tag{32}$$

As before we write

$$g_\mathbf{k}(\mathbf{r}) = f_\mathbf{k}(\mathbf{r}) - f_\mathbf{k}^0$$

and, in addition, assume that the thermal gradient is small enough that it is meaningful to talk about a local temperature and a local chemical potential. With these assumptions, the Boltzmann equation becomes

$$-\mathbf{v}_\mathbf{k} \cdot \frac{\partial f_\mathbf{k}}{\partial \mathbf{r}} - \frac{e}{\hbar} \mathbf{E} \cdot \frac{\partial f_\mathbf{k}}{\partial \mathbf{k}} + \frac{\partial f_\mathbf{k}}{\partial t} \bigg|_{\text{coll}} = 0 \tag{33}$$

We take $f_\mathbf{k}^0(\mathbf{r})$ to be the equilibrium distribution with the local temperature $T(\mathbf{r})$ and the local chemical potential $\mu'(\mathbf{r})$ controlling the density at point $\mathbf{r}$. Noting that $\epsilon_\mathbf{k} + e\phi(\mathbf{r}) - \mu'(\mathbf{r}) = \epsilon_\mathbf{k} - \mu(\mathbf{r})$, we have

$$f_\mathbf{k}^0(\mathbf{r}) = f^0\{\epsilon_\mathbf{k}, \mu(\mathbf{r}), T(\mathbf{r})\} = \left[\exp \left\{ \frac{\epsilon_\mathbf{k} - \mu(\mathbf{r})}{k_B T(\mathbf{r})} \right\} + 1 \right]^{-1} \tag{34}$$

and hence

$$\frac{\partial f_k^0}{\partial \mathbf{r}} = \frac{\partial f_k^0}{\partial T}\frac{\partial T}{\partial \mathbf{r}} + \frac{\partial f_k^0}{\partial \mu}\frac{\partial \mu}{\partial \mathbf{r}} \tag{35}$$

We next make the relaxation-time approximation (7) and in the spirit of the linearized Boltzmann equation, neglect terms such as

$$\frac{\partial g}{\partial \mathbf{r}} \quad \text{and} \quad \frac{e}{\hbar}\mathbf{E}\cdot\frac{\partial g}{\partial \mathbf{k}}$$

Using

$$\frac{\partial f_k^0}{\partial T} = -\frac{\epsilon_k - \mu}{T}\frac{\partial f_k^0}{\partial \epsilon_k} \qquad \frac{\partial f_k^0}{\partial \mu} = -\frac{\partial f_k^0}{\partial \epsilon_k}$$

and collecting terms, we obtain

$$\frac{1}{\tau}g_k = -\frac{\partial f_k^0}{\partial \epsilon_k}\mathbf{v}_k\cdot\left[-\frac{\epsilon_k - \mu}{T}\frac{\partial T}{\partial \mathbf{r}} + \left(e\,\mathbf{E} - \frac{\partial \mu}{\partial \mathbf{r}}\right)\right] \tag{36}$$

The potential difference measured by, say, a voltmeter is not

$$\int \mathbf{E}\cdot d\mathbf{s}$$

but rather the quantity

$$\Psi = \int \left(\mathbf{E} - \frac{1}{e}\nabla\mu\right)\cdot d\mathbf{s}$$

We therefore introduce the "electromotive field" or "observed" field

$$\mathscr{E} = \mathbf{E} - \frac{1}{e}\nabla\mu = -\frac{1}{e}\nabla\mu' \tag{37}$$

Clearly, $\mathscr{E}$ is of more interest than the electric field $\mathbf{E}$ itself. We now define the kinetic coefficients L_{CC}, L_{CQ}, L_{QC}, and L_{QQ} through

$$\mathbf{j}_C = L_{CC}\,\mathscr{E} + L_{CQ}\,\nabla\!\left(\frac{1}{T}\right)$$

$$\mathbf{j}_Q = L_{QC}\,\mathscr{E} + L_{QQ}\,\nabla\!\left(\frac{1}{T}\right) \tag{38}$$

Using (31) and (32), we see that the kinetic coefficients can all be expressed in terms of the integral

$$I_\alpha = \int d\epsilon\left(-\frac{\partial f^0}{\partial \epsilon}\right)(\epsilon - \mu)^\alpha \sigma(\epsilon) \tag{39}$$

where

$$\sigma(\epsilon) = e^2\tau\int \frac{d^3k}{4\pi^3}\delta(\epsilon - \epsilon_k)\mathbf{v}_k : \mathbf{v}_k \tag{40}$$

is the generalized energy-dependent form of the conductivity tensor (15). We shall evaluate I_α for conditions appropriate to a metal and in this case

$$-\frac{\partial f^0}{\partial \epsilon} = \frac{\beta \exp\{\beta(\epsilon - \mu)\}}{[\exp\{\beta(\epsilon - \mu)\} + 1]^2} \tag{41}$$

can be taken to be nonzero only in a narrow energy range of order $k_B T$ around ϵ_F. We introduce the new variable $z = \beta(\epsilon - \mu)$ and expand

$$\sigma(k_B Tz + \mu) = \sigma(\mu) + k_B Tz\frac{\partial \sigma}{\partial \mu} + \cdots$$

Substituting in (40), we then have the transport coefficients expressed in terms of

$$I_\alpha \approx (k_B T)^\alpha \int_{-\infty}^\infty dz \frac{z^\alpha e^z}{(1 + e^z)^2}\left[\sigma(\mu) + k_B Tz\frac{\partial \sigma}{\partial \mu}\right] \tag{42}$$

Defining

$$Q_j = \int_{-\infty}^\infty dz \frac{z^j}{(e^z + 1)(e^{-z} + 1)}$$

we have $Q_0 = 1$, $Q_1 = 0$, $Q_2 = \pi^2/3$, and $Q_3 = 0$. Taking $\mu \approx \epsilon_F$, we thus obtain

$$L_{CC} = \sigma(\epsilon_F) = \sigma \tag{43a}$$

$$L_{CQ} = TL_{QC} = \frac{\pi^2}{3e}k_B^2 T^3 \frac{\partial \sigma(\epsilon)}{\partial \epsilon}\bigg|_{\epsilon = \epsilon_F} \tag{43b}$$

$$L_{QQ} = \frac{\pi^2}{3e^2}k_B^2 T^3 \sigma \tag{43c}$$

We see that in our approximate treatment we obtain $L_{CQ} = TL_{QC}$, which is an Onsager relation, as can easily be shown (Problem 8.9). We note also that for electrons ($e = -|e|$), L_{QC} and L_{CQ} are negative. If these coefficients are found experimentally to be positive, it is an indication that the charge carriers are holes.

To obtain the thermal conductivity we require that there be no electric current, or from (38),

$$\mathscr{E} = -L_{CC}^{-1}L_{CQ}\nabla\left(\frac{1}{T}\right) \tag{44}$$

Substituting into the second equation of (38), we have for the thermal conductivity

$$\kappa = \frac{L_{QQ} - L_{QC}L_{CC}^{-1}L_{CQ}}{T^2} \tag{45}$$

We now argue that in a metal the second term in (45) is small compared to the first. We first note that the second law of thermodynamics implies that κ

is positive or $L_{CC}L_{QQ} > L_{CQ}L_{QC}$. To obtain an order-of-magnitude estimate, we make the approximation (on dimensional grounds) $\partial\sigma/\partial\epsilon \sim \sigma/\epsilon$ (for free electrons $\partial\sigma/\partial\epsilon = 3\sigma/2\epsilon$) and thus have

$$\frac{L_{CQ}L_{QC}}{L_{CC}L_{QQ}} \approx \frac{\pi^2}{3}\left(\frac{k_B T}{\epsilon_F}\right)^2 \sim 10^{-4}$$

for a typical metal at room temperature. Neglecting the second term in (45), we therefore find

$$\kappa = \frac{L_{QQ}}{T^2} = \frac{\pi^2}{3e^2}k_B^2 T\sigma \tag{46}$$

This result is known as the Wiedemann–Franz law and is often derived from more elementary considerations (see, e.g., Ashcroft and Mermin, 1976). However, the present derivation suggests that this law should hold under quite general circumstances. The main assumption made above was the use of the relaxation-time approximation, which relies, in turn, on the assumption of elastic scattering. The dominant inelastic processes are emission and absorption of phonons and it is interesting that the most significant deviations from the Wiedemann–Franz law occur near the Debye temperature. At lower temperatures most electron–phonon processes are frozen out, while at room temperature or higher, phonon energies are small compared to $k_B T$, and the scattering processes can be taken to be "quasi elastic."

The coupled transport equations (38) suggest other thermoelectric effects such as the Seebeck and Peltier effects. We refer to the literature for a discussion of these.

PROBLEMS

8.1. *Thomas–Fermi and Debye Screening.* If an external potential energy $\phi(\mathbf{r})$ varies sufficiently slowly the chemical potential must satisfy

$$\mu = \mu_0(n(\mathbf{r})) + \phi(\mathbf{r}) = \text{constant}$$

where μ_0 is the chemical potential of a system of particles of density $n(\mathbf{r})$ in the absence of the external field.

(a) Consider an electron gas at $T = 0$ with electron states filled up to a Fermi level $\epsilon_F = \hbar^2 k_F^2/2m$, where k_F is the Fermi wave vector. Show that in these circumstances the free particle static susceptibility is given by

$$\chi_0(q) = -\frac{\partial n_0}{\partial \epsilon_F} = -\frac{1}{2\pi^2}\left(\frac{2m}{\hbar^2}\right)^{3/2}\epsilon_F^{1/2}$$

(b) Show that the mean field dielectric function can be expressed in the form

$$\epsilon(\mathbf{q}) = 1 + \frac{k_{TF}^2}{q^2}$$

where $k_{TF} = [4me^2 k_F/(\pi\hbar^2)]^{1/2}$ is the Thomas–Fermi wave vector.

(c) Show that if $\phi_{ext} = -e^2/r$ is the potential energy due to an external point charge, the mean field effective potential becomes

$$\phi_{tot}(\mathbf{r}) = -\frac{e^2}{r} e^{-k_{TF}r}$$

(d) Consider a classical gas of charged particles at temperature T. Show that with the same approximations

$$\phi_{tot}(\mathbf{r}) = -\frac{e^2}{r} e^{-r/\lambda_D}$$

where λ_D is the Debye screening length given by $\lambda_D = (4\pi n_0 e^2/k_B T)^{1/2}$.

8.2. *Pair Distribution Function of an Ideal Bose Gas.*
   (a) Compute the dynamic structure factor $S(\mathbf{q}, \omega)$ for an ideal Bose gas at $T \neq 0$.
   (b) Find the expression for the geometric structure factor (i) by integrating the expression for $S(\mathbf{q}, \omega)$ over frequency and (ii) by evaluating the expression for the geometric structure factor directly.
   (c) Evaluate numerically the pair distribution function $g(r)$ for an ideal Bose gas for a temperature (i) above and (ii) below the Bose condensation temperature (see Section 7.A). Show that in the latter case $g(r)$ does not approach unity as $r \to \infty$.

8.3. *Mean Field Approximation for the Bose Gas.*
   (a) Show that the mean field response function (8.B.19) satisfies the f-sum rule.
   (b) The mean field theory for the weakly interacting Bose gas gives rise to a pair distribution function which is quite unphysical at short distances. Show that (8.B.20) yields a form for $g(r)$ which diverges as $r \to 0$. This result clearly suggests the need to introduce local field corrections of the type (8.B.11) in a more realistic theory.

8.4. *Dispersion of Plasma Oscillations.* Extend the calculation of the mean field response of an electron gas at $T = 0$ for $\omega >> \hbar^2 q^2/2m$ to next order in q and show that

$$\Omega_{pl}(q) = \Omega_{pl}(0)\left[1 + \frac{3q^2 v_F^2}{10\Omega_{pl}^2(0)} + \cdots \right]$$

where $v_F = \hbar k_F/m$ is the Fermi velocity and $\Omega_{pl}^2(0) = 4\pi e^2 n/m$.

8.5. *Screening in Two Dimensions.* Consider a two-dimensional system of electrons distributed on a surface of area A at $T = 0$ and with a compensating uniform positive background ensuring overall neutrality.
   (a) Show that the two-dimensional Fourier transform of the Coulomb potential is $v_q = 2\pi e^2/q$.
   (b) Show that the static free-particle susceptibility is

$$\chi_0(\mathbf{q}, 0) = -\frac{m}{\hbar^2 \pi} \qquad\qquad q < 2k_F$$

$$= -\frac{m}{\hbar^2 \pi}\left[1 - \left(1 - \frac{4k_F^2}{q^2}\right)^{1/2}\right] \qquad q > 2k_F$$

   (c) Consider a point charge e a distance z above the layer of electrons. Show that the two-dimensional Fourier transform of the bare Coulomb potential from this charge in the electron gas is $2\pi e^2 \exp\{-qz\}/q$.

(d) Find expressions for the mean field screened potential and for the screening charge distribution in the electron layer.

8.6. *Low-Temperature Properties of the Heisenberg Model.* Consider a spin system described by the isotropic Heisenberg model with spin wave excitation spectrum $\epsilon(\mathbf{q}) = J\hbar Sa^2q^2$. Assume that the mean number of spin waves with wave vector $\mathbf{q}$ obeys the Bose–Einstein distribution function. The total magnetization is reduced from its saturation value $N\hbar S$ by $\hbar$ for each spin wave that is excited.

(a) Show that with these assumptions the low-temperature spontaneous magnetization is of the form

$$M(T) = M(0)(1 - \text{const. } T^{3/2})$$

in three dimensions. Find the value of the multiplicative constant.

(b) Show that the integral which gives the deviation of the spontaneous magnetization from the saturation value diverges in one and two dimensions. This result is an indication of the absence of long-range order in the Heisenberg model in one and two dimensions. For a proof not based on spin wave theory, see Mermin and Wagner [1966].

8.7. *Symmetry of Transport Coefficients.* Consider the thermal conductivity tensor L defined in (8.C.20).

(a) Show that in the case of a cubic crystal the symmetry of the lattice requires that this tensor be diagonal.

(b) Show that for a hexagonal crystal $L_{12} = -L_{21}$, $L_{11} = L_{22}$, and $L_{i3} = L_{3i} = 0$ for $i = 1, 2$, where the 3 axis is perpendicular to the plane of hexagonal symmetry.

8.8. *Estimate of Relaxation Time for Impurity Scattering.* Consider a situation in which the transition rate for impurity scattering can be described by plane-wave matrix elements of a spherically symmetric impurity potential $u(\mathbf{r})$:

$$W(\mathbf{k}, \mathbf{k}') = \frac{2\pi}{\hbar} |\langle \mathbf{k}' | u(\mathbf{r}) | \mathbf{k} \rangle|^2 \, \delta(\epsilon_{\mathbf{k}'} - \epsilon_{\mathbf{k}})$$

and let

$$u_{\mathbf{q}} = \int d^3r \, e^{-i\mathbf{q}\cdot\mathbf{r}} u(\mathbf{r})$$

(a) Consider a free electron gas at $T = 0$ with n_i impurities per unit volume. Show that substitution of

$$g_{\mathbf{k}} = -\tau e \, \mathbf{E} \cdot \mathbf{v}_{\mathbf{k}} \frac{\partial f^0}{\partial \epsilon}$$

into the linearized Boltzmann equation yields

$$\frac{1}{\tau} = \frac{n_i}{4\pi^2 v_F \hbar^2} \int d^2S_F \, u_{\mathbf{k}-\mathbf{k}'}^2 \left(1 - \frac{\mathbf{k} \cdot \mathbf{k}'}{k_F^2}\right)$$

where $\mathbf{k}$ lies on the Fermi surface, the integration over $\mathbf{k}'$ extends over the Fermi surface, and $v_F = \hbar k_F/m$ is the Fermi velocity.

(b) Estimate the low-temperature resistivity in ohms/meter of a sample of 0.1 % Mg in Na. Since Mg has a valence $Z = 2$ (one more than Na), the impurity potential can be taken to be that of a Thomas–Fermi screened point charge e (use Lindhard screening if you have easy access to a computer). The Fermi wave vector of Na is $0.92 \, \text{Å}^{-1}$.

8.9. *Onsager Relation for the Thermoelectric Effect.* To obtain the Onsager relation (8.D.43b) in the situation in which there is a thermal gradient as well as an electric field, one writes, as in (8.D.28),

$$T\,ds = du - \mu'\,dn$$

where $\mu' = \mu + e\phi(\mathbf{r})$ and where μ is the chemical potential in the absence of an electric field. The energy and particle currents, defined as in Section 8.C, then obey the phenomenological equations

$$\mathbf{j}_E = L_{EE}\,\nabla\!\left(\frac{1}{T}\right) + L_{EN}\,\nabla\left(-\frac{\mu'}{T}\right)$$

$$\mathbf{j}_N = L_{NE}\,\nabla\!\left(\frac{1}{T}\right) + L_{NN}\,\nabla\left(-\frac{\mu'}{T}\right)$$

with an Onsager relation $L_{NE} = L_{EN}$. Show that the relation (8.D.43b) $L_{CQ} = TL_{QC}$ follows.

9

Disordered Systems

Real materials are seldom, if ever, the idealized pure systems that we have discussed to this point. Magnetic crystals invariably contain defects and nonmagnetic impurities. Liquids, which we have generally taken to be composed of a single component, invariably have impurities dissolved in them. Even liquid helium will have a certain amount of isotopic disorder. Thus it is important to understand the effect of disorder on the properties of materials and to what extent the theoretical framework that we have constructed remains valid when we attempt to describe real materials.

The physics of disordered systems is a vast subject with an extensive literature. In one chapter we will hardly be able to give a comprehensive treatment of the material, and the reader may wish to consult, for example, the book by Ziman [1979] or one of the review articles that we shall mention when we discuss specific topics.

The effects of disorder begin at the microscopic level. The energy levels and wave functions of a particle in a disordered medium can be substantially different from those in a pure material. In our discussion of electrons in metals we have, to this point, assumed translational invariance. This assumption is, for the statistical mechanics of a pure metal, quite adequate. The periodic potential due to the ions changes the shape of the Fermi surface and is responsible for the detailed properties of individual materials. The incorporation of such effects into our statistical formalism presents us only with minor technical problems; the basic physical picture is the same as that of the idealized system. Similarly, the use of realistic force constants, rather than springs between nearest neighbors, in the vibrational energy of a crystal is a conceptually trivial change. The translational symmetry of the Hamiltonian makes both of these problems in principle very simple, although technical difficulties may arise in actual computations.

The situation is dramatically different in a disordered material. One can show, for example, that in a one-dimensional disordered material the electronic wave functions are *localized* rather than *extended* as they are in a periodic potential. In two and three dimensions we believe that some, if not all, of the states of a disordered system are localized. This has important implications for transport coefficients. Intuitively, one expects that if the Fermi level falls into a range of localized states, the conductivity will decrease as the temperature is lowered—in contrast to the behavior of a pure material in which the *resistivity* decreases due to the freezing out of phonon scattering. We discuss the properties of single-particle states in Section 9.A.

In our discussion of phase transitions we have noted that a true phase transition can occur only in the thermodynamic limit. Any finite system has a partition function which is a smooth function of its variables. Consider now a crystal that has a certain concentration of magnetic atoms interacting with each other through short-range exchange interactions. Clearly, if the concentration of magnetic atoms is too small, they will be for the most part isolated, and even in the thermodynamic limit, no infinite cluster of interacting magnetic atoms will exist. Thus one of the important aspects of disordered systems is the geometry or connectivity of random clusters. The question as to when a system of randomly occupied lattice sites forms an infinite connected cluster, what the dimensionality and other properties of this cluster are, is the subject of *percolation* theory, which is discussed in Section 9.B.

Another question that one can consider is the effect of randomness on the critical behavior of a system. Are the critical exponents the same as in a pure material? Does the phase transition remain well defined (i.e., occur at a unique critical temperature T_c), or is it smeared out over some temperature interval? We will briefly discuss these questions in Section 9.C.

To this point we have mentioned effects of disorder which are partially understood and for which a certain number of analytical results exist, at least for simplified models. A subject that is far less well understood is the statistical mechanics of glassy or amorphous materials. Only phenomenological and partially successful theories for the glass transition exist. This is also the case for spin glasses—a set of materials that display singularities in some thermodynamic properties at well-defined temperatures but without simply ordered low-temperature phases. We discuss some aspects of such systems in Section 9.D.

Before beginning our treatment of these topics we note that, from a statistical point of view, disorder is often classified as *annealed* or *quenched*. By quenched disorder we mean that randomly distributed impurities or defects do not equilibrate in the temperature range or time interval in which the properties of the system which are of direct interest are changing. An example of this would be a solid solution of magnetic and nonmagnetic atoms at very low temperature. The diffusion of the two atomic species is then a very slow process, sometimes involving time constants comparable to the age of the universe, and usually slower than other processes, such as the demagnetization of the sample due, for example, to excitation of spin waves as the temperature is increased. Annealed disorder refers to the converse situation in which the distribution that describes the disorder is also changing in the temperature or time interval of interest. An example of this might be the case of β-brass (Section 3.C) near the order–disorder transition. If one is interested in the electronic properties of CuZn near 740 K, one must take the variation of the atomic distribution into account, and in a fundamental approach to this system, one would of course attempt to derive the effective interactions that drive the order–disorder transition from the band structure of the system. In the following sections we take disorder to be quenched unless we state explicitly that it is annealed.

9.A SINGLE-PARTICLE STATES IN DISORDERED SYSTEMS

To have a concrete model to work with, we consider the following simple tight-binding Hamiltonian (a description of the tight-binding method can be found in most solid-state physics texts, see, e.g., Ashcroft and Mermin [1976]).

$$H = - \sum_{\langle ij \rangle} t_{ij}(c_i^+ c_j + c_j^+ c_i) + \sum_j \epsilon_j c_j^+ c_j \tag{1}$$

where the c's are the usual fermion creation and annihilation operators (see the Appendix) and where we have ignored the electron spin. The sum in (1) is over the sites of a perfect lattice and disorder is introduced by taking either, or both, of the coefficients ϵ_j or t_{ij} to be random variables subject to some distribution. The second term in (1) represents a set of "atomic" levels with energies ϵ_j and corresponding Wannier functions centered on sites j. The first term represents a covalent lowering of the energy due to overlap of atomic orbitals on neighboring sites. In what follows we take the disorder to be "site diagonal" and take $t_{ij} = t$ to be a constant for i, j nearest neighbors and zero otherwise. The on-site energies ϵ_j will be taken to be ϵ_A with probability p_A and ϵ_B with probability p_B, independent of the site j. In this case (1) is a crude model for a *substitutionally* disordered binary alloy. In situations in which the Wannier representation is appropriate, the independent electron approximation is generally unrealistic. In particular, electron correlations are essential for an understanding of the origin of magnetic effects. However, we wish to concentrate on the effects of disorder and ignore these complications. Also, in a more realistic model the hopping matrix element t_{ij} would depend on the nature of the atoms occupying sites i and j and we would also expect to see a certain amount of clustering of atoms in solid solutions (see Section 3.C). Hence at least the probabilities p_{AA}, p_{AB}, and p_{BB} for the occurrence of AA, AB, and BB pairs should be specified in order to make the model reasonably realistic. However, even with our simplified version we will be able to demonstrate some of the effects of disorder.

(a) Electron States in One Dimension

Consider the two limiting cases $p_A = 1$ and $p_B = 1$. In this case the Hamiltonian (1) can be written in the form

$$H = -t \sum_i (c_{i+1}^+ c_i + c_i^+ c_{i+1}) + \epsilon_{A,B} \sum_i c_i^+ c_i \tag{2}$$

We easily obtain a diagonal form by making the canonical transformation

$$c_j^+ = \frac{1}{\sqrt{N}} \sum_k c_k^+ e^{ikja}$$

$$c_j = \frac{1}{\sqrt{N}} \sum_k c_k e^{-ikja} \tag{3}$$

with $k = 2\pi m/(\mathrm{Na})$, $m = 0, \pm 1, \pm 2, \ldots, \pm (N/2 - 1), N/2$ for a chain of length Na and with periodic boundary conditions. Substituting, we find

$$H = \sum_k \epsilon(k) c_k^+ c_k = \sum_k \epsilon(k) n_k \tag{4}$$

where n_k is the occupation number (0 or 1) of state k and with energy levels given by

$$\epsilon(k) = \begin{cases} \epsilon_A - 2t \cos ka & \text{for } p_A = 1 \\ \epsilon_B - 2t \cos ka & \text{for } p_B = 1 \end{cases} \tag{5}$$

The eigenstate corresponding to the eigenvalue $\epsilon(k)$ is simply

$$|\psi(k)\rangle = c_k^+ |0\rangle = \frac{1}{\sqrt{N}} \sum_j c_j^+ e^{-ikja} |0\rangle \tag{6}$$

Defining $|j\rangle = c_j^+ |0\rangle$, we see that the probability amplitude for the electron to be on site j is

$$\langle j | \psi_k \rangle = \frac{1}{\sqrt{N}} e^{-ijka} \tag{7}$$

and hence the probability of finding the electron on site j is simply $1/N$ independent of j and k. The eigenstates are therefore *extended*. This property is, of course, independent of dimensionality, as is the diagonalization procedure. In three dimensions, for the simple cubic lattice, one simply has

$$\epsilon_{A,B}(\mathbf{k}) = \epsilon_{A,B} - 2t(\cos k_x a + \cos k_y a + \cos k_z a) \tag{8}$$

In one dimension it is also possible to find an analytic expression for the density of states per site. The result for spinless fermions is

$$n_{A,B}(E) = \frac{1}{\pi} \frac{1}{\sqrt{4t^2 - (E - \epsilon_{A,B})^2}} \tag{9}$$

for $\epsilon_{A,B} - 2t \leq E \leq \epsilon_{A,B} + 2t$ and zero outside this interval.

We note that in the pure system the Hamiltonian has the property $H(x + ja) = H(x)$, where j is any integer. This immediately implies Bloch's theorem and the classification of eigenstates in terms of wave vectors k (Ashcroft and Mermin, 1976).

(b) Transfer Matrix

We now consider the more general case of a one-dimensional disordered chain and assume that $\epsilon_m = \epsilon_A$ with probability p_A and $\epsilon_m = \epsilon_B$ with probability p_B for all m. Since the system is not periodic we no longer have Bloch's theorem to help us classify the eigenstates. Intuitively, we expect that the energy levels, for any configuration, will be confined to the range $\epsilon_A - 2t \leq E \leq \epsilon_B + 2t$ for $\epsilon_A < \epsilon_B$, and this is indeed the case (see Ziman [1979] for further discussion of this point). We can formulate the one-electron problem in terms of a transfer matrix in a manner similar to our treatment of the one-dimensional Ising model (Section 3.F). We assume that

$$|\psi\rangle = \sum_{m=1}^{N} A_m |m\rangle = \sum_{m=1}^{N} A_m c_m^+ |0\rangle \tag{10}$$

is an eigenstate of (1) with eigenvalue E. Then

$$\langle j | H | \psi \rangle = EA_j = \epsilon_j A_j - t(A_{j+1} + A_{j-1}) \tag{11}$$

for $j = 1, 2, \ldots, N$ and where $A_{N+1} = A_1, A_0 = A_N$ in the case of periodic boundary conditions. Defining the vector

$$\phi_m = \begin{bmatrix} A_{m+1} \\ A_m \end{bmatrix}$$

we immediately have the recursion relation

$$\phi_j = \mathbf{T}_j \phi_{j-1} \tag{12}$$

where

$$\mathbf{T}_j(\epsilon_j, E) = \begin{bmatrix} \dfrac{\epsilon_j - E}{t} & -1 \\ 1 & 0 \end{bmatrix} \tag{13}$$

is the transfer matrix. Thus the solution of the Schrödinger equation is reduced to finding E such that

$$\phi_N = \mathbf{T}_N(\epsilon_N, E)\mathbf{T}_{N-1}(\epsilon_{N-1}, E) \cdots \mathbf{T}_1(\epsilon_1, E)\phi_N \tag{14}$$

or

$$\prod_{j=1}^{N} \mathbf{T}_j(\epsilon_j, E) = 1 \tag{15}$$

We first show that this equation reproduces the eigenstates of the pure chain. If $\epsilon_j = \epsilon_A$ for all j, we can easily show (Problem 9.1) that (15) reduces to the requirement that the two eigenvalues of $\mathbf{T}$, λ_+ and λ_-, be complex ($\lambda_- = \lambda_+^*$), and since $\lambda_+ \lambda_- = 1$ we have

$$\lambda_+ = \frac{\epsilon_A - E}{2t} + i \sqrt{1 - \left(\frac{\epsilon_A - E}{2t}\right)^2} = e^{i\theta}$$

$$\lambda_- = \frac{\epsilon_A - E}{2t} - i \sqrt{1 - \left(\frac{\epsilon_A - E}{2t}\right)^2} = e^{-i\theta} \tag{16}$$

which yields

$$E = \epsilon_A - 2t \cos \theta$$

The periodic boundary condition ($\lambda^N = 1$) produces $\theta = 2\pi j/N$. In the case of real eigenvalues ($\lambda \neq \pm 1$) (i.e., $|\epsilon_A - E| > 2t$) it is impossible to satisfy the periodic boundary conditions, and this corresponds to an energy gap in the spectrum of the translationally invariant chain.

In the disordered situation we have, in (15), a product of transfer matrices which are either of the form $\mathbf{T}_j = \mathbf{T}_A$ or $\mathbf{T}_j = \mathbf{T}_B$ depending on whether site j is occupied by an A or B atom. One would expect that if E lies in an energy gap of *both* the pure A and pure B system, it will be impossible to satisfy (15). Similarly, if E lies in the forbidden region of one of the materials, say A, then each time the transfer matrix $\mathbf{T}_A$ appears in the product (15), the vector ϕ will in general increase in modulus since the larger of the two eigenvalues of $\mathbf{T}_A$ is greater than 1. Thus it seems unlikely that (15) could be satisfied for any E that lies outside the region of allowed energies common to the two pure materials. However, extensive numerical work (see Ishii [1973] for a review) has shown that there are, in fact, eigenstates of H in this region but that these wave functions are *localized*. We discuss localization further below.

The determination of energy levels in the disordered case requires the solution of the eigenvalue problem (15) for a particular realization of a chain of length N. If we are interested only in a rough determination of the location of the eigenvalues or of the average density of states, we can use the following simple method, which is applicable only in one dimension. Quite generally, the eigenfunction corresponding to the mth eigenvalue has $m - 1$ nodes (this is strictly true only for nonperiodic chains). A node between sites j and $j + 1$ in our case corresponds to the ratio A_{j+1}/A_j being negative and this ratio can easily be expressed in terms of the matrix elements of $\mathbf{T}_j$ and the ratio A_j/A_{j-1}. Thus if we fix E and the initial ratio A_2/A_1 and simply count the number of negative ratios of successive coefficients A_j for a particular configuration of the

N potentials, we have, equivalently, a count of the number of energy levels below E. Averaging over many realizations of the random chain, a process easily
carried out on a computer, we find the average integrated density of states:

$$\mathcal{N}(E) = \int_{-\infty}^{E} dE'\, n(E') = S_N(E)$$

where $S_N(E)$ is the total number of nodes in the wave function at energy E. In
Figures 9.1 and 9.2 we display the density of states of a 500-atom chain ob-
tained by numerical differentiation of $\mathcal{N}(E)$ for $\epsilon_A = 2$, $\epsilon_B = 3$, $t = 1$ and two
different concentrations of the constituents. The density of states is plotted as a

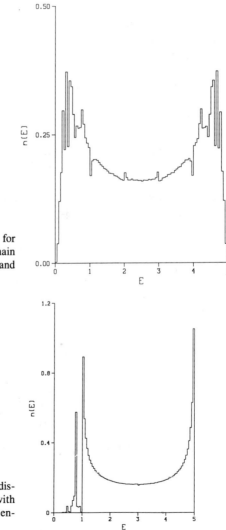

Figure 9.1 Density of states $n(E)$ for
the disordered one-dimensional chain
with $\epsilon_A/t = 2$, $\epsilon_B/t = 3$,   and
$p_A = p_B = 0.5$.

Figure 9.2 Density of states for the disordered one-dimensional alloy with
$p_A = 0.05$, $p_B = 0.95$, and the same energy parameters as in Figure 9.1.

histogram instead of as a smooth curve since we have no information as to how $n(E)$ varies between the points at which the nodes were counted.

A number of features of Figures 9.1 and 9.2 are of interest. In the strongly disordered alloy (Figure 9.1, $p_A = p_B = \frac{1}{2}$) no trace of the square-root singularities (9) near the upper and lower band edges remains. On the other hand, $n(E)$ is quite noisy in these regions. This is not a statistical artifact. A measure of statistical error is the degree of asymmetry of the density of states around the average energy $E = 2.5$ and we see that this is negligible. The noisy structure near the band edges is due to the existence of gaps in the spectrum at arbitrarily fine energy resolution (see Gubernatis and Taylor [1973] for a graphic demonstration of this fine structure). We also note that the density of states seems to extend essentially to the pure system band edges $\epsilon_A - 2$ and $\epsilon_B + 2$, although it drops rather precipitously near these points. These low-density regions are known as Lifshitz tails. One can understand that states with energy very close to the lower limit $\epsilon_A - 2$ will exist. In a random system there is a finite probability that a very long island, say of length L, of pure A material will occur. At least one eigenstate of the complete system should therefore be very close to the lowest eigenstate of a pure A system of length L. A similar argument applies at the upper band edge.

Figure 9.2 shows the density of states for a weakly disordered chain ($p_A = 0.05$). In this case we see almost the pure B density of states with a noisy impurity band essentially separated from the main part of the spectrum at lower energies.

We now return to the question of localization of the eigenstates. It has been rigorously proven that *all* eigenstates (except possibly a set of measure zero) of a one-dimensional disordered system are localized. The proof requires theorems on the properties of random matrices which are beyond the scope of this book and the reader is referred to Matsuda and Ishii [1970] for the rigorous mathematical arguments. We present here a weaker demonstration of localization which should at least make the result plausible.

We consider the quantity

$$\phi_{N+1}^+ \phi_{N+1} = A_{N+1}^2 + A_N^2 = \phi_0^+ (T_1^+ T_2^+ \cdots T_N^+ T_N T_{N-1} \cdots T_1) \phi_0 \qquad (17)$$

The Hermitian matrix $M_N = T_1^+ T_2^+ \cdots T_N^+ T_N \cdots T_1$ will be of the form

$$M_N = \begin{bmatrix} a_N & b_N \\ b_N & c_N \end{bmatrix} \qquad (18)$$

We easily derive a set of recursion relations for the matrix elements. We have

$$M_N = T_1^+ M_{N-1} T_1 = \begin{bmatrix} \dfrac{\epsilon_1 - E}{t} & 1 \\ -1 & 0 \end{bmatrix} \begin{bmatrix} a_{N-1} & b_{N-1} \\ b_{N-1} & c_{N-1} \end{bmatrix} \begin{bmatrix} \dfrac{\epsilon_1 - E}{t} & -1 \\ 1 & 0 \end{bmatrix} \qquad (19)$$

and find

$$a_N = \left(\frac{\epsilon_1 - E}{t}\right)^2 a_{N-1} + 2\left(\frac{\epsilon_1 - E}{t}\right) b_{N-1} + c_{N-1}$$

$$b_N = -\left(\frac{\epsilon_1 - E}{t}\right) a_{N-1} - b_{N-1} \tag{20}$$

$$c_N = a_{N-1}$$

We now average these recursion relations over the probability distribution of the atomic potentials. This yields the expectation value of the quantity $\phi_{N+1}^+ \phi_{N+1}$, rather than the probability distribution of this quantity which is needed for a proper proof of localization. Once the average of (20) is obtained, the difference equations can be solved by the ansatz

$$\begin{bmatrix} a_N \\ b_N \\ c_N \end{bmatrix} = \begin{bmatrix} x_1 \\ x_2 \\ x_3 \end{bmatrix} \lambda^N \tag{21}$$

which then yields a 3×3 eigenvalue problem. An extended eigenstate of the system corresponds to an eigenvalue λ of magnitude 1 and it is easy to show that in the pure case ($p_A = 1$ or $p_B = 1$) we recover the usual dispersion relations for the energy levels.

Since the secular equation for this eigenvalue problem is cubic and rather unenlightening, we simply plot, in Figure 9.3, the quantity

$$L^{-1} = \lim_{N \to \infty} \frac{1}{N} \ln \frac{\phi_{N+1}^+ \phi_{N+1}}{\phi_0^+ \phi_0} \tag{22}$$

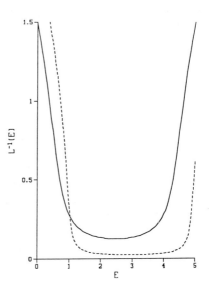

Figure 9.3 Plot of the inverse localization length (22) as a function of E for the disordered one-dimensional alloys whose densities of state are plotted in Figures 9.1 and 9.2. Solid curve, $p_A = 0.5$; dashed curve, $p_A = 0.05$.

as a function of E for the two sets of parameters for which we have plotted the average density of states in Figures 9.1 and 9.2, namely $\epsilon_A = 2$, $\epsilon_B = 3$ and $p_A = 0.5$ and 0.05. We also take as boundary condition $A_0 = 0$. The quantity L clearly has a physical interpretation as a mean localization length. Referring to Figure 9.3, we see that in both cases the localization length is a smooth function of the energy E and that it decreases sharply near the band edges but remains finite throughout the entire energy range. The feature of strong localization near band edges is thought to hold in three dimensions as well as in one dimension. The three-dimensional case is distinguished from the one-dimensional one by the existence of a *mobility edge* which separates a region of extended states from localized states.

The foregoing discussion for the case of a disordered tight binding model holds as well for the vibrational properties of an isotopically disordered chain (random masses, fixed spring constants), for a disordered Kronig–Penney model (Problem 9.2) and, in general, for any disordered one-dimensional system. An extensive discussion of both numerical and analytic results can be found in the review by Ishii [1973].

(c) Localization in Three Dimensions

The notion that wave functions in a disordered three-dimensional system might be localized is due to Anderson [1958], who, in a classic paper, considered the Hamiltonian (1) with a continuous distribution of energies ϵ_j characterized by a width W. Anderson attempted a calculation of the probability $a_j(\infty)$ for an electron to be at site j at $t = \infty$, given that it is there at $t = 0$. The mathematical methods he used are complicated and we shall not attempt to repeat his arguments. His conclusion was that given sufficiently strong disorder (W large enough), the electronic eigenstates are localized.

One can partially understand this conclusion, at least in the pathological limit in which the energy levels of the atoms are vastly different. Suppose that we have only two possible energies ϵ_A and ϵ_B and that $|\epsilon_A - \epsilon_B| >> t$ and, moreover, that the concentration of B atoms is small. An electron, initially on a B site, will be unable to tunnel through A sites and can escape the vicinity of its initial location only if there exists a continuous connected path of B atoms which extends to infinity. As we shall see in the next section, such a path exists only if the concentration of B atoms is greater than a critical value called the percolation concentration. Thus we have a situation in which at least some of the electronic states are localized. Once we accept this possibility, it is not hard to see that there might be a transition, as a function of electron energy, from localized to extended wave functions.

The picture that we now believe to be correct is shown schematically in Figure 9.4 and is supported by extensive numerical work. For a given band the localized states extend from the band edges to mobility edges at energies E_1 and E_2. The localized states have wave functions with an exponential envelope.

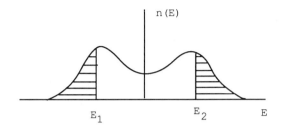

Figure 9.4 Density of states of a disordered alloy with localized states in the shaded regions ($E < E_1, E > E_2$), extended states for $E_1 < E < E_2$.

There is some evidence (Last and Thouless, 1973: Licciardello and Thouless, 1975) that there may be a region in which the "localized" states of a three-dimensional system have a power law envelope rather than an exponential envelope as well. The notion of sharp mobility edges separating localized and extended states seems to be due to Mott [1967]. Much of the numerical work on this topic is discussed by Thouless [1979].

It is quite clear that the effect of localized states on the conductivity will be quite interesting and theoretically complicated. At zero temperature we expect that the conductivity will be zero. Wave functions of localized states will in general not overlap with other localized states of the same energy and hence there can be no tunneling between different states without thermal activation (hopping). At small but finite temperature this argument no longer holds, but there are now two length scales in the problem—the extent of the localized wave function L and the inelastic mean free path L_i. If $L_i < < L$ the electron is scattered many times while traversing the distance L, and hence the information as to which localized state it started in is lost. Localization is irrelevant in this case and resistivity is dominated by the usual scattering processes. Conversely, if $L << L_i$, the effect of localization becomes important and this manifests itself in added dependence of the resistivity of thin wires and films on the physical dimensions of the sample. These ideas have been used by Thouless and co-workers and Wegner to develop scaling theories of localization which are reviewed, together with the experimental situation, by Lee and Ramakrishnan [1985].

(d) Density of States

The equilibrium thermodynamics of a disordered system can be expressed entirely as a functional of the density of states or spectral density. The response to external perturbations (conductivity, thermal conductivity, etc.) depends in greater detail on the nature of the wave functions and presents more difficult problems. The density of states is understood in much more detail and reliable approximations have been developed which allow us to determine this function in many cases. The simplest method for calculating the spectral density is the *rigid band* or *virtual crystal* approximation; the most successful the *coherent potential* approximation. We briefly discuss them below.

(i) Virtual crystal approximation. Consider, again, the Hamiltonian (1) with $\epsilon_j = \epsilon_A$ with probability $p_j = p_A$ and $\epsilon_j = \epsilon_B$ with probability $p_j = p_B$. If we average the Hamiltonian over the impurity distribution function, before attempting to diagonalize it, we obtain the translationally invariant effective Hamiltonian

$$\overline{H} = -t \sum_{\langle i,j \rangle} (c_i^+ c_j + c_j^+ c_i) + \overline{\epsilon} \sum_i c_i^+ c_i \tag{23}$$

with $\epsilon = p_A \epsilon_A + p_B \epsilon_B$. Using Bloch's theorem, we easily find the energy levels to be

$$\epsilon(\mathbf{k}) = \overline{\epsilon} - t \sum_\delta e^{i\mathbf{k} \cdot \delta} \equiv \overline{\epsilon} + w(\mathbf{k}) \tag{24}$$

where the sum over δ extends over nearest-neighbor vectors and where $\mathbf{k}$ lies in the first Brillouin zone. The entire band is thus displaced uniformly without change in shape. The approximation (23) has the virtue of simplicity but is not realistic, especially when the two atomic potentials ϵ_A and ϵ_B are substantially different.

(ii) Coherent potential approximation. A more sophisticated and far more accurate approximation is the coherent potential approximation (Soven, 1967). The essential idea is to replace each atom by an "effective" atom so that on the average no scattering takes place on each site. To be more precise, consider a single impurity of type B at site i in an otherwise type A crystal. The Hamiltonian then is

$$H = H_A + (\epsilon_B - \epsilon_A)c_i^+ c_i = H_A + U \tag{25}$$

We now define the resolvent operator (or Green's function) $G(z)$:

$$\begin{aligned} G(z) &= (z - H)^{-1} = (z - H_A - U)^{-1} \\ &= (z - H_A)^{-1} + (z - H_A)^{-1}U(z - H)^{-1} \end{aligned} \tag{26}$$

The last equation can easily be shown to be correct by premultiplying by $(z - H_A)$ and postmultiplying by $(z - H)$. Taking matrix elements in the atomic basis and defining

$$G_{mj}(z) = \langle m | (z - H)^{-1} | j \rangle \qquad G_{mj}^0(z) = \langle m | (z - H_A)^{-1} | j \rangle$$

we obtain, on iterating (26),

$$\begin{aligned} G_{mj}(z) &= G_{mj}^0(z) + G_{mi}^0(z)U_{ii}G_{ij}(z) \\ &= G_{mj}^0(z) + G_{mi}^0(z)U_{ii}G_{ij}^0(z) + G_{mi}^0(z)U_{ii}G_{ii}^0(z)U_{ii}G_{ij}^0(z) + \cdots \\ &= G_{mj}^0(z) + G_{mi}^0 U_{ii}(1 - G_{ii}^0 U_{ii})^{-1}G_{ij}^0 \end{aligned} \tag{27}$$

The operator $T = U(1 - G^0 U)^{-1}$ is known as the T-matrix of the potential U and has, in the particular case of a single impurity (25), only diagonal matrix elements. The generalization of (26) and (27) for an arbitrary perturbation U is, in operator form,

$$G(z) = G^0(z) + G^0(z)UG(z) = G^0(z) + G^0(z)T(z)G^0(z) \qquad (28)$$

The Green's function $G(z)$ yields the density of states, as we now show. Suppose that the eigenstates of H are $|\phi_m\rangle$ with energies E_m and consider

$$\text{Tr } G(E + i\eta) = \sum_m \langle \phi_m | (E - H - i\eta)^{-1} | \phi_m \rangle$$

$$= \sum_m (E - E_m + i\eta)^{-1} \qquad (29)$$

Using

$$\lim_{\eta \to 0} \frac{1}{E - E_m + i\eta} = P\frac{1}{E - E_m} - i\pi\delta(E - E_m) \qquad (30)$$

we see that

$$n(E) = \sum_m \delta(E - E_m) = -\frac{1}{\pi} \text{Im Tr } G(E + i0^+) \qquad (31)$$

Since the trace (31) can be evaluated in any basis, we are free to use our Wannier states $|m\rangle$ or the Bloch states $|k\rangle$ to calculate the density of states. It is only necessary to find the diagonal matrix elements of the operator $G(z)$.

In the coherent potential approximation one determines the Green's function G by assuming that there exists an effective complex translationally invariant potential such that the averaged T-matrix vanishes for each site. Thus one writes the Hamiltonian in the form

$$H = \sum_k [w(k) + \Sigma(E)] c_k^+ c_k + \sum_j [\epsilon_j - \Sigma(E)]c_j^+ c_j$$

$$= H_0 + \sum_j U_j(E) \qquad (32)$$

where we have used a mixed k-space and real space representation and where $w(k)$ is defined in (24). Using $\langle k | j \rangle = N^{-1/2} \exp\{-i k \cdot r_j\}$, we have, for the diagonal matrix elements of the Green's function G^0,

$$G_{ii}^0(z) = \langle i | [z - H_0]^{-1} | i \rangle = \frac{1}{N} \sum_k \frac{1}{z - \Sigma(z) - w(k)} \qquad (33)$$

The T matrix of a single "impurity" potential at site i is then diagonal and is given by

$$T_i(z) = [\epsilon_i - \Sigma(z)]\{1 - [\epsilon_i - \Sigma(z)]G_{ii}^0(z)\}^{-1} \qquad (34)$$

Averaging over the atomic distribution, we have

$$\langle T_i(z) \rangle = p_A[\epsilon_A - \Sigma(z)]\{1 - [\epsilon_A - \Sigma(z)]G_{ii}^0(z)\}^{-1}$$
$$+ p_B[\epsilon_B - \Sigma(z)]\{1 - [\epsilon_B - \Sigma(z)]G_{ii}^0(z)\}^{-1} \equiv 0 \tag{35}$$

which determines the unknown function $\Sigma(z)$. Rearranging, we find the simpler form

$$\Sigma(z) = p_A\epsilon_A + p_B\epsilon_B - [\Sigma(z) - \epsilon_A][\Sigma(z) - \epsilon_B]G_{ii}^0(z) \tag{36}$$

which can be solved numerically for the complex function $\Sigma(z)$. The CPA approximation for the *average* density of states is then

$$n(E) = -\frac{1}{\pi} \text{Im Tr } G^0(E + i0)$$
$$= -\frac{1}{\pi N} \text{Im} \sum_{\mathbf{k}} \frac{1}{E + i0 - w(\mathbf{k}) - \Sigma(E + i0)} \tag{37}$$

In Figures 9.5 and 9.6 we plot this function for the one-dimensional disordered alloy for the same parameters as used to obtain the average density of states by the exact node counting method (Figures 9.1 and 9.2). As we see, the gross features of the density of states are well reproduced by the coherent potential approximation, and indeed, near the center of the band, the two functions are essentially identical. What is missing in the CPA density of states is the fine structure near the band edges and the low-density tails, which in the exact calculation, extend to the edges of the pure system bands. This is not surprising since the CPA deals with the impurity distribution in a strictly local manner. The structure in the Lifshitz tails is due to islands of one species and the probability of such islands never enters into the CPA formalism.

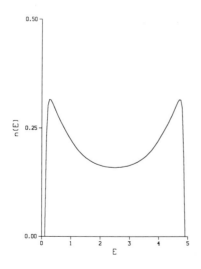

Figure 9.5 Density of states from the coherent potential approximation for the one-dimensional disordered binary alloy. Parameters are the same as in Figure 9.1.

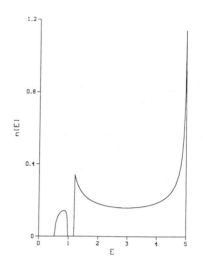

Figure 9.6 Density of states from the coherent potential approximation for the one-dimensional disordered binary alloy. Parameters as in Figure 9.2.

One can understand the success of the coherent potential approximation in the weak scattering limit, which we characterize by $|G_{jj}^0(z)\,T_j(z)| << 1$. Using (28) and expanding the full T matrix, we have

$$G_{mj} = G_{mj}^0 + \sum_i G_{mi}^0 T_i G_{ij}^0 + \sum_{k \neq i} G_{mi}^0 T_i G_{ik}^0 T_k G_{kj}^0$$
$$+ \sum_{k \neq i,\, n \neq k} G_{mi}^0 T_i G_{ik}^0 T_k G_{kn}^0 T_n G_{nj}^0 \tag{38}$$
$$+ \sum_{k \neq i,\, n \neq k,\, s \neq n} G_{mi}^0 T_i G_{ik}^0 T_k G_{kn}^0 T_n G_{ns}^0 T_s G_{sj}^0 + \cdots$$

Taking G^0 to be the CPA Green's function and averaging over the impurity distribution, we see that because of the restrictions on successive indices in the summations in (38), the first few terms vanish and

$$\langle G_{mj} \rangle = G_{mj}^0 + \sum_{k \neq i} G_{mi}^0 \langle T_i G_{ik}^0 T_k G_{ki}^0 T_i G_{ik}^0 T_k \rangle G_{kj}^0 + O(T^6) \tag{39}$$

Thus the corrections to the coherent potential approximation are of order T^4 and higher.

For a much more extensive discussion of the coherent potential approximation for the case of the disordered binary alloy, the reader is referred to Velicky et al. [1968]. The review of Elliott et al. [1974] provides a thorough discussion of applications of the CPA to vibrational properties of disordered materials, spin waves in disordered magnets, and other physical situations.

As shown by Matsubara and Yonezawa [1967], the expansion (27) can be used to construct a diagrammatic expansion for the density of states. The resulting moment expansion is similar in form to the high-temperature expansions discussed in Section 5.B.

9.B PERCOLATION

(a) Introduction

In our brief discussion of localization in three dimensions, we have noted that the nature of connected clusters of equivalent atoms in a disordered binary alloy may be important in determining the nature of electronic states. In dilute magnetic alloys where the fraction of magnetic atoms is p, the existence of a finite temperature phase transition depends on whether or not there exists an infinite connected cluster (of suitably high dimensionality) of interacting magnetic atoms. A random resistor network formed of elements with finite resistance (probability p) and infinite resistance (probability $1 - p$) will conduct only if the network of conducting elements is continuous across the sample. Many other physical situations depend in an essential way on the geometric properties of random clusters (see Stauffer [1985] for a discussion of forest fires) and, in particular, on the existence of an infinite connected cluster which spans the system in question. The study of such clusters is the subject of percolation theory.

There are two basic percolation problems: *site* percolation and *bond* percolation. In the site percolation problem the vertices of a lattice are occupied with a given probability p and occupied sites are considered to be connected if they are nearest neighbors. In Figure 9.7 we show a 20 × 20 segment of a

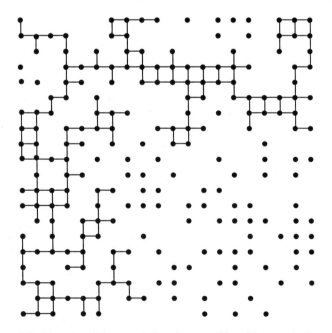

Figure 9.7   Example of site percolation cluster on 20 × 20 square lattice at concentration 0.5575. The largest cluster has been connected.

square lattice occupied with probability 0.5575 (223 particles). The particles in the largest cluster have been connected and we see that it spans the lattice in both the horizontal and vertical directions but that the connections are tenuous. In fact, the existence of a spanning cluster at this probability of occupation is a finite-size effect. The site percolation probability (i.e., the probability at which the infinite cluster forms) on the square lattice is known to be $p_c \approx 0.59$.

Bond percolation, on the other hand, is defined in terms of the probability of occupation of nearest-neighbor bonds on a lattice, and this model is clearly relevant to conduction in inhomogeneous media. The site and bond problems are closely related and one can show, for example, that on any lattice the critical probability for bond percolation $p_c^B \leq p_c^S$, the site percolation critical probability (see, e.g., Essam, 1972).

We now define the quantities of interest in a percolation problem. Let $n_s(p)$ be the number of clusters per lattice site of size s. Then the probability that a given site is occupied and part of a cluster of size s is simply $s \cdot n_s(p)$. Now let $P(p)$ be the fraction of *occupied* sites that belong to the infinite cluster. Clearly, $P(p) = 1$ for $p = 1$ and $P(p) = 0$ for $p < p_c$. In this sense $P(p)$ is similar to the order parameter of a system which has an ordinary thermal phase transition. Indeed, percolation is often referred to as a geometric phase transition. Clearly, we have the relation

$$\sum_s sn_s(p) + pP(p) = p \qquad (1)$$

where the summation extends over all finite clusters.

Another quantity of interest is the mean size of finite clusters, which we denote by $S(p)$. This quantity can also be related to $n_s(p)$ through the relation

$$S(p) = \frac{\sum_s s^2 n_s(p)}{\sum_s sn_s(p)} \qquad (2)$$

where, once again, the summation is over all finite clusters. Finally, one can define the "pair connectedness" $C(p, r)$ to be the probability that occupied sites a distance r apart are part of the same cluster, and one can then show that the quantity $S(p)$ can be expressed in terms of $C(p, r)$ (Essam, 1972); Lubensky, 1979).

We also note that $S(p)$ can be related to the low-temperature susceptibility of a dilute Ising model. For $k_B T << J$ all nearest-neighbor magnetic atoms will be in the same state, either up or down. Thus in a magnetic field h the magnetization per site is given by

$$\langle \sigma \rangle = pP(p) + \sum_s sn_s(p) \tanh \frac{sh}{k_B T} \qquad (3)$$

where the first term on the right is the contribution from the infinite cluster. Differentiating, we find for the zero-field susceptibility:

$$\chi(0, T) = \frac{1}{k_B T} \sum_s s^2 n_s(p) = \frac{p[1 - P(p)]}{k_B T} S(p) \tag{4}$$

The functions $P(p)$, $S(p)$, and $C(p, r)$ can be calculated by most of the methods that we have developed for other statistical problems, such as series expansions, Monte Carlo simulations, and renormalization group methods. To date the percolation problem in two and three dimensions remains unsolved, although the percolation probability for the triangular lattice is known to be $p_c^S = 0.5$ exactly. The one-dimensional percolation problem is, of course, trivial and both the bond and site percolation probability can be calculated exactly for a Bethe lattice (see, e.g., Thouless, 1979). We will first exploit the analogy between percolation and phase transitions to postulate a simple scaling theory.

(b) Scaling Theory of Percolation

The analogy between a thermal phase transition and percolation alluded to above can be put on a rigorous basis by means of an argument due to Kasteleyn and Fortuin [1969], who showed that the bond percolation problem is isomorphic to the q-state Potts model in the limit $q \to 1$ (see also Lubensky [1979] for a discussion of this point). The probability of occupation of a bond is related to a Boltzmann weight in the Potts model and is therefore like the temperature variable in the scaling theory of Chapter 5 (the analog of the magnetic field can be introduced as well but does not have as direct an interpretation in the percolation problem). The correspondence between thermodynamic functions and the geometric quantities defined above is as follows. The analog of the free energy per site is the total number of clusters per site, that is,

$$G(p) = \sum_s n_s(p) \tag{5}$$

The role of an order parameter is played by the probability $P(p)$, as is intuitively obvious. Similarly, the mean size of finite clusters $S(p)$ is equivalent to the susceptibility and the pair connectedness $C(p, r)$ is equivalent to the thermal pair correlation function.

We therefore expect these quantities to have power law singularities at the percolation probability and can define the usual set of exponents:

$$G(p) \sim |p - p_c|^{2-\alpha} \tag{6}$$

$$P(p) \sim (p - p_c)^{\beta} \tag{7}$$

$$S(p) \sim |p - p_c|^{-\gamma} \tag{8}$$

$$C(p, r) \sim \frac{\exp\{-r/\xi(p)\}}{r^{d-2+\eta}} \tag{9}$$

where, moreover, the correlation length ξ is expected to diverge as

$$\xi(p) \sim |p - p_c|^{-\nu} \tag{10}$$

If universality (Section 5.D) holds, the percolation exponents α, β, γ, ...
will depend on dimensionality but not on details such as the type of lattice or
whether it is bond or site percolation that is being considered.

The basic assumption of the scaling theory (Stauffer, 1979) is that for p
near p_c, or for a given ξ in (9), there is a typical cluster size s_ξ which yields the
dominant contribution to the functions (6)–(8). This cluster size must diverge
as $p \to p_c$ and we assume that the divergence is of the power law form:

$$s_\xi \sim |p - p_c|^{-1/\sigma} \tag{11}$$

which defines the exponent σ. We further assume that the number of clusters
of size s at probability p can be related to s/s_ξ and to $n_s(p_c)$ through

$$n_s(p) = n_s(p_c) f\left(\frac{s}{s_\xi}\right) \tag{12}$$

where $f(x) \to 0$ as $x \to \infty$ and $f(x) \to 1$ as $x \to 0$ but is otherwise unspecified.
It is known from Monte Carlo simulations (Stauffer, 1979) that for large s,
$n_s(p_c) \sim s^{-\tau}$, where τ is an exponent that depends on the dimensionality d.
Using this form and (12), we have

$$n_s(p) = s^{-\tau}\phi(s\,|p - p_c|^{1/\sigma}) \tag{13}$$

which, as in the case of thermodynamic scaling (Section 5.C), allows us to express the percolation exponents in terms of the two independent exponents σ
and τ. For example,

$$G(p) = \sum_s n_s(p) = \sum_s s^{-\tau}\phi(s\,|p - p_c|^{1/\sigma})$$
$$\approx |p - p_c|^{\frac{\tau-1}{\sigma}} \int dx\, \chi^{-\tau}\phi(x) \tag{14}$$

where $x = s\,|p - p_c|^{1/\sigma}$. Assuming that the integral over x converges, we
have $\alpha = 2 - \frac{\tau-1}{\sigma}$. Similarly, we obtain $\gamma = (3 - \tau)/\sigma$ and $\beta = (\tau - 2)/\sigma$
and we find the familiar scaling relation,

$$\alpha + 2\beta + \gamma = 2$$

The pair connectedness can also be included in the scaling formalism. If we integrate $C(p, r)$ over the volume of the sample, we should obtain the mean
cluster size $S(p)$, that is,

$$S(p) \sim |p - p_c|^{-\gamma} \sim \int dr\, r^{d-1}\frac{e^{-r/\xi(p)}}{r^{d-2+\eta}} \sim |p - p_c|^{-\nu(2-\eta)} \tag{15}$$

and hence $\gamma = \nu(2 - \eta)$. The hyperscaling equation $d\nu = 2 - \alpha$ can also be
obtained with one further assumption. The dominant cluster size at a given p,

from (11), obeys the relation $s \sim |p - p_c|^{-1/\sigma}$ and the concentration of these clusters (13) is $n_s \sim |p - p_c|^{\tau/\sigma}$. Assuming that this concentration is inversely proportional to the volume $\xi^d(p)$ occupied by these clusters, we find that

$$|p - p_c|^{\tau/\sigma} = |p - p_c|^{d\nu}$$

or $d\nu = 2 - \alpha$.

In the review of Stauffer [1979], the reader will find a critical discussion of numerical tests of the scaling theory. Although the exponents obtained from Monte Carlo simulations and series expansions are not extremely accurate, they are consistent with the scaling theory. There is also no evidence for violation of the universality hypothesis–percolation exponents seem to depend only on the dimensionality. In two dimensions there is evidence (see Binder and Stauffer, 1984) that the critical exponents are given by the rational numbers $\alpha = -\frac{2}{3}$, $\beta = \frac{5}{36}$, $\gamma = \frac{43}{18}$, $\nu = \frac{4}{3}$, and $\eta = \frac{5}{24}$. In three dimensions $\beta \approx 0.4$, $\gamma \approx 1.7$, and $\nu \approx 0.82$.

(c) Approximation Methods: Series Expansions and Renormalization Group

The application of Monte Carlo methods to the percolation problem is straightforward and we shall not discuss it here. Series expansions can be constructed for all the relevant functions discussed in Section 9.B(b) and analyzed by the standard techniques (Section 5.B) used for thermodynamic functions. Consider, for example, the quantity $S(p)$ (2), the mean size of finite clusters. We wish to obtain a power series in p for this function, and since we are expanding around $p = 0$, the denominator in (2) is simply p from (1). We write

$$S(p) = \frac{1}{p} \sum_s s^2 n_s(p) = \sum_{j=0}^{\infty} a_j p^j \tag{16}$$

One can systematically calculate the coefficients a_j by enumerating clusters of increasing size and calculating the probability of occurrence. To be specific, consider the square lattice. The probability that a given site is occupied and isolated is simply $p(1 - p)^4 = n_1(p)$. Consider now the next few clusters shown below. Cluster (a) will contribute terms of order p^2 and higher, (b) and (c) terms of order p^3, and the remaining five graphs terms of order p^4 and higher.

It is easiest to calculate the quantity $sn_s(j, p)$, where j refers to the label of the graph. Taking a site, say 0, in the lattice and associating it with each of the inequivalent vertices of the graph j, we count the number of configurations in which site 0 is part of cluster j. For graph (c) in the table below, we have, for example,

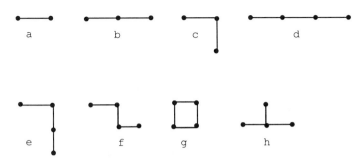

where the crossed vertex is located at site 0 and the numbers 8 and 4 are the number of different configurations. Thus

$$3n_3(c, p) = 12p^3(1 - p)^7$$

Carrying out this calculation for the remaining graphs and expanding the factor $(1 - p)^t$, where t is the number of perimeter sites, we find, through order p^3,

$$S(p) = 1 + 4p + 12p^2 + 24p^3 + \cdots \tag{17}$$

It is clear that as in the case of high-temperature series for the Ising model (Section 5.B), the calculation of high-order terms becomes a formidable problem of graph enumeration. In the review of Essam [1972] the coefficients a_j, in (16), are tabulated for a number of two- and three-dimensional lattices (to $j = 11$ for the square lattice). The reader may wish to derive one or two more terms to add to (17) or to analyze the longer series by applying the ratio methods of Section 5.B. As is the case with high-temperature expansions, one can obtain well-converged estimates of p_c and to a lesser extent of the critical exponents by the series expansion method (Dunn et al., 1975).

We now briefly discuss a renormalization group approach to percolation. The renormalization group has been used primarily in two different ways to attack the percolation problem. The first method makes use of the formal equivalence between percolation and the $q \rightarrow 1$ limit of the q-state Potts model, which we have already mentioned. One can carry out an ϵ expansion (around $d = 6$, the upper critical dimension for percolation) and attempt to evaluate the critical exponents at $\epsilon = 3$ (Harris et al., 1975; Priest and Lubensky, 1976).

A technically simpler approach is to apply real space renormalization group techniques directly. The essential idea is that at the percolation probability, the system should be invariant under rescaling by an arbitrary length. Thus, in a simple case (Figure 9.8), we divide the triangular lattice into blocks of three sites and attempt to calculate the block occupation probability in terms of the site occupation probability. We assume that the block is occupied if the majority of its sites are occupied. Thus the probability for occupation of the blocks is given by

$$p' = R_{\sqrt{3}}(p) = p^3 + 3p^2(1 - p) \tag{18}$$

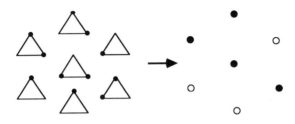

Figure 9.8 Example of one step in the renormalization procedure described for the triangular lattice. The occupied blocks are shown as heavy dots on the right.

This recursion relation has trivial fixed points at $p = 0$ and $p = 1$ and a nontrivial fixed point at $p^* = 0.5$ which we identify with the percolation concentration and which is, fortuitously, exact. The correlation length exponent ν can be determined immediately from the recursion relation. Recall that $\xi(p) \sim |p - p_c|^{-\nu}$ and that $\xi'(p') = \xi(p)/L \sim |p' - p_c|^{-\nu}$, where $L = 3^{1/2}$ is the rescaling length. Thus

$$p' - p_c = L^{1/\nu}(p - p_c)$$

near the fixed point. From (18) we therefore obtain $\nu = 1.35$, which is in excellent agreement with the accepted result $\nu = \frac{4}{3}$.

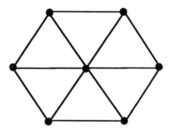

One can easily generalize these results to larger clusters. For the seven-site block shown above, the recursion relation is

$$p' = R_{\sqrt{7}}(p) = 35p^4 - 84p^5 + 70p^6 - 20p^7 \tag{19}$$

which, once again, has a nontrivial fixed point at $p^* = p_c = 0.5$. In this case the correlation length exponent is given by

$$\nu = \frac{\ln 7}{2 \ln \frac{7}{2}} = 1.243$$

in respectable agreement with $\nu = \frac{4}{3}$.

It is clear how to generalize this procedure to arbitrary lattices and that the Monte Carlo renormalization group methods described in Section 6.D can also be easily adapted to the percolation problem. We refer to the reviews of Binder and Stauffer [1984], Stanley et al. [1982], and Kirkpatrick [1979] for further discussion.

(d) Conclusion

In this section we have concentrated on the aspects of percolation which are formally equivalent to the statistical mechanics of phase transitions. A subject that we have ignored completely is the internal structure of the percolating cluster near the critical concentration. A glance at Figure 9.7 will convince the reader that this structure is indeed interesting and complicated—the word "ramified" is often used to describe percolation clusters. The elementary excitations and transport properties (conduction, heat conduction, etc.) will depend in detail on the nature of this structure. In recent years the concepts of self-similarity and fractal geometry (Mandelbrot, 1982) have been applied to percolation and other cluster growth models. For more details, the reader is referred to Aharony [1986] or Stanley and Ostrowsky [1986].

9.C PHASE TRANSITIONS IN DISORDERED MATERIALS

In this section we discuss some aspects of phase transitions in disordered materials. We use as our primary model a crystalline ferromagnet randomly diluted with nonmagnetic atoms. We assume that the magnetic atoms interact through short-range exchange interactions (nearest neighbors) and that the interactions are all ferromagnetic. This is an important simplification, as it at least allows us to determine the ground state. If the system has a mixture of ferromagnetic and antiferromagnetic interactions, even the determination of the ground state may be a difficult if not unsolvable problem. Such systems are discussed in Section 9.D.

To be specific, we consider primarily the Hamiltonian

$$H = -J\sum_{\langle i, j \rangle} \epsilon_i \epsilon_j \mathbf{S}_i \cdot \mathbf{S}_j - h\sum_i \epsilon_i S_{iz} \tag{1}$$

where $\mathbf{S}_i = (S_{i1}, S_{i2}, \ldots, S_{in})$ is an n-component spin and where the variables ϵ_i are either 1, if the site i is occupied by a magnetic atom, or 0, if site i is occupied by a nonmagnetic impurity. We assume that the random variables ϵ_i are uncorrelated (i.e., $\langle \epsilon_i \epsilon_j \rangle = \langle \epsilon_i \rangle \langle \epsilon_j \rangle = p^2$, where p is the concentration of magnetic atoms). We also assume that the disorder is quenched, namely that there is no change in the distribution of magnetic atoms as a function of temperature. The Hamiltonian (1) thus encompasses the disordered version of the standard models for magnetism that we have studied in previous chapters ($n = 1$, Ising; $n = 2$, XY; $n = 3$, Heisenberg, etc.).

Physical realizations of the n-vector model can be found in several series of compounds (mostly antiferromagnets, with short-range interactions) such as $Co_pZn_{1-p}Cs_3Cl_5$, which is a diluted three-dimensional Ising model, $Co_pZn_{1-p}(C_5H_5NO)_6(ClO_4)_2$ (XY model), and $KMn_pMg_{1-p}F_2$ (Heisenberg model). A review of experimental work and comparison with theory can be found in Stinchcombe [1983].

The ground state of (1) is clearly the state with all spins aligned with the magnetic field and with a magnetization per site given by

$$m(h = 0^+, T = 0) = \frac{1}{N} \sum_{i=1}^{N} \epsilon_i \langle S_{iz} \rangle \big|_{h=0^+, T=0} = pS \tag{2}$$

We will primarily be interested in the question of whether or not a sharp phase transition (i.e., a unique well-defined critical temperature) exists in the presence of disorder; if so, what is the dependence of the critical temperature $T_c(p)$ on concentration, and what, if any, changes are to be expected in the critical exponents? It is clear that with all interactions ferromagnetic, dilution of the system with nonmagnetic impurities can only lower the critical temperature. A crude generalization of mean field theory (Section 3.A) can be constructed immediately. The mean effective field, averaging over both impurity distribution and the Boltzmann distribution, acting on a magnetic atom in the case of an Ising model is

$$h_{\text{eff}} = pqJm + h \tag{3}$$

where q is the number of nearest neighbors. Hence

$$\langle S_{iz} \rangle = \tanh\left[\beta(pqJm + h)\right] \tag{4}$$

which yields $T_c(p)/T_c(1) = p$ for the critical temperature. This result is obviously wrong in view of the discussion of Section 9.B. There cannot be a finite-temperature phase transition for $p < p_c$, the percolation concentration, and the critical temperature, as a function of p, must be qualitatively like the solid curve in Figure 9.9 rather than the dashed mean field approximation. Before dealing further with these topics, we digress briefly on the statistical formalism necessary for a description of disordered systems.

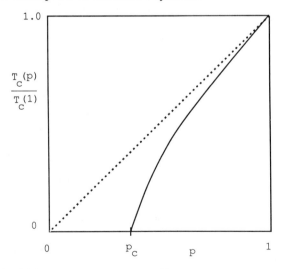

Figure 9.9   Critical temperature of model (1) as function of concentration p in mean field theory (dashed curve). The solid curve is qualitative but recognizes that $T_c = 0$ for $p < p_c$.

(a) Statistical Formalism and the Replica Trick

As mentioned above, we will be interested primarily in the case of quenched disorder. Thus the partition function is a function of the variables $\{\epsilon_j\}$,

$$Z = \text{Tr} \exp\{-\beta H[\epsilon_i, S_i]\} = Z(\epsilon_1, \epsilon_2, \ldots, \epsilon_N) \tag{5}$$

as is the free energy,

$$G(h, T, \epsilon_1, \epsilon_2, \ldots, \epsilon_N) = -k_B T \ln Z(h, T, \epsilon_1, \epsilon_2, \ldots, \epsilon_N) \tag{6}$$

In practice, it is impossible to calculate the free energy (6) for an arbitrary configuration of impurities. To simplify the problem somewhat, we assume that the system can be divided into a large number of subvolumes, each much larger than the correlation length. These subvolumes then constitute an ensemble of statistically independent systems, each with an impurity configuration governed by the same distribution. The result of an experiment thus will yield an average of the observable with respect to the impurity distribution. For example, the measured free energy per site is given by

$$g(h, T, p) = -\frac{k_B T}{N} \langle \ln Z(h, T, \epsilon_1, \epsilon_2, \ldots, \epsilon_N) \rangle \tag{7}$$

where the concentration p is a parameter, not a thermodynamic variable. Other thermodynamic functions are derived from (7) in the usual way.

On the other hand, if the impurities are annealed rather than quenched, the variables ϵ_i must be treated as dynamical variables on the same footing as the spin variables. The impurity concentration can be fixed through the introduction of a chemical potential and the appropriate partition function is

$$Z(h, T, \mu) = \sum_{\{\epsilon_i = 0, 1\}} \text{Tr} \exp\left\{-\beta H[\epsilon_i, S_i] + \beta\mu \sum_{i=1}^{N} \epsilon_i\right\} \tag{8}$$

with

$$Np = k_B T \frac{\partial}{\partial \mu} \ln Z(h, T, \mu) \tag{9}$$

It is clear that the two cases are fundamentally different.

The calculation of the free energy (7) is a very difficult problem, one for which universally applicable methods have not yet been developed. We now describe briefly one of the more intriguing and popular methods used to carry out the average in (7) known as the "replica trick" (Grinstein, 1974; see also Stinchcombe [1983], Lubensky [1979], Anderson [1979] and Binder and Young [1986] for discussions). This method makes use of the formal identity

$$\left\langle \lim_{n \to 0} \frac{x^n - 1}{n} \right\rangle = \langle \ln x \rangle \tag{10}$$

Taking x to be the partition function (5) with a given set of variables ϵ_1, ϵ_2, ..., ϵ_N, we see that

$$Z^n = [\text{Tr}_{\mathbf{S}_i} e^{-\beta H\{\mathbf{S}_i, \epsilon_i\}}]^n = \text{Tr}_{\{s_{i,1}\}} \text{Tr}_{\{s_{i,2}\}} \cdots \text{Tr}_{\{s_{i,n}\}} \exp\left\{-\beta \sum_{\alpha=1}^{n} H[\mathbf{S}_{i,\alpha}, \epsilon_i]\right\} \quad (11)$$

The spin variables $S_{i,\alpha}$, for different α, are independent dynamical variables and we essentially have n copies, or replicas, of the system, all with the same configuration of magnetic atoms specified by the set of variables $\{\epsilon_j\}$. One now interchanges the calculation of the average over the impurity distribution and the limit $n \to 0$ in (10) and, as well, attempts to analytically continue to $n = 0$ from integer values of n. This is, in fact, the delicate step in the procedure and, at least in one well-documented situation, (Anderson, 1979) leads to unphysical results at low temperature. Even in this case the replica method proved invaluable in leading to the correct solution (Mezard, et al. 1987 and references therein).

However, if one nevertheless proceeds, the average over $\{\epsilon_j\}$ can be carried out (in closed form if the distribution is simple enough) and the resulting effective Hamiltonian will be translationally invariant but couple spins in different replicas, that is,

$$\int d\epsilon_1 d\epsilon_2, \ldots d\epsilon_N P(\epsilon_1, \epsilon_2, \ldots, \epsilon_N) \text{Tr} \exp\left\{-\beta \sum_{\alpha=1}^{n} H[\mathbf{S}_{i,\alpha}, \epsilon_i]\right\}$$
$$= \text{Tr} \exp\{-\beta H'[\mathbf{S}_{i,1}, \mathbf{S}_{i,2}, \ldots, \mathbf{S}_{i,n}]\} = Z'(h, T, n) \quad (12)$$

where the functional form of H' depends on the impurity distribution function $P(\epsilon_1, \epsilon_2, \ldots, \epsilon_N)$. The translationally invariant effective Hamiltonian can then be treated in the usual way in mean field theory (Chapter 3) or by renormalization group methods (Chapter 6). In most cases the results obtained by the replica methods are consistent with those arrived at by other means.

(b) Nature of Phase Transitions in Disordered Systems

To this point few exact results on the statistical mechanics of disordered materials for dimensionality $d > 1$ exist. One of the exceptions is a calculation by McCoy and Wu (reviewed by McCoy [1972] and by McCoy and Wu [1973]) for a two-dimensional Ising model. They considered the case of a constant nearest-neighbor interaction in the horizontal direction on a square lattice and random interactions between spins in different rows. McCoy and Wu found that a well-defined transition temperature exists but that the critical behavior is modified by disorder. For example, the zero-field specific heat is infinitely differentiable but nonanalytic at T_c in contrast to the logarithmic divergence found in the pure model (Section 5.A).

A second interesting analytic result is due to Griffiths [1969]. He showed that the magnetization $m(h, T)$ of a disordered Ising model is not an analytic function of h at $h = 0$ for *any* $T < T_c(p = 1)$. Thus the phase diagram in the h–T plane of the disordered Ising model is as shown in Figure 9.10. The *Griffiths singularities,* in the region $T_c(p) < T < T_c(1)$, are weak essential singularities which may in fact be unobservable experimentally. They arise, even for $p < p_c$, the percolation concentration, from arbitrarily large clusters of connected spins which occur with finite probability for any nonzero p. This effect of large connected islands is reminiscent of the singular structure (Lifshitz tails) of the density of states near band edges discussed in Section 9.A.

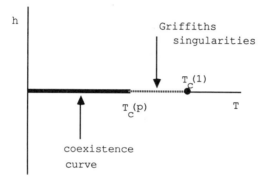

Figure 9.10   Phase diagram in the h–T plane of a disordered Ising model.

Finally, we mention a heuristic argument due to Harris [1974] regarding the nature of phase transitions in disordered materials. We have seen, in Chapters 5 and 6 that the singular part of the free energy obeys a homogeneity or scaling relation of the form

$$g(t, h, u) = |t|^{d/y_1} g\left(\frac{t}{|t|}, h\,|t|^{-y_h/y_1}, u\,|t|^{-y_u/y_1}\right) \qquad (13)$$

In the case of a disordered system [such as (1)], we identify the quantity u with the fluctuation in nearest-neighbor coupling constants and the fixed point $t = h = u = 0$ with the pure system fixed point. The quantities y_1, y_h, and y_u are the exponents determined from the linearized recursion relations near the pure system fixed point. If the exponent y_u is greater than zero, the perturbation u is relevant and the fixed point is unstable. The renormalization group flow is then usually to another, stable, fixed point and it is conventional to call the exponent $\phi_u = y_u/y_1$ the *crossover* exponent. Since the fluctuations in the dimensionless coupling constants are temperature-like variables, we expect that $\phi_u = 1$.

It is clear, from (13), that in general the variable u becomes important at a characteristic temperature

$$u\,|t|^{-\phi_u} \approx 1 \qquad (14)$$

We assume that the system at temperature t has correlation length ξ. A measure of the magnitude of u is therefore

$$u \approx \frac{[\int_{\xi^d} d^d r \langle (J(\mathbf{r}) - \langle J \rangle)^2 \rangle]^{1/2}}{\xi^d \langle J \rangle} \approx \xi^{-d/2} \tag{15}$$

where $J(r)$ is the coupling constant at point r and where the angular brackets indicate averaging with respect to the impurity distribution function. Using (15) in (14), we have

$$u \, |t|^{-\phi_u} \approx u \, |t|^{-1} \approx |t|^{d\nu/2-1} = |t|^{-\alpha/2}$$

where, in the last step, we have used the hyperscaling relation $d\nu = 2 - \alpha$, where α is the specific heat exponent of the pure system. We conclude, therefore, that disorder is relevant if the specific heat exponent α is greater than zero (three-dimensional Ising model), irrelevant if it is less than zero (three-dimensional Heisenberg model, possibly XY model).

A slightly different way of phrasing the argument above is to think of the system as a collection of domains of linear dimension ξ. The degree of order in each domain is characteristic of a system a distance ΔT from the critical point, and this deviation is determined by t and by the quantity u defined in (15). As $t \to 0$, the entire system will approach the critical point simultaneously only if $\Delta T(t, u) \to 0$. It is tempting to conclude that if this does not occur, the transition will not be sharp but will rather be smeared over some finite temperature interval and that the sharp power law singularities characteristic of pure systems will be replaced by smooth functions characterized by some width ΔT. However, the evidence (see, e.g., Stinchcombe, 1983) from renormalization group and Monte Carlo calculations is that a sharp transition exists nevertheless, but with modified critical exponents, namely those of a disorder fixed point.

Another interesting case that we can consider is the model (1) with constant exchange interactions but with a random magnetic field with mean value $\langle h_i \rangle = 0$ (Imry and Ma, 1975). In this case one can show that the system may be unstable, at low temperature, against domain formation. The argument goes as follows. The energy cost of introducing a domain of volume L^d into the sample is of order L^{d-1} for an Ising model and of order L^{d-2} for models with continuous symmetry. (The domain wall is diffuse in this case. The spins can rotate continuously over the distance L. Writing $\mathbf{S}_i \cdot \mathbf{S}_j = S^2 \cos(\hat{n}_i \cdot \hat{n}_j) \approx S^2[1 - \frac{1}{2}(\pi/L)^2]$, for i, j nearest neighbors, we immediately arrive at the result that the domain wall energy scales as L^{d-2}.) The sum of the random fields, over a finite region of size L^d, has a finite expectation value which is randomly positive or negative but typically has a magnitude given by $\langle (\Sigma_i h_i)^2 \rangle^{1/2} \approx L^{d/2}$. Thus the energy gain due to alignment with the local field is of order $L^{d/2}$. Therefore, for large L, domain formation is energetically favorable if

$$L^{d/2} > L^{d-1} \quad \text{(Ising model)}$$

$$L^{d/2} > L^{d-2} \quad (n \geq 2)$$

which implies that the Ising model in a random field will break up into domains for $d < 2$ and models with continuous symmetry for $d < 4$. Note that the entropy which has not been considered can only help this process. Domain formation of course implies that the order parameter is zero. The situation of a random magnetic field does not turn out to be of purely academic interest. Imry and Ma [1975] discuss a number of possible physical realizations of this case.

We now briefly discuss the dependence of the critical temperature on p, the concentration of magnetic atoms in the sample, assuming that the system has a sharp continuous transition with exponents which may or may not be those of the pure system. Our discussion closely follows that of Lubensky [1979]. For p close to 1 it is intuitively clear that T_c will decrease linearly with p. More interesting is the behavior of T_c near the percolation concentration p_c. A scaling argument can be used to find the dependence of T_c on p. We assume that at $T = 0$, $p = p_c$, a crossover occurs between the thermal fixed point (not necessarily the pure system fixed point) and the percolation fixed point. Thus, as in (13), we may write the free energy or other functions of interest in the scaled form

$$f(T, p - p_c) = |p - p_c|^{d/y_p} F\left[\frac{g(T)}{|p - p_c|^\phi}\right] \qquad (16)$$

where $g(T)$ and the crossover exponent ϕ remain to be identified, and where y_p is the appropriate percolation exponent equivalent to the thermal exponent y_1. In (9.B.4) we have seen that for an Ising model at low temperature, $T\chi(0, T)$ is related to the percolation function $S(p)$. Thus we write

$$T\chi(0, T, p) = |p - p_c|^{-\gamma_p} \Psi\left[\frac{g(T)}{|p - p_c|^\phi}\right] \qquad (17)$$

where γ_p again is the percolation exponent.

To make further progress we use a conceptual picture of percolation clusters due to Skal and Shklovskii [1975] and de Gennes [1976]. We assume that after all dangling bonds (irrelevant for phase transitions) are removed from the infinite cluster, the resulting "backbone" is schematically of the form shown in Figure 9.11. In this picture relatively dense regions (the black "blobs") are connected by long, essentially one-dimensional strands. We assume that the average distance between the blobs is the percolation correlation length ξ_p and that the number of segments in the connecting strand is $L \geq \xi_p$. The spin-spin correlation length $\xi_M(T)$ along one of the one-dimensional segments can be determined analytically. For the one-dimensional Ising model we have (Section 3.F)

$$\xi_M(T) = \frac{-1}{\ln\left[\tanh\left(J/k_B T\right)\right]} \approx \frac{1}{2} e^{2J/k_B T} \qquad (18)$$

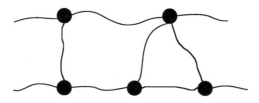

Figure 9.11   Schematic picture of the backbone of the infinite cluster near the percolation threshold.

at low temperatures. For the classical n component Heisenberg model Stanley [1969] has shown that

$$\xi_M(T) \approx \frac{2n}{n-1} \frac{J}{k_B T} \tag{19}$$

This length ξ_M is characteristic of the magnetic order as long as $\xi_M << L$. Once the magnetic correlation length becomes comparable to L, the blobs form a strongly interacting d-dimensional network and the one-dimensional correlations are no longer relevant. We assume, therefore, that

$$T\chi(0, T, p) = |p - p_c|^{-\gamma_p} \Psi\left[\frac{\xi_M(T)}{L(p)}\right] \tag{20}$$

and, further, that $L(p) \sim |p - p_c|^{-\zeta}$, where $\zeta \geq \nu_p$ since $L \geq \xi_p$. The crossover from percolative to magnetic behavior then occurs at $\xi_M(T) \approx L \approx |p - p_c|^{-\zeta}$, and for the Ising model we therefore find

$$\frac{k_B T_c(p)}{J} \approx -\frac{2}{\zeta \ln|p - p_c|} \tag{21}$$

while

$$\frac{k_B T_c(p)}{J} \approx \frac{2n}{n-1} |p - p_c|^{\zeta} \tag{22}$$

for the n vector model. The exponent ζ is thought to be exactly 1 in all dimensions on the basis of renormalization group calculations (Lubensky, 1979), and if this holds, T_c approaches zero linearly as a function of $p - p_c$ for the n vector models but, independent of ζ, with an infinite slope for the Ising model. High temperature series and Monte Carlo calculations support these conclusions (Stinchcombe, 1983).

We now turn briefly to the calculation of critical exponents by renormalization group methods. The application of the position space renormalization group methods described in Chapter 6 to disordered systems is considerably more difficult than in the case of pure materials. Under a rescaling transformation, not only the coupling constants but also the probability distribution of the random interactions is modified and at the fixed points is represented by a *fixed distribution* rather than by a finite set of numbers. We will not discuss the various methods that have been developed to deal with these difficulties but rather, refer the reader to the reviews of Kirkpatrick [1979] and Stinchcombe [1983].

Field-theoretic and momentum space renormalization group methods have also been applied to the problem of dilute magnets. We again refer to the aforementioned reviews and references therein. The physical picture that has emerged from this work is consistent with the Harris criterion. In pure systems with specific heat exponents greater than zero the pure system fixed point is unstable with respect to disorder, and a new disorder fixed point with its own set of critical exponents governs the critical behavior of the system. Conversely, the exponents of systems with negative specific heat exponents are unaffected by disorder.

9.D STRONGLY DISORDERED SYSTEMS: AMORPHOUS MATERIALS AND SPIN GLASSES

We now turn to the most difficult and least understood subject, which we have included under the general heading of disordered systems, namely the physics of amorphous or glassy materials and spin glasses. As in the previous sections, we have leaned heavily on recent reviews, in particular those of Anderson [1979], Lubensky [1979], Jäckle [1986], and Binder and Young [1986]. Because the theory of "strongly disordered" or "topologically disordered" systems, as such materials are sometimes referred to, is very much under development, we have kept this section short and have only attempted to discuss a few of the more important concepts.

A large number of quite different materials form glasses as they are cooled from the melt. Films of monatomic materials such as Ge and Si can be readily prepared in an amorphous form. Metallic glasses, often a mixture of a metal such as Ni or Co with a metalloid such as P can be prepared by rapid cooling techniques at least in certain composition ranges. Intermetallic compounds ($TbFe_2$ and others) will form glasses if sputtered onto a cold substrate.

The common feature in these preparation techniques is the rapid extraction of thermal energy, and it is possible that even the simplest monatomic liquids would form glasses if heat could be extracted rapidly enough. Phenomenologically, the glass transition is characterized by a rapid freezing out of transport processes at the transition temperature, T_g—the system becomes rigid, diffusion decreases rapidly, the shear viscosity increases. The transition temperature T_g depends on the cooling rate, as do the low-temperature physical properties such as the specific volume. This is already an indication that in contrast to the phase transitions that we have described previously, the glass transition may not be an equilibrium phenomenon. Further evidence for this point of view comes from the observation that below the transition temperature the properties of the glass continue to change as a function of time, although very slowly, and it is believed that given sufficient time the sample will eventually reach the equilibrium low-temperature configuration, namely a crystal.

One of the unusual features of amorphous materials is the low-temperature behavior of the specific heat, which is linear in T, even for insulators, rather than cubic as one expects for the phonon specific heat in three-dimensional crystals. This feature can be understood on the basis of the following conceptual picture, which also qualitatively accounts for the observed kinetic effects alluded to above. A simplified version is shown below in Figure 9.12. At high temperature the free energy $A(x)$ of the system, as a function of a number of configurational parameters x, has a unique minimum which is accessible from any point in the configuration space. At lower temperatures it is assumed that this free energy is the more complicated function, with many local minima shown in the lower graph and an absolute minimum (presumably a crystalline state) some distance in configuration space from the high-temperature minimum. It is clear that if the system is rapidly cooled—more rapidly than the characteristic relaxation times—it may become trapped at one of the local minima. Once the free-energy barriers have formed, the probability of a transition is $0(\exp\{-\beta\Delta A\})$, where ΔA is the height of the free-energy barrier, and if ΔA is roughly of magnitude 1 eV, the transition rate can become vanishingly small. On the other hand, if the cooling process is slow, the system may be able to follow the absolute free-energy minimum to the ground state.

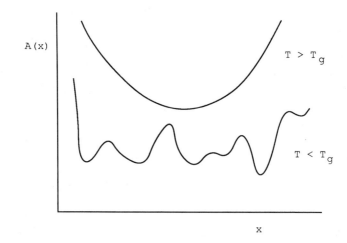

Figure 9.12   Schematic picture of the free energy of a system above and below the glass transition temperature (see the text).

Returning to the specific heat, we now assume that there are states very close in energy to the local free energy minima and that the density $n(E)$ of these states approaches a constant, $n(0)$, as the energy difference between the excited and local minimum states approaches zero. Note that this dependence is in sharp contrast to the phonon density of states in a crystal which has the functional form $n_{ph}(E) \sim E^3$ for small E. The original picture of these "tunneling states" is due to Anderson et al. [1971]. Assuming that this picture is correct, we have

$$\langle E(T) \rangle = \int dE' n(E') \frac{E'}{\exp\{\beta E'\} - 1} \approx \frac{\pi^2}{6} n(0)(k_B T)^2 \tag{1}$$

and for the specific heat,

$$C(T) \approx \frac{\pi^2}{3} n(0)k_B^2 T \tag{2}$$

This formula, without further justification, is purely phenomenological. Anderson et al., however, predicted a number of other effects on the basis of the tunneling model, which seem to be consistent with experiment and which lend support to this picture. In a later calculation of the low-lying states of a model Hamiltonian for a spin glass, Walker and Walstedt [1977] found that the density of low-lying excitations indeed approaches a constant as the energy approaches zero. In spin glasses the magnetic specific heat is, as in physical glasses, linear in the temperature.

A convincing microscopically based theory of the glass transition has not to this point been developed. For a review of much of the more recent work, the reader is referred to the article by Jäckle [1986] and references therein.

We now briefly discuss spin glasses, which, in some ways, are the magnetic counterpart of molecular glasses. The first examples of the spin glass transition (see, e.g., Canella and Mydosh, 1972) were found in dilute alloy systems such as $Au_{1-x}Fe_x$ with x very small. Experimentally, one sees a rather sharp maximum in the zero-field susceptibility, a broad maximum in the specific heat, and an absence of any long-range order below this spin glass transition temperature, although there is both hysteresis and remanence. Many other materials (including nonmagnetic ones) have since been identified as having a transition of the spin glass type (see Binder and Young, 1986).

As in the case of molecular glasses, one interprets the spin glass transition as a freezing out of fluctuations—in the molecular glasses, the freezing out of large-scale structural rearrangements, in the spin glass the freezing out of magnetic transitions. The existence of hysteresis and remanence is already evidence that this occurs. There is, however, one important conceptual difference between the two types of systems. In the case of molecular glasses, one usually knows that the true equilibrium state at low temperatures is a crystal. In spin glasses there may be no unique ground state. The magnetic atoms in dilute alloys, such as $Au_{1-x}Fe_x$ or $Cu_{1-x}Mn_x$, do not diffuse appreciably at low temperatures. The interaction between magnetic atoms is therefore also frozen or quenched and, in a metallic environment, this interaction is most simply modeled by the Ruderman–Kittel–Kasuya–Yosida (RKKY) interaction, which is of the form (Mattis, 1981)

$$H = \sum_{i,j} J(R_{ij})\mathbf{S}_i \cdot \mathbf{S}_j = \sum_{i,j} J_0 \frac{\cos (2k_F R_{ij} + \phi)}{R_{ij}^3}\mathbf{S}_i \cdot \mathbf{S}_j \tag{3}$$

(i.e., *long range* and *oscillatory*). The RKKY oscillations are caused by the sharpness of the Fermi surface in the same way as the Friedel oscillations discussed in Chapter 8. A given magnetic atom thus interacts both via ferromagnetic and antiferromagnetic interactions with others, and it is conceivable that no simple ground state analogous to the crystalline state of molecular glasses exists.

This idea can be made more precise and is generally referred to as *frustration*. Consider a simpler model Hamiltonian than (3), namely the nearest-neighbor Ising model on a square lattice but with interactions which are randomly either $+J$ (ferromagnetic) or $-J$ (antiferromagnetic). In the diagram below we show two elementary *plaquettes* (elementary square units of four spins), one of which is frustrated, one which is not. It is easy to see that no arrangement of spins on the frustrated plaquette will leave all bonds satisfied. A choice of $\sigma = 1$ in the lower right-hand corner leaves the lower horizontal bond in an unfavorable configuration, the opposite choice leaves the right vertical bond in a high-energy state. A plaquette is frustrated if the product of the four coupling constants is negative, not frustrated if it is positive, as can easily be verified explicitly. We note in passing that the nonrandom Ising antiferromagnet on the triangular lattice is frustrated and this model has a finite ground-state entropy per spin and no phase transition.

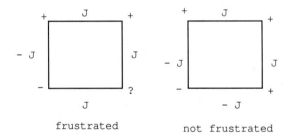

frustrated   not frustrated

Consider now the segment of the square lattice shown below, which contains four frustrated plaquettes labelled by the open circles. The $+$ and $-$ signs between the lattice sites indicate the sign of the interaction between spins. In the absence of a magnetic field we have spin-flip symmetry and can choose the orientation of one of the spins (upper left-hand corner) freely. We next choose the remaining spins, in turn, satisfying all bonds until we reach a frustrated plaquette in which we are forced to make one unfavorable assignment. Since the bond in the unfavorable configuration is shared between two plaquettes, the unfavorable assignment propagates until we arrive at another frustrated plaquette at which the process terminates. This is indicated in Figure 9.13 by the dotted lines perpendicular to the bonds in unfavorable configurations. The length of these lines or "strings" is a measure of the energy of the spin configuration relative to a state with all bonds satisfied. The problem of determining the ground state is thus equivalent to finding the set of strings, connecting frustrated plaquettes, with shortest total length. This is a nontrivial problem and we can already see from this simple example how a high degree of degeneracy can arise. In Figure 9.14 we show two string configurations with

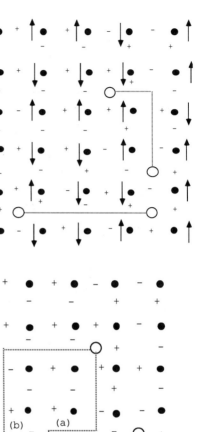

Figure 9.13 Section of square lattice with four frustrated plaquettes.

Figure 9.14 String configurations (a) and (b) together with the nearest-neighbor string are each degenerate with that of Figure 9.13. The assignment of coupling constants is, of course, the same.

the same energy as the state of Figure 9.13, which, however, imply different orientations of the spins.

Thus a frustrated system is, we believe, characterized by a high degree of ground-state degeneracy. In more realistic models we would also expect to find a high density of low-lying excited states and dynamical effects due to string motion. Further support for the idea that frustration is an important aspect of spin glass ordering comes from the Mattis model (Mattis, 1976), which, although random in nature, undergoes an ordinary continuous phase transition. The Hamiltonian of the Mattis model is

$$H = -\sum_{i,j} J(R_{ij})\epsilon_i\epsilon_j \mathbf{S}_i \cdot \mathbf{S}_j - h\sum_i S_{iz} \qquad (4)$$

where $\epsilon_i = +1$ or -1 are quenched random variables. Defining new spins

$$\boldsymbol{\tau}_i = \epsilon_i \mathbf{S}_i$$

we obtain

$$H = - \sum_{i,j} J(R_{ij}) \boldsymbol{\tau}_i \cdot \boldsymbol{\tau}_j - h \sum_i \epsilon_i \tau_{iz} \tag{5}$$

which represents a translationally invariant "pseudo-spin" interaction and a Zeeman term with a random magnetic field. For $h = 0$, randomness is completely absent and the partition function for this model with quenched disorder is the same as the partition function of the pure Heisenberg, XY, or Ising model, depending on the number of spin components. The specific heat, internal energy, and other zero-field thermodynamic functions thus have simply the singularities of the pure model.

It is clear that the ground state of (5) in zero magnetic field is the state with the $\boldsymbol{\tau}$ spins fully aligned; hence there is no frustration. The original spins $\{S\}$ are randomly oriented, but as far as the critical behavior of the system is concerned, the model is equivalent to a ferromagnet and it is the absence of frustration that makes it unsuitable as a model for spin glasses.

A model that does seem to have the essential features necessary to describe the spin glass transition is the Edwards–Anderson model (Edwards and Anderson, 1975), which has the Hamiltonian

$$H = - \sum_{i,j} J(R_{ij}) \mathbf{S}_i \cdot \mathbf{S}_j - h \sum_i S_{iz} \tag{6}$$

where the interaction $J(R_{ij})$ is a random variable subject to a Gaussian distribution

$$P\{J(R_{ij})\} = \frac{1}{\sqrt{2\pi[\Delta J(r_{ij})]^2}} \exp\left\{ -\frac{[J(R_{ij}) - \bar{J}(R_{ij})]^2}{2[\Delta J(R_{ij})^2]} \right\} \tag{7}$$

and where the range of the interaction can be varied. We shall not review the various methods that have been devised to deal with this model (Binder and Young, 1986) but do wish to discuss briefly the characterization of the spin glass phase in terms of an order parameter.

It is believed that for small values of $\bar{J}$ there is no spontaneous magnetization or other form of long-range order. It is clear then that expectation values such as

$$\mathbf{m_q} = \frac{1}{N} \sum_i \langle \mathbf{S}_i \rangle \, e^{i\mathbf{q} \cdot \mathbf{r}_i}$$

where the angular brackets indicate thermal averages, must all be zero. On the other hand, if the spin configuration freezes, the local expectation value $\langle \mathbf{S}_i \rangle$ is well defined but will be oriented in different directions, depending on the site. One choice of order parameter which distinguishes between a high-temperature paramagnetic phase and a low-temperature frozen configuration is the Edwards–Anderson order parameter which arises naturally from the replica approach.

$$q \equiv \langle \mathbf{S}_i \rangle^2 = \frac{1}{N} \sum_i \langle \mathbf{S}_i \rangle^2 \tag{8}$$

which is zero at high temperature, since each individual spin has zero expectation value in the high-temperature phase, and positive in the spin glass state. When the interactions are long range the order parameter (8) is inadequate and the situation is more complex (Mezard, et al. [1987]).

The study of glasses and spin glasses is a fascinating but difficult subject which promises to be an active area of research for many years to come. We hope that we have given the reader at least an introduction to the subject.

PROBLEMS

9.1. *Electron States in One Dimension.* Consider the transfer matrix **T** (9.A.13) for the case of a pure material: $\epsilon_j = \epsilon_A$ for all j.

(a) Show that equation (9.A.15) can only be satisfied if the eigenvalues of **T** are complex or ± 1.

(b) Formulate the one-electron problem with fixed end boundary conditions $A_1 = A_N = 0$ and derive the eigenvalues and eigenfunctions using the transfer matrix approach for a pure type A material.

9.2. *One-Dimensional Liquid: Disordered Kronig–Penney Model.* Consider the problem of an electron of momentum $\hbar k$ incident from the left on a set of $N + 1$ scatterers with potential

$$V(x) = \sum_{i=0}^{N} \frac{\hbar^2}{2m} u\delta(x - x_i)$$

where $u > 0$ and where the location of the δ-function scatterers is for the moment unspecified except for $x_{i+1} > x_i$. The Schrödinger equation is therefore

$$-\frac{d^2\psi(x)}{dx^2} + \sum_{i=0}^{N} u\delta(x - x_i)\psi(x) = k^2\psi(x)$$

Let $\Delta_j = x_j - x_{j-1}, j = 1, 2, \ldots, N$, and write

$$\psi(x) = A_0 e^{ik(x-x_0)} + B_0 e^{-ik(x-x_0)} \qquad x < x_0$$
$$= A_j' e^{ik(x-x_{j-1})} + B_j' e^{-ik(x-x_{j-1})} \qquad x_{j-1} < x < x_j$$
$$= A_j e^{ik(x-x_j)} + B_j e^{-ik(x-x_j)} \qquad x_{j-1} < x < x_j$$
$$= A_{N+1}' e^{ik(x-x_N)} \qquad x \geq x_N$$

where $A_j' = A_j \exp\{-ik\Delta_j\}$ and $B_j' = B_j \exp\{ik\Delta_j\}$.

(a) Apply the usual continuity conditions to the wave function and show that

$$\begin{bmatrix} A_j \\ B_j \end{bmatrix} = \begin{bmatrix} \dfrac{1}{t} & \dfrac{r^*}{t^*} \\ \dfrac{r}{t} & \dfrac{1}{t^*} \end{bmatrix} \begin{bmatrix} A_{j+1}' \\ B_{j+1}' \end{bmatrix} = \mathbf{Q} \begin{bmatrix} A_{j+1}' \\ B_{j+1}' \end{bmatrix}$$

where $t = |t| \, e^{i\delta} = [1 - u/2ik]^{-1}$ and $r = -i \, |r| \, e^{i\delta} = u/2ik \, [1 - u/2ik]^{-1}$ are the complex transmission and reflection coefficients of each scatterer. Using the relation between A', B' and A, B given above, we have

$$\begin{bmatrix} A'_j \\ B'_j \end{bmatrix} = \mathbf{M}_j \mathbf{Q} \begin{bmatrix} A'_{j+1} \\ B'_{j+1} \end{bmatrix} \qquad \text{where } \mathbf{M}_j = \begin{bmatrix} e^{-ik\Delta_j} & 0 \\ 0 & e^{ik\Delta_j} \end{bmatrix}$$

contains the spacing between neighboring scatterers.

(b) Define $T_N = |A'_{N+1}|/|A_0|$ and $R_N = |B_0|/|A_0|$ to be the transmission and reflection amplitudes of the array of $N + 1$ potentials and show that

$$\frac{1 + R_N^2}{T_N^2} = \frac{|A_0|^2 + |B_0|^2}{|A'_{N+1}|^2} = a_N$$

where

$$\begin{bmatrix} a_N & b_N \\ b_N^* & a_N \end{bmatrix} = [\mathbf{Q}^+ \mathbf{M}_N^+ \mathbf{Q}^+ \mathbf{M}_{N-1}^+ \cdots \mathbf{Q}^+ \mathbf{M}_1^+ \mathbf{Q}^+ \mathbf{Q} \mathbf{M}_1 \mathbf{Q} \cdots \mathbf{M}_N \mathbf{Q}]$$

(c) Show that the matrix elements a_N, b_N obey the recursion relation

$$a_N = a_{N-1} \frac{1 + |r|^2}{|t|^2} + 2 \,\text{Re} \, \frac{b_{N-1} r e^{2ik\Delta_N}}{|t|^2}$$

$$b_N = 2a_{N-1} \frac{r^*}{(t^*)^2} + \frac{b_{N-1}^* (r^*)^2 e^{-2ik\Delta_N} + b_{N-1} e^{2ik\Delta_N}}{(t^*)^2}$$

(d) Show that if the spacings Δ_j are independent random variables governed by a distribution $P(\Delta)$ with the property

$$\int_0^\infty d\Delta \, P(\Delta) e^{2ik\Delta} = 0$$

the transmission coefficient is zero.

(e) More generally, assume that $P(\Delta) = 1/W$ for $0 < \Delta < W$. Show numerically, by iteration of the recursion relations for a_N and b_N for $u = 1$, $W = 1$ and a range of k, that $T_N(k) \to 0$ as N becomes large.

9.3. *Real Space Renormalization Group for Percolation on the Square Lattice.* Consider the following nine-site block on the square lattice:

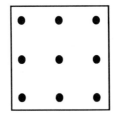

(a) Adapt the real space renormalization group treatment of Section 9.B to this cluster. Define the block to be occupied if the majority of the sites are occupied, empty otherwise. Find the fixed point and the correlation length exponent ν.

(b) As an alternative, define the cluster to be occupied if a path of connected sites from the left side of the block to the right side exists. Find p_c and ν.

A

Occupation Number Representation

An important aspect of quantum many-particle systems is the symmetry requirement that identical particles obey either Bose or Fermi statistics. This requirement can be conveniently incorporated into the formalism by making use of what is known as the occupation number or Fock representation, sometimes also referred to as second quantization. The underlying concept is that when dealing with a system of identical particles, one cannot specify which particle is in which state, only which states are occupied, and how many particles there are in each.

Let $|i\rangle$ be a complete set of one-particle states with wave functions

$$\varphi_i(\mathbf{r}) = \langle \mathbf{r} \,|\, i \rangle \tag{1}$$

We assume that these states are orthonormal, that is,

$$\langle i \,|\, j \rangle = \int d^3r \, \varphi_i^*(\mathbf{r})\varphi_j(\mathbf{r}) = \delta_{ij} \tag{2}$$

A state in which particle 1 is in state 1, particle 2 is in state 2, and so on, is represented by the product wave function

$$(\mathbf{r}_1, \mathbf{r}_2, \ldots, \mathbf{r}_N \,|\, 1, 2, 3, \ldots, N) = \varphi_1(\mathbf{r}_1)\varphi_2(\mathbf{r}_2) \cdots \varphi_N(\mathbf{r}_N) \tag{3}$$

If the particles are identical, this state is indistinguishable from one in which the labels are *permuted,* for example

$$\varphi_{P1}(\mathbf{r}_1)\varphi_{P2}(\mathbf{r}_2) \cdots \varphi_{PN}(\mathbf{r}_N) \tag{4}$$

where P is a permutation operator, or a mapping, of the numbers 1, 2, 3, . . . , N on themselves, for example,

$$P = \begin{pmatrix} 1 & 2 & 3 & 4 & 5 & 6 & 7 & 8 & 9 \\ 9 & 2 & 4 & 6 & 1 & 7 & 3 & 5 & 8 \end{pmatrix} \tag{5}$$

In this example $P1 = 9$, $P2 = 2$, $P3 = 4$, and so on. Each such permutation can be written as a product of *transpositions*. A transposition is a permutation in which only two labels are interchanged, for example,

$$T_{36} = \begin{pmatrix} 1 & 2 & 3 & 4 & 5 & 6 & 7 & 8 & 9 \\ 1 & 2 & 6 & 4 & 5 & 3 & 7 & 8 & 9 \end{pmatrix}$$

and the permutation (5) can be written

$$P = T_{89} T_{67} T_{59} T_{46} T_{34} T_{19}$$

The way a permutation is factorized into transpositions is clearly not unique, but it is possible to classify all permutations as being either even or odd, depending on whether an even or odd number of transpositions are required to achieve it. Altogether, the numbers 1, 2, 3, . . . , N can be permuted in $N!$ different ways. The quantum-mechanical symmetry requirement for a system of identical particles is then that a state must be a linear combination of all possible ways the single-particle state labels can be attached to the particles. Let

$$S_P = \begin{cases} 1 & \text{for } P \text{ even} \\ -1 & \text{odd} \end{cases} \tag{6}$$

For a system of *fermions* the *Pauli exclusion principle* states that the wave function must be odd under transpositions. Consider the situation which one of N fermions is in single-particle state 1, one in state 2, and so on. The corresponding many-particle state can be expressed as

$$|1, 2, \ldots, N\rangle = \nu \sum_P S_P |P1, P2, \ldots, PN) \tag{7}$$

where ν is a normalization factor, and where we use the notation that kets $|1, 2, \ldots\rangle$ have been properly symmetrized, while $|1, 2, \ldots)$ is a many-particle state in which particle 1 is in state 1, particle 2 in state 2, and so on. As an example, consider a state with three particles,

$$(\mathbf{r}_1, \mathbf{r}_2, \mathbf{r}_3 \,|\, 1, 2, 3\rangle = \nu \sum_P S_P(\mathbf{r}_1, \mathbf{r}_2, \mathbf{r}_3 \,|\, P1, P2, P3)$$

$$\begin{aligned} &= \nu \,(\varphi_1(\mathbf{r}_1)\varphi_2(\mathbf{r}_2)\varphi_3(\mathbf{r}_3) - \varphi_1(\mathbf{r}_1)\varphi_3(\mathbf{r}_2)\varphi_2(\mathbf{r}_3) \\ &\quad + \varphi_2(\mathbf{r}_1)\varphi_3(\mathbf{r}_2)\varphi_1(\mathbf{r}_3) - \varphi_2(\mathbf{r}_1)\varphi_1(\mathbf{r}_2)\varphi_3(\mathbf{r}_3) \\ &\quad + \varphi_3(\mathbf{r}_1)\varphi_1(\mathbf{r}_2)\varphi_2(\mathbf{r}_3) - \varphi_3(\mathbf{r}_1)\varphi_2(\mathbf{r}_2)\varphi_1(\mathbf{r}_3)) \\ &= \langle \mathbf{r}_1, \mathbf{r}_2, \mathbf{r}_3 \,|\, 1, 2, 3) = \langle \mathbf{r}_1, \mathbf{r}_2, \mathbf{r}_3 \,|\, 1, 2, 3\rangle \end{aligned} \tag{8}$$

if we require that the states are normalized. In the case of the three-particle system we thus have $\nu = (3!)^{-1/2}$ and in general,

$$\nu = (N!)^{-1/2} \tag{9}$$

Equations (7) and (8) can be expressed as determinants known as *Slater determinants*.

$$\langle \mathbf{r}_1, \mathbf{r}_2, \ldots, \mathbf{r}_N \,|\, 1, 2, \ldots \rangle$$

$$= \frac{1}{(N!)^{1/2}} \begin{vmatrix} <\mathbf{r}_1|1> & <\mathbf{r}_2|1> & \cdots & <\mathbf{r}_N|1> \\ <\mathbf{r}_1|2> & <\mathbf{r}_2|2> & \cdots & <\mathbf{r}_N|2> \\ \vdots & \vdots & & \vdots \\ <\mathbf{r}_1|N> & <\mathbf{r}_2|N> & \cdots & <\mathbf{r}_N|N> \end{vmatrix} \tag{10}$$

Since a determinant is zero when two rows or two columns are equal, it follows that no two fermions can occupy the same state.

In the case of *Bose particles* the symmetry requirement is that the wavefunction should be even under the transposition of two particle or state labels. Consider a state in which n_1 bosons are in state 1, n_2 are in state 2, and so on. Let N be the total number of particles

$$\sum_i n_i = N$$

We obtain properly normalized wave functions if we write

$$|n_1, n_2, \ldots \rangle = \left(\frac{\prod (n_i!)}{N!} \right)^{1/2} \sum_{\text{distinct } P}{}' |1, \ldots 1, 2, \ldots, , \ldots) \tag{11}$$

In (11) the notation Σ' indicates that only those permutations in which the particles do not remain in the same state are included.

We wish to consider a system of particles that interact through two-body interactions

$$H = \sum_{i=1}^{N} \frac{p_i^2}{2m} + \sum_{i=1}^{N} U(\mathbf{r}_i) + \frac{1}{2} \sum_{i \neq j} v(\mathbf{r}_i - \mathbf{r}_j) \tag{12}$$

Only when the interparticle interaction is zero are the eigenstates of the Hamiltonian the simple product states that we used above. In general, a many-particle eigenstate cannot be expressed in terms of a single symmetrized product wave function. Instead, linear combinations of such wave functions are required. It is then no longer practical to make use of wave functions that make explicit reference as to which particles are in which states. We must also rewrite the Hamiltonian without reference to particle labels. In this process the emphasis is shifted from the wave functions to the operators.

The fundamental operators in the occupation number representation are the *creation* and *annihilation* operators. Consider first the case of *fermions* and let $|0\rangle$ be the vacuum (no particle) state. We define the creation operator $a_1^\dagger$ by the effect it has on a state in which the single-particle states 2, 3, . . . , N are occupied:

$$|1, 2, 3, \ldots, N\rangle = a_1^\dagger |2, 3, \ldots, N\rangle$$
$$= a_1^\dagger a_2^\dagger a_3^\dagger \cdots a_N^\dagger |0\rangle \tag{13}$$

Remembering that this state must be odd under transposition of labels, we must have, for example,

$$|1, 2, 3, 4, \ldots, N\rangle = -|1, 2, 4, 3, \ldots, N\rangle \tag{14}$$

We conclude that the creation operators formally satisfy the algebra

$$a_i^\dagger a_j^\dagger = -a_j^\dagger a_i^\dagger \tag{15}$$

when operating on any properly symmetrized state. From (15) we see that

$$(a_i^\dagger)^2 = 0 \tag{16}$$

implying that no two fermions can occupy the same state.

We write for the Hermitian conjugate of (13)

$$\langle 0 | a_N a_{N-1} \cdots a_2 a_1 = (a_1^\dagger a_2^\dagger \cdots a_{N-1}^\dagger a_N^+ | 0\rangle)^\dagger \tag{17}$$

Since

$$|i\rangle = a_i^\dagger |0\rangle \tag{18}$$

we have

$$\langle 0 | a_i | i\rangle = 1 \tag{19}$$

If we consider a_i as an operator acting to the right, we see that it has the effect of removing a particle from state i. For this reason we refer to a_i as the annihilation (or destruction) operator, that is,

$$a_i |i\rangle = |0\rangle \tag{20}$$
$$a_i a_j = -a_j a_i \tag{21}$$

Using all this information we find for $i \neq j$

$$a_i^\dagger a_j |j, 1, 2, \ldots\rangle = |i, 1, 2, \ldots\rangle = a_j |j, i, 1, 2, \ldots\rangle$$
$$= -a_j |i, j, 1, 2, \ldots\rangle = -a_j a_i^\dagger |j, 1, 2, \ldots\rangle \tag{22}$$

We see that when acting on a symmetrized state and for $i \neq j$,

$$a_i^\dagger a_j = -a_j a_i^\dagger \tag{23}$$

Next consider a many-particle configuration in which the single-particle state i is occupied. In this case

$$a_i^\dagger a_i |\cdots i \cdots\rangle = |\cdots i \cdots\rangle$$
$$a_i a_i^\dagger |\cdots i \cdots\rangle = 0 \tag{24}$$

Similarly, if i is empty,

$$a_i^\dagger a_i |\cdots\rangle = 0$$
$$a_i a_i^\dagger |\cdots\rangle = |\cdots\rangle \tag{25}$$

Since the single-particle state is either empty or occupied, we must have

$$a_i^\dagger a_i + a_i a_i^\dagger = 1 \tag{26}$$

Introducing the *anticommutator*

$$[A, B]_+ = AB + BA \tag{27}$$

we can summarize the commutation relations for the fermion creation and annihilation operators as

$$[a_i^\dagger, a_j^\dagger]_+ = [a_i, a_j]_+ = 0$$
$$[a_i^\dagger, a_j]_+ = \delta_{ij} \tag{28}$$

We next define the corresponding operators for a system of particles
obeying Bose statistics. Consider a many-particle state in which there are n_i
particles in the ith single-particle state. We define the annihilation operator b_i
through

$$b_i |n_1 \cdots n_i \cdots\rangle = (n_i)^{1/2} |n_1 \cdots (n_i - 1) \cdots\rangle \tag{29}$$

In the special case where a single-particle state is unoccupied, the annihilation
operators returns zero. The Hermitian conjugate operator $b_i^\dagger$ must satisfy

$$b_i^\dagger |n_1 \cdots (n_i - 1) \cdots\rangle = (n_i)^{1/2} |n_1 \cdots n_i \cdots\rangle \tag{30}$$

or

$$b_i^\dagger |n_1 \cdots n_i \cdots\rangle = (n_i + 1)^{1/2} |n_1 \cdots (n_i + 1) \cdots\rangle \tag{31}$$

and we see that $b_i^\dagger$ has the interpretation of creation operator. From (29) and
(30) we see that

$$b_i^\dagger b_i |\cdots n_i \cdots\rangle = n_i |\cdots n_i \cdots\rangle \tag{32}$$

that is, $b_i^\dagger b_i$ can be interpreted as the *number operator*, while from (31) and
(29) we see that

$$b_i b_i^\dagger |\cdots n_i \cdots\rangle = (n_i + 1) |\cdots n_i \cdots\rangle \tag{33}$$

We now define the commutator

$$[A, B]_- = AB - BA \tag{34}$$

Since commutators occur much more frequently than anticommutators we will
henceforth drop the subscript for the commutator. With this notation we find
that the boson creation and annihilation operators satisfy

$$[b_i, b_j] = [b_i^\dagger, b_j^\dagger] = 0 \tag{35}$$

since the many-particle states are symmetric under transposition of particle labels. Similarly, from (32) and (33) we find

$$[b_i, b_j^\dagger] = \delta_{ij} \tag{36}$$

We have implicitly assumed that the representation (1) for the single-particle states is complete, that is,

$$\sum_i |i\rangle\langle i| = 1 \tag{37}$$

In addition to the spatial variable $\mathbf{r}$, there is often an internal degree of freedom such as the spin. We must then consider the wave functions to be spinors; for example, for spin$-1/2$ particles,

$$\langle \mathbf{r} | i \rangle = \varphi_i(\mathbf{r}, \uparrow)\binom{1}{0} + \varphi_i(\mathbf{r}, \downarrow)\binom{0}{1} \tag{38}$$

and we write

$$\langle \mathbf{r}\sigma | i \rangle = \varphi_i(\mathbf{r}, \sigma)$$

It is often convenient to work with creation and annihilation operators that are not tied to any particular representation $|i\rangle$ for the single-particle states. This leads us to define *field operators*. In the case of fermions we define

$$\psi(\mathbf{r}, \sigma) = \sum_i \langle \mathbf{r}, \sigma | i \rangle a_i = \sum_i \varphi_i(\mathbf{r}, \sigma)a_i$$
$$\psi^\dagger(\mathbf{r}, \sigma) = \sum_i \langle i | \mathbf{r}, \sigma \rangle a_i^\dagger = \sum_i \varphi_i^*(\mathbf{r}, \sigma)a_i^\dagger \tag{39}$$

Using the orthonormality of the wave functions, we invert (39) to obtain

$$a_i = \int d^3r \sum_\sigma \varphi_i^*(\mathbf{r}, \sigma)\psi(\mathbf{r}, \sigma)$$
$$a_i^\dagger = \int d^3r \sum_\sigma \varphi_i(\mathbf{r}, \sigma)\psi^\dagger(\mathbf{r}, \sigma) \tag{40}$$

The completeness requirement (37) can be reexpressed as

$$\sum_i \varphi_i^*(\mathbf{r}', \sigma')\varphi_i(\mathbf{r}, \sigma) = \delta(\mathbf{r} - \mathbf{r}')\delta_{\sigma\sigma'} \tag{41}$$

and we see that the field operators satisfy the anticommutation relations

$$[\psi(\mathbf{r}, \sigma), \psi(\mathbf{r}', \sigma')]_+ = [\psi^\dagger(\mathbf{r}, \sigma), \psi^\dagger(\mathbf{r}', \sigma')]_+ = 0$$
$$[\psi(\mathbf{r}, \sigma), \psi^\dagger(\mathbf{r}', \sigma')]_+ = \delta(\mathbf{r} - \mathbf{r}')\delta_{\sigma\sigma'} \tag{42}$$

The analogous operators for (spinless) Bose particles can be written

$$\chi(\mathbf{r}) = \sum_i \varphi_i(\mathbf{r})b_i \qquad \chi^\dagger(\mathbf{r}) = \sum_i \varphi_i^*(\mathbf{r})b_i^\dagger \tag{43}$$

and the commutation relations become

$$[\chi(\mathbf{r}), \chi(\mathbf{r}')] = [\chi^\dagger(\mathbf{r}), \chi^\dagger(\mathbf{r}')] = 0$$
$$[\chi(\mathbf{r}), \chi^\dagger(\mathbf{r}')] = \delta(\mathbf{r} - \mathbf{r}') \tag{44}$$

We may also express the operator associated with total particle number in terms of field operators

$$N = \sum_i b_i^\dagger b_i = \int d^3r \, \chi^\dagger(\mathbf{r})\chi(\mathbf{r}) \tag{45}$$

for spinless bosons, and

$$N = \sum_i a_i^\dagger a_i = \int d^3r \sum_\sigma \psi^\dagger(\mathbf{r}, \sigma)\psi(\mathbf{r}, \sigma) \tag{46}$$

for fermions. By considering a set of states $|j\rangle$ which are localized within a subvolume Ω and taking the limit $\Omega \to 0$, we identify the particle density operator as

$$n(\mathbf{r}) \equiv \sum_{\text{particles } i} \delta(\mathbf{r} - \mathbf{r}_i) = \chi^\dagger(\mathbf{r})\chi(\mathbf{r}) \tag{47}$$

for spinless bosons, and

$$n(\mathbf{r}) = \sum_\sigma \psi^\dagger(\mathbf{r}, \sigma)\psi(\mathbf{r}, \sigma) \tag{48}$$

for fermions. A particularly useful single-particle representation is the momentum representation. In this case the operators are related to the field operators through

$$\psi(\mathbf{r}, \sigma) = V^{-1/2} \sum_{\mathbf{k}} e^{i\mathbf{k}\cdot\mathbf{r}} a_{\mathbf{k}, \sigma}$$
$$\psi^\dagger(\mathbf{r}, \sigma) = V^{-1/2} \sum_{\mathbf{k}} e^{-i\mathbf{k}\cdot\mathbf{r}} a_{\mathbf{k}, \sigma}^\dagger \tag{49}$$

and

$$a_{\mathbf{k}, \sigma} = V^{-1/2} \int d^3r \, e^{-i\mathbf{k}\cdot\mathbf{r}} \psi(\mathbf{r}, \sigma)$$
$$a_{\mathbf{k}, \sigma}^\dagger = V^{-1/2} \int d^3r \, e^{i\mathbf{k}\cdot\mathbf{r}} \psi^\dagger(\mathbf{r}, \sigma) \tag{50}$$

in the case of fermion operators. The expression for Bose operators are analogous.

The Fourier transform of the density operator is

$$\rho(\mathbf{q}) = \int d^3r \, e^{-i\mathbf{q}\cdot\mathbf{r}} n(\mathbf{r}) = \sum_{\text{particles } i} e^{-i\mathbf{q}\cdot\mathbf{r}_i} \tag{51}$$

In the occupation number representation we see that

$$\rho(\mathbf{q}) = \sum_{\mathbf{k}\sigma} a^\dagger_{\mathbf{k}-\mathbf{q},\,\sigma} a_{\mathbf{k},\,\sigma} \tag{52}$$

We next wish to express the Hamiltonian (12) in terms of the field operators. We first note that (12) contains two types of terms. We refer to expressions such as

$$\sum_i \frac{p_i^2}{2m} \qquad \sum_i U(\mathbf{r}_i)$$

as *one-body operators,* since they can be evaluated by adding up single-particle contributions. The last term in (12),

$$\tfrac{1}{2} \sum_{i \neq j} v(\mathbf{r}_{ij}) \tag{53}$$

is an example of a *two-body operator.* As before, let $|i\rangle$ be a complete set of single-particle states, characterized by wave functions $\varphi_i(\mathbf{r}) = \langle \mathbf{r}\,|\,i\rangle$. A one-body operator T can be completely specified through its matrix elements

$$\langle i\,|\,T\,|\,j\rangle = \int d^3 r\; \varphi_i^*(\mathbf{r})T(\mathbf{r})\varphi_j(\mathbf{r}) \tag{54}$$

Similarly, a two-body operator V is fully determined by the matrix elements

$$(ij\,|\,V\,|\,kl) = \int d^3 r_1\, d^3 r_2\; \varphi_i^*(r_1)\varphi_j^*(r_2)V(\mathbf{r}_1,\mathbf{r}_2)\varphi_k(r_2)\varphi_l(r_1) \tag{55}$$

Let

$$|n_1, \ldots, n_i, \ldots, n_j, \ldots\rangle \tag{56}$$

be a properly symmetrized many-particle state (13) or (7) in which there are n_1 particles in state 1, n_i particles in states i, and so on (for fermions n_i can, of course, only take on the values 0 and 1). Also, let $c_i^\dagger$, c_i be the appropriate creation and annihilation operators ($a_i^\dagger$, a_i, or $b_i^\dagger$, b_i).

$$c_i\,|n_1, \ldots, n_i, \ldots, n_j, \ldots\rangle$$
$$= (n_i)^{1/2}\,|n_1, \ldots, (n_i - 1), \ldots, n_j, \ldots\rangle$$
$$c_i^\dagger\,|n_1, \ldots, (n_i - 1), \ldots, n_j, \ldots\rangle$$
$$= (n_i)^{1/2}\,|n_1, \ldots, n_i, \ldots, n_j, \ldots\rangle \tag{57}$$

A little reflection should convince the reader that

$$\sum_{\text{particles } p} T(\mathbf{r}_p)\,|n_1, \ldots, n_i, \ldots, n_j, \ldots\rangle \tag{58}$$

has nonzero matrix elements only between states in which at most one particle is in a different state than (56) and that the matrix elements of (58) are identical to those of

$$\sum_{\text{states } i,j} \langle i\,|\,T\,|\,j\rangle c_i^\dagger c_j\,|n_1,\,\ldots,\,n_i,\,\ldots,\,n_j,\,\ldots\rangle \tag{59}$$

This allows us to write formally

$$\sum_{\text{particles } p} T(\mathbf{r}_p) = \sum_{\text{states } i,j} \langle i\,|\,T\,|\,j\rangle c_i^\dagger c_j \tag{60}$$

As a check, consider an operator V which is diagonal in the $\mathbf{r}$-representation. We have, using (47) (similar results obtain with the analogous expression for fermions),

$$\sum_p V(\mathbf{r}_p) = \int d^3r\, V(\mathbf{r})n(\mathbf{r}) = \int d^3r\, V(\mathbf{r})\chi^\dagger(\mathbf{r})\chi(\mathbf{r}) \tag{61}$$

Substitution of (43) then yields (60). Let T be an arbitrary Hermitian one-body operator. One can always express T in terms of a generalized density, in the representation in which T is diagonal, multiplied by the appropriate eigenvalue. By use of completeness and orthogonality relations, one recovers (60).

In the case of two-body operators we note that

$$\sum_{p \neq p'} V(\mathbf{r}_p, \mathbf{r}_{p'})\,|n_1,\,\ldots,\,n_i,\,\ldots,\,n_j,\,\ldots\rangle \tag{62}$$

can only have nonzero matrix elements between states where at most two particles have changed state. Equation (62) is therefore indistinguishable from

$$\sum_{i,j,k,l} (ij\,|\,V\,|\,kl)c_i^\dagger c_j^\dagger c_k c_l\,|n_1,\,\ldots,\,n_i,\,\ldots,\,n_j,\,\ldots\rangle \tag{63}$$

and we can formally write

$$\sum_{p \neq p'} V(\mathbf{r}_p, \mathbf{r}_{p'}) = \sum_{i,j,k,l} (ij\,|\,V\,|\,kl)c_i^\dagger c_j^\dagger c_k c_l \tag{64}$$

It is instructive to verify (64) by reexpressing the two-body operator in terms of density operators. We have

$$\sum_{p \neq p'} V(\mathbf{r}_p, \mathbf{r}_{p'}) = \sum_{p,p'} V(\mathbf{r}_p, \mathbf{r}_{p'}) - \sum_p V(\mathbf{r}_p, \mathbf{r}_p)$$

$$= \int d^3r \left(\int d^3r'\, V(\mathbf{r}, \mathbf{r}')n(\mathbf{r}') - V(\mathbf{r}, \mathbf{r}) \right) n(\mathbf{r})$$

$$= \int d^3r \int d^3r'\, V(\mathbf{r}, \mathbf{r}')(\chi^\dagger(\mathbf{r})\chi(\mathbf{r}')\chi^\dagger(\mathbf{r})\chi(\mathbf{r}) \tag{65}$$

$$-\, \delta(\mathbf{r} - \mathbf{r}')\chi^\dagger(\mathbf{r})\chi(r))$$

$$= \int d^3r \int d^3r'\, V(\mathbf{r}, \mathbf{r}')\chi^\dagger(\mathbf{r}')\chi^\dagger(\mathbf{r})\chi(\mathbf{r})\chi(\mathbf{r}')$$

where we have made use of the commutation relations (44) of the field operators. Note that since an even number of transpositions are involved, the argument leading to (65) holds equally well for fermions. Substituting the definitions of the field operators we recover (64).

We can now reexpress the Hamiltonian (12), assuming that the potentials have Fourier transforms given by

$$U(\mathbf{q}) = \int d^3r \, e^{-i\mathbf{q}\cdot\mathbf{r}} U(\mathbf{r})$$

$$v(\mathbf{q}) = \int d^3r_{ij} e^{-i\mathbf{q}(\mathbf{r}_i - \mathbf{r}_j)} v(\mathbf{r}_i - \mathbf{r}_j) \tag{66}$$

Then

$$H = \sum_{k,\sigma} \frac{\hbar^2 k^2}{2m} c^\dagger_{\mathbf{k},\sigma} c_{\mathbf{k},\sigma} + \frac{1}{V} \sum_{\mathbf{k},\mathbf{q},\sigma} U(\mathbf{q}) c^\dagger_{\mathbf{k}+\mathbf{q},\sigma} c_{\mathbf{k}\sigma}$$

$$+ \frac{1}{V} \sum_{\mathbf{k}\mathbf{k}'\mathbf{q},\sigma\sigma'} v(\mathbf{q}) c^\dagger_{\mathbf{k}+\mathbf{q},\sigma} c^\dagger_{\mathbf{k}'-\mathbf{q},\sigma'} c_{\mathbf{k}',\sigma'} c_{\mathbf{k},\sigma} \tag{67}$$

As a final example, we express the pair distribution function (4.B.8) and structure factor (4.B.11) as expectation values of products of annihilation and creation operators, and evaluate the expectation values for the special case of a noninteracting Fermi gas at $T = 0$. We have from (4.B.3) and (4.B.8)

$$g(\mathbf{r}) = \frac{V}{\langle N \rangle^2} \left\langle \sum_{i \neq j} \int d^3x \, \delta(\mathbf{r}_i - \mathbf{x}) \delta(\mathbf{r}_j - \mathbf{x} - \mathbf{r}) \right\rangle \tag{68}$$

or

$$g(\mathbf{r}) = \frac{V}{\langle N \rangle^2} \left\langle \int d^3x \left(\sum_i \delta(\mathbf{r}_i - \mathbf{x}) \sum_j \delta(\mathbf{r}_j - \mathbf{x} - \mathbf{r}) \right.\right.$$

$$\left.\left. - \sum_i \delta(\mathbf{r}_i - \mathbf{x}) \delta(\mathbf{r}_i - \mathbf{x} - \mathbf{r}) \right) \right\rangle$$

From (47) and (48) we obtain

$$g(\mathbf{r}) = \frac{V}{\langle N \rangle^2} \left\langle \int d^3x \, n(\mathbf{x}) n(\mathbf{x} + \mathbf{r}) - \frac{V}{\langle N \rangle} \delta(\mathbf{r}) \right\rangle$$

$$= \frac{V}{\langle N \rangle^2} \left\langle \int d^3x \sum_{\sigma,\sigma'} \psi^\dagger(\mathbf{x}, \sigma') \psi(\mathbf{x}, \sigma') \psi^\dagger(\mathbf{x}+\mathbf{r}, \sigma) \psi(\mathbf{x} + \mathbf{r}, \sigma) \right\rangle$$

$$- \frac{V}{\langle N \rangle} \delta(\mathbf{r})$$

Using the commutation relations (42), this expression reduces to

$$g(\mathbf{r}) = \frac{V}{\langle N \rangle^2} \left\langle \int d^3x \sum_{\sigma,\sigma'} \psi^\dagger(\mathbf{x} + \mathbf{r}, \sigma) \psi^\dagger(\mathbf{x}, \sigma') \psi(\mathbf{x}, \sigma') \psi(\mathbf{x} + \mathbf{r}, \sigma) \right\rangle \tag{69}$$

In the momentum representation (49) and (50) the corresponding equation becomes

$$g(\mathbf{x}) = \frac{1}{\langle N \rangle^2} \sum_{\mathbf{k,p,q},\sigma,\sigma'} e^{i\mathbf{q}\cdot\mathbf{x}} \langle a^\dagger_{\mathbf{p}+\mathbf{q},\sigma} a^\dagger_{\mathbf{k}-\mathbf{q},\sigma'} a_{\mathbf{k},\sigma'} a_{\mathbf{p},\sigma} \rangle \tag{70}$$

An expression for the structure factor can be obtained from (4.B.11) and (51):

$$S(q) = \frac{1}{\langle N \rangle} \langle \rho(\mathbf{q})\rho(-\mathbf{q}) \rangle - \langle N \rangle \delta_{\mathbf{q},0}$$

Substitution of (52) yields

$$S(\mathbf{q}) = \frac{1}{\langle N \rangle} \left\langle \sum_{\mathbf{k,p},\sigma,\sigma'} a^\dagger_{\mathbf{k}-\mathbf{q},\sigma'} a_{\mathbf{k},\sigma} a^\dagger_{\mathbf{p}+\mathbf{q},\sigma} a_{\mathbf{p},\sigma} \right\rangle - \langle N \rangle \delta_{\mathbf{q},0}$$

By making use of the commutation relations this expression can be rewritten as

$$S(\mathbf{q}) - 1 = \frac{1}{\langle N \rangle} \left\langle \sum_{\mathbf{k,p},\sigma,\sigma'} a^\dagger_{\mathbf{p}+\mathbf{q},\sigma} a^\dagger_{\mathbf{k}-\mathbf{q},\sigma} a_{\mathbf{k},\sigma'} a_{\mathbf{p},\sigma} \right\rangle - \langle N \rangle \delta_{\mathbf{q},0} \tag{71}$$

which in terms of the field operators becomes

$$S(q) - 1 = \frac{1}{\langle N \rangle} \left\langle \sum_{\sigma,\sigma'} \int d^3r \int d^3x\, e^{i\mathbf{q}\cdot\mathbf{r}} \psi^\dagger(\mathbf{x} + \mathbf{r}, \sigma) \right.$$
$$\left. \psi^\dagger(\mathbf{x}, \sigma')\psi(\mathbf{x}, \sigma')\psi(\mathbf{x} + \mathbf{r}, \sigma) \right\rangle - \langle N \rangle \delta_{\mathbf{q},0} \tag{72}$$

We now turn to the problem of evaluating the expressions for $g(\mathbf{r})$ and $S(\mathbf{q})$ in the special case of a noninteracting gas of spin-$\frac{1}{2}$ fermions. The ground state has $N/2$ particles in each spin state, filling up the lowest momentum states up to a Fermi wave vector k_F. We write

$$g(\mathbf{x}) = \sum_{\sigma\sigma'} g_{\sigma\sigma'}(\mathbf{x}) = g_{\uparrow\uparrow}(\mathbf{x}) + g_{\uparrow\downarrow}(\mathbf{x}) + g_{\downarrow\uparrow}(\mathbf{x}) + g_{\downarrow\downarrow}(\mathbf{x}) \tag{73}$$

where

$$g_{\uparrow\uparrow}(\mathbf{x}) = \frac{1}{\langle N \rangle^2} \sum_{\mathbf{k,p,q}} e^{i\mathbf{q}\cdot\mathbf{x}} \langle 0| a^\dagger_{\mathbf{p}+\mathbf{q},\uparrow} a^\dagger_{\mathbf{k}-\mathbf{q},\uparrow} a_{\mathbf{k},\uparrow} a_{\mathbf{p},\uparrow}) |0\rangle$$

and $|0\rangle$ represents the ground state. To obtain nonzero matrix elements we must have $k, p < k_F$ and either $q = 0$ or $\mathbf{k} - \mathbf{q} = \mathbf{p}$. In the former case the matrix element is 1 in the latter case -1. We obtain, assuming that the ground state is an eigenstate of the number operator with eigenvalue N,

$$g_{\uparrow\uparrow}(\mathbf{x}) = \frac{1}{N^2}\left\{ \frac{N}{2}\left[\frac{N}{2} - 1 \right] \right\} - \frac{1}{N^2} \sum_{k<k_F,\, p<k_F,\, \mathbf{p}\neq\mathbf{k}} e^{i(\mathbf{k}-\mathbf{p})\cdot\mathbf{x}}$$

This expression can be cast into the form

$$g_{\uparrow\uparrow}(\mathbf{x}) = \frac{1}{4} - \left| \frac{1}{N} \sum_{p<k_F} e^{i\mathbf{p}\cdot\mathbf{x}} \right|^2 \tag{74}$$

Note that $g_{\uparrow\uparrow}(0) = 0$ in agreement with the Pauli exclusion principle. Clearly, $g_{\uparrow\uparrow}(\mathbf{x}) = g_{\downarrow\downarrow}(\mathbf{x})$. Next

$$g_{\uparrow\downarrow}(\mathbf{x}) = \frac{1}{N^2} \sum_{\mathbf{k},\mathbf{p},\mathbf{q}} e^{i\mathbf{q}\cdot\mathbf{x}} \langle 0| a^\dagger_{\mathbf{p}+\mathbf{q},\uparrow} a^\dagger_{\mathbf{k}-\mathbf{q},\downarrow} a_{\mathbf{k},\downarrow} a_{\mathbf{p},\uparrow} |0\rangle$$

Here the only possibility is $q = 0$ and

$$g_{\uparrow\downarrow}(\mathbf{x}) = g_{\downarrow\uparrow}(\mathbf{x}) = \frac{1}{N^2}\left(\frac{N}{2}\right)^2 = \frac{1}{4} \tag{75}$$

Collecting terms, we find

$$g(\mathbf{x}) = 1 - 2\left| \frac{1}{N} \sum_{p<k_F} e^{i\mathbf{p}\cdot\mathbf{x}} \right|^2 \tag{76}$$

The sum in (76) can be evaluated analytically to give

$$\frac{1}{N} \sum_{p<k_F} e^{i\mathbf{p}\cdot\mathbf{x}} = \frac{V}{N} \int_{p<p_F} e^{i\mathbf{p}\cdot\mathbf{x}} \frac{d^3p}{(2\pi)^3} = \frac{V}{2\pi^2 N}\left(\frac{1}{x^3} \sin k_F x - \frac{k_F}{x^2} \cos k_F x \right) \tag{77}$$

Similar arguments yield for the structure factor

$$S(q) = 1 - \frac{2}{N} \sum_{p<k_F,\, |\mathbf{p}+\mathbf{q}|<k_F} 1 = 1 - \frac{2V}{N(2\pi)^3} \int_{p<k_F,\, |\mathbf{p}+\mathbf{q}|<k_F} d^3p$$

The reader may wish to verify that this expression holds both for $q = 0$ and $q \neq 0$. Evaluating the integral, we find that

$$S(q) = \frac{3q}{4k_F} - \frac{q^3}{16k_F^3} \quad \text{for } q < 2k_F \qquad S(q) = 1 \quad \text{for } q \geq 2k_F \tag{78}$$

Note that the result $S(0) = 0$ for $T = 0$ is expected from (4.B.16) and (4.B.10).

References

ABRAHAM, F. F. [1986]. *Adv. Phys.* **35**:1.

ABRAHAM, F. F., RUDGE, W. E., AUERBACH, D. J. and KOCH, S. W. [1984]. *Phys. Lett.* **52**:445.

ABRIKOSOV, A. A., GORKOV, L. P., and DZYALOSHINSKY, I. E. [1963]. *Methods of Quantum Field Theory in Statistical Physics.* Englewood Cliffs, N. J.: Prentice-Hall.

ADLER, S. L. [1962]. *Phys. Rev.* **126**:413.

AHARONY, A. [1976]. In *Phase Transitions and Critical Phenomena,* Vol. 6, ed. C. Domb and M. S. Green. New York: Academic Press.

AHARONY, A. [1986]. In *Directions in Condensed Matter Physics,* Vol. 1, ed. G. Grinstein and G. F. Mazenko. Singapore: World Scientific.

ALDER, B. J., and HECHT, C. E. [1969]. *J. Chem. Phys.* **50**:2032.

ALS-NIELSEN, J. [1985]. *Z. Phys.* **B61**:411.

ANDERSON, P. W. [1958]. *Phys. Rev.* **109**:1492.

ANDERSON, P. W. [1979]. In *Ill Condensed Matter,* ed. R. Balian, R. Maynard, and G. Toulouse, Amsterdam: North-Holland.

ANDERSON, P. W. [1984]. *Basic Notions in Condensed Matter Physics.* Menlo Park, Calif.: Benjamin-Cummings.

ANDERSON. P. W., HALPERIN, B. I., and VARMA, C. M. [1971]. *Philos. Mag.* **25**:1.

ANDRONIKASHVILI, E. L., and MAMALADZE. YU. G. [1966]. *Rev. Mod. Phys.* **38**:567.

ASHCROFT, N., and MERMIN, N. D. [1976]. *Solid State Physics.* New York: Holt, Rinehart and Winston.

BAKER, G. A., NICKEL, B. G., and MEIRON, D. I. [1976]. *Phys. Rev. Lett.* **36**:1351.

BALESCU, R. [1975]. *Equilibrium and Non-equilibrium Statistical Mechanics.* New York: Wiley.

BARDEEN, J., COOPER, L. N., and SCHRIEFFER, J. R. [1957]. *Phys. Rev.* **108**:1175.

BARKER, J. A., and HENDERSON, D. [1967]. *J. Chem. Phys.* **43**:4714.

BARKER, J. A., and HENDERSON, D. [1971]. *Mol. Phys.* **21**:187.

BARKER, J. A., and HENDERSON, D. [1976]. *Rev. Mod. Phys.* **48**:587.

BAXTER, R. J. [1973]. *J. Phys.* **C6**:L445.

BAXTER, R. J. [1982]. *Exactly Solved Models in Statistical Mechanics.* New York: Academic Press.

BEAGLEHOLE, D. [1979]. *Phys. Rev. Lett.* **43**:2016.

BERRY, M. V. [1981]. *Ann. Phys.* **131**:163.

BETHE, H. [1935]. *Proc. R. Soc. London* **A150**:552.

BETTS, D. D. [1974]. In *Phase Transitions and Critical Phenomena,* Vol. 3, ed. C. Domb and M. S. Green. New York: Academic Press.

BIJL, A. [1940]. *Physica* **7**:860.

BINDER, K. [1986]. *Monte Carlo Methods in Statistical Mechanics,* 2nd ed. Berlin: Springer-Verlag.

BINDER, K., and STAUFFER, D. [1984]. In *Applications of the Monte Carlo Method in Statistical Physics,* ed. K. Binder. Berlin: Springer-Verlag.

BINDER, K., and YOUNG, A. P. [1986]. *Rev. Mod. Phys.* **58**:801.

BISHOP, D. J., and REPPY, J. D. [1978]. *Phys. Rev. Lett.* **40**:1727.

BLOTE, H. W. J., and SWENDSEN, R. [1979]. *Phys. Rev.* **B20**:2077.

BLUME, H., EMERY, V. J., and GRIFFITHS, R. B. [1971]. *Phys. Rev.* **A4**:1071.

BOGOLIUBOV, N. N. [1947]. *J. Phys. USSR* **11**:23.

BREZIN, E., LE GUILLOU, J. C., and ZINN-JUSTIN, J. [1976]. In *Phase Transitions and Critical Phenomena,* Vol. 6, ed. C. Domb and M. S. Green. New York: Academic Press.

CAHN, J. W., and HILLIARD, J. E. [1958]. *J. Chem. Phys.* **28**:258.

CALLAWAY, J. [1974]. *Quantum Theory of the Solid State.* Part B. New York: Academic Press.

CALLEN, H. B. [1985]. *Thermodynamics and an Introduction to Thermostatistics,* 2nd ed. New York: Wiley.

CAMP, W. J., and VAN DYKE, J. P. [1976]. *J. Phys.* **A9**:731.

CANELLA, V., and MYDOSH, J. A. [1972]. *Phys. Rev.* **B6**:4220.

CHAPELA, G., SAVILLE, G., THOMPSON, S. M., and ROWLINSON, J. S. [1977]. *J. Chem. Soc. Faraday Trans. 2* **73**:1133.

COOPER, L. N. [1956]. *Phys. Rev.* **104**:1189.

COWLEY, R. A., and WOODS, A. D. B. [1971]. *Can. J. Phys.* **49**:177.

DE GENNES, P. G. [1976]. *J. Phys. Paris. Lett.* **37**:L1.

DE GENNES P. G. [1974], *The Physics of Liquid Crystals,* Oxford: Clarendon.

DOLGOV, O. V., KIRZHNITZ, D. A., and MAKSIMOV, E. G. [1981]. *Rev. Mod. Phys.* **53**:81.

DOMB, C. [1974]. In *Phase Transitions and Critical Phenomena.* Vol. 3, ed. C. Domb and M. S. Green. New York: Academic Press.

DOMB, C., and HUNTER, D. L. [1965]. *Proc. Phys. Soc.* **86**:1147.

DUNN, A. G., ESSAM, J. W., and RITCHIE, D. S. [1975]. *J. Phys.* **C8**:4219.

DYSON, F. J. [1958]. *Phys. Rev.* **102**:1217, 1230.

EDWARDS, S. F., and ANDERSON, P. W. [1975]. *J. Phys.* **F5**:965.

ELLIOTT, R. J., KRUMHANSL, J. A., and LEATH, P. L. [1974]. *Rev. Mod. Phys.* **46**:465.

ESSAM, J. W. [1972]. In *Phase Transitions and Critical Phenomena,* Vol. 2, ed. C. Domb and M. S. Green. New York: Academic Press.

FELLER, W. [1957]. *An Introduction to Probability Theory and Its Applications,* 2nd ed. New York: Wiley.

FETTER, A. L., and WALECKA, J. D. [1971]. *Quantum Theory of Many Particle Systems.* New York: McGraw-Hill.

FEYNMAN, R. P. [1954]. *Phys. Rev.* **94**:262.

FISHER, M. E. [1974]. *Rev. Mod. Phys.* **46**:597.

FISK, S., and WIDOM, B. [1968]. *J. Chem. Phys.* **50**:3219.

FLAPPER, D. P., and VERTOGEN, G. [1981a]. *Phys. Rev.* **A24**:2089.

FLAPPER, D. P., and VERTOGEN, G. [1981b]. *J. Chem. Phys.* **75**:3599.

FOWLER, R. H., and GUGGENHEIM, E. A. [1940]. *Proc. R. Soc. London* **A174**:189.

FRISKEN, B. J., BERGERSEN, B., and PALFFY-MUHORAY, P. [1987]. *Mol. Cryst. Liq. Cryst.* **148**:45.

FROHLICH, H. [1950]. *Phys. Rev.* **79**:845.

GAUNT, D. S., and GUTTMANN, A. J. [1974]. In *Phase Transitions and Critical Phenomena,* Vol. 3, ed. C. Domb and M. S. Green. New York: Academic Press.

GELL-MANN, M., and BRUECKNER, K. [1957]. *Phys. Rev.* **139**:407.

GOLDSTEIN, H. [1980]. *Classical Mechanics,* 2nd ed. Reading, Mass.: Addison-Wesley.

GORISHNY, S. G., LARIN, S. A., and TKACHOV, F. V. [1984]. *Phys. Lett.* **A101**:120.

GREYTAK, T. J., and KLEPPNER, D. [1984]. In *New Trends in Atomic Physics,* Les Houches Summer School, 1982, ed. G. Greenberg and R. Stora. Amsterdam: North-Holland.

GRIFFITHS, R. B. [1967]. *Phys. Rev.* **158**:176.

GRIFFITHS, R. B. [1969]. *Phys. Rev. Lett.* **23**:17.

GRIFFITHS, R. B. [1970]. *Phys. Rev. Lett.* **24**:1479.

GRIFFITHS, R. B., and PEARCE, P. A. [1978]. *Phys. Rev. Lett.* **41**:917.

GRINSTEIN, G. [1974]. *AIP Conf. Proc.* **24**:313.

GUBERNATIS, J. E., and TAYLOR, P. L. [1973]. *J. Phys.* **C6**:1889.

GUGGENHEIM, E. A. [1965]. *Mol. Phys.* **9**:199.

GUGGENHEIM, E. A. [1967]. *Thermodynamics. An Advanced Treatment for Chemists and Physicists.* 5th ed. Amsterdam: North-Holland.

HANSEN, J. P. and MCDONALD, I. R. [1986]. *Theory of Simple Liquids.* 2nd ed. London: Academic Press.

HARRIS, A. B. [1974]. *J. Phys.* **C7**:1671.

HARRIS, A. B., LUBENSKY, T. C., HOLCOMB, W. K., and DASGUPTA, C. [1975]. *Phys. Rev. Lett.* **35**:327.

HICKS, C. P., and YOUNG, C. L. [1977]. *J. Chem. Soc. Faraday Trans. 2* **73**:597.

HIRSCHFELDER, J. O., CURTISS, C. F., and BIRD, R. B. [1954]. *Molecular Theory of Gases and Liquids.* New York: Wiley.

HO, J. T., and LITSTER, J. D. [1969]. *Phys. Rev. Lett.* **22**:603.

HOHENBERG, P. C. [1967]. *Phys. Rev.* **158**:383.

HOUTAPPEL, R. M. F. [1950]. *Physica* **16**:425.

HUANG, K. [1987]. *Statistical Mechanics,* 2nd ed. New York: Wiley.

HUSIMI, K. and SYOZI, I. [1950]. *Prog. Theor. Phys.* **5**:177, 341.

IMRY, Y., and MA, S. K. [1975]. *Phys. Rev. Lett.* **35**:1399.

ISHII, K. [1973]. *Prog. Theor. Phys. Suppl.* **53**:77.

JACKLE, J. [1986]. *Rep. Prog. Phys.* **49**:171.

JASNOW, D., and WORTIS, M. [1968]. *Phys. Rev.* **176**:739.

JAYNES, E. T. [1957]. *Phys. Rev.* **106**:620.

JOSE, J. V., KADANOFF, L. P., KIRKPATRICK, S., and NELSON, D. R. [1977]. *Phys. Rev.* **B16**:1217.

KADANOFF, L. P. [1971]. In *Proceedings of 1970 Varenna Summer School on Critical Phenomena,* ed. M. S. Green. New York: Academic Press.

KADANOFF, L. P., GÖTZE, W., HAMBLEN, D., HECHT, R., LEWIS, E. A. S., PALCIAUKAS, V. V., RAYL, M., SWIFT, J., ASPNES, D. and KANE, J. [1967]. *Rev. Mod. Phys.* **39**:395.

KASTELEYN, P. W., and FORTUIN, C. M. [1969]. *J Phys. Soc. Japan Suppl.* **16**:11.

KIRKPATRICK, S. [1979]. In *Ill Condensed Matter,* ed. by R. Balian, R. Maynard and G. Toulouse. Amsterdam: North Holland.

KIRZHNITZ, D. A. [1976]. *Usp. Fiz. Nauk* **119**:357 (Engl. transl. Sov. Phys. Usp.) **19**:530.

KITTEL, C. [1976]. *Introduction to Solid State Physics,* 5th Edition. New York: Wiley.

KOSTERLITZ, J. M., and THOULESS, D. J. [1973]. *J. Phys.* **C6**:1181.

KUBO, R. [1957]. *J. Phys. Soc.* Japan **12**:570.

KUBO, R., ICHIMURA, H., USUI, T., and HASHIZUME, N. [1965]. *Statistical Mechanics, An Advanced Course.* Amsterdam: North Holland.

LANDAU, L. D. [1941]. *J. Phys.* USSR **5**:71.

LANDAU, L. D. [1947]. *J. Phys.* USSR **11**:91.

LANDAU, L. D., and LIFSHITZ, E. M. [1980]. *Statistical Physics,* parts 1 and 2. Oxford: Pergamon Press.

LAST, B. J., and THOULESS, D. J. [1973]. *J. Phys.* **C7**:715.

LAWRIE, I. D., and SARBACH, S. [1984]. In *Phase Transitions and Critical Phenomena,* Vol. 9, ed. C. Domb and J. L. Lebowitz. New York: Academic Press.

LEBOWITZ, J. L., and LIEB, E. H. [1969]. *Phys. Rev. Lett.* **22**:631.

LE GUILLOU, J. C., and ZINN-JUSTIN, J. [1987]. *J. Phys. Paris.* **48**:19.

LEE, P. A., and RAMAKRISHNAN, T. V. [1985]. *Rev. Mod. Phys.* **57**:287.

LICCIARDELLO, D. C., and THOULESS, D. J. [1975]. *J. Phys.* **C8**:4157.

LIFSHITZ, I. M. [1964]. *Adv. Phys.* **13**:483.

LONGUET-HIGGINS, H. C., and WIDOM, B. [1965]. *Molecular Physics.* **8**:549.

LUBENSKY, T. [1979]. In *Ill Condensed Matter,* ed. by R. Balian, R. Maynard and G. Toulouse. Amsterdam: North-Holland.

MA, S.-K. [1976a]. *Modern Theory of Critical Phenomena.* New York: Benjamin.

MA, S.-K. [1976b]. *Phys. Rev. Lett.* **37**:461.

MA, S.-K. [1985]. *Statistical Mechanics.* Philadelphia: World Scientific.

MAHAN, G. D. [1981]. *Many Particle Physics.* New York: Plenum.

MAIER, W., and SAUPE, A. [1959]. *Z. Naturforsch.* **A14**:882.

MAIER, W., and SAUPE, A. [1960]. *Z. Naturforsch.* **A15**:287.

MAITLAND, G. C., RIGBY, M., SMITH, E. B., and WAKEHAM, W. A. [1981]. *Intermolecular Forces, Their Origin and Determination.* Oxford: Clarendon Press.

MANDELBROT, B. [1982]. *The Fractal Geometry of Nature.* San Francisco: W. H. Freeman.

MATSUBARA, T., and YONEZAWA, F. [1967]. *Prog. Theor. Phys.* **37**:1346.

MATSUDA, H., and ISHII, K. [1970]. *Prog. Theor. Phys. Suppl.* **45**:56.

MATTIS, D. C. [1976]. *Phys. Lett.* **A56**:421.

MATTIS, D. C. [1981]. *The Theory of Magnetism I.* Berlin: Springer-Verlag.

MAYER, J. E., and MAYER, M. G. [1940]. *Statistical Mechanics.* New York: Wiley.

McCoy, B. [1972]. In *Phase Transitions and Critical Phenomena.* Vol. 2, ed. C. Domb and M. S. Green. New York: Academic Press.

McCoy, B., and Wu, T. T. [1973]. *The Two-Dimensional Ising Model.* Cambridge, Mass: Harvard University Press.

McKenzie, S. [1975]. *J. Phys.* **A8:**L102.

McNelly, T. F., Rogers, S. J., Channin, D. J., Rollefson, R. J., Goubau, W. M., Schmidt, G. E., Krumhansl, J. A. and Pohl, R. O. [1970]. *Phys. Rev. Lett.* **24:**100.

Mermin, N. D. [1968]. *Phys. Rev.* **176:**250.

Mermin, N. D., and Wagner, H. [1966]. *Phys. Rev. Lett.* **17:**1133.

Mezard, M., Parisi, G. and Virasoro, M.A., [1987]. *Spin Glass Theory and Beyond.* Singapore: World Scientific.

Mott, N. F. [1967]. *Adv. Phys.* **16:**49.

Murray, C. A., and Van Winkle, D. H. [1987]. *Phys. Rev. Lett.* **58:**1200.

Nauenberg, M., and Nienhuis, B. [1974]. *Phys. Rev. Lett.* **33:**944.

Nelson, D. R. [1983]. In *Phase Transitions and Critical Phenomena,* Vol. 7, ed. C. Domb and T. L. Lebowitz. New York: Academic Press.

Nelson, D. R., and Halperin, B. I. [1979]. *Phys. Rev.* **B19:**2457.

Nickel, B. G. [1982]. *Phase Transitions, Cargese 1980,* ed. M. Levy, J.-C. Le Gouillou, and J. Zinn-Justin. New York: Plenum, p. 291.

Niemeijer, T. H., and van Leeuwen, J. M. J. [1976]. *Phase Transitions and Critical Phenomena.* Vol. 6, ed. C. Domb and M. S. Green. New York: Academic Press.

Onsager, L. [1931]. *Phys. Rev.* **37:**405; **38:**2265.

Onsager, L. [1944]. *Phys. Rev.* **65:**117.

Onsager, L. [1949]. *Nuovo Cimento Suppl.* **2:**249.

Opechowski, W. [1937]. *Physica* **4:**181.

Osborne, D. V. [1950]. *Proc. Phys. Soc. London* **A63:**909.

Palffy-Muhoray, P., and Bergersen, B. [1987]. *Phys. Rev.* **A35:**2704.

Palffy-Muhoray, P., and Dunmur, D. A. [1983]. *Mol. Cryst. Liq. Cryst.* **97:**337.

Pathria, R. K. [1983]. *Can. J. Phys.* **61:**228.

Pawley, G. S., Swendsen, R. H., Wallace, D. J., and Wilson, K. G. [1984]. *Phys. Rev.* **B29:**4030.

Peierls, R. E. [1936]. *Proc. Cambridge Philos. Soc.* **32:**477.

Penrose, O. [1951]. *Philos. Mag.* **42:**1373.

Penrose, O., and Onsager, L. [1956]. *Phys. Rev.* **104:**576.

Pfeuty, P., and Toulouse, G. [1977]. *Introduction to the Renormalization Group and to Critical Phenomena.* London: Wiley.

Pines, D., and Nozieres, P. [1966]. *The Theory of Quantum Liquids.* New York: Benjamin.

Plischke, M., Henderson, D., and Sharma, S. R. [1985]. *Physics and Chemistry of Disordered Systems,* ed. D. Adler, H. Fritsche, and S. R. Ovshinsky. New York: Plenum.

Potts, R. B. [1952]. *Proc. Cambridge Philos. Soc.* **48:**106.

Priest, R. G., and Lubensky, T. C. [1976]. *Phys. Rev.* **B13:**4159.

Priestley, E. B., Wojtowicz, P. J., and Sheng Ping. [1974]. *Introduction to Liquid Crystals.* New York: Plenum.

Rasetti, M. [1986]. *Modern Methods in Statistical Mechanics.* Philadelphia: World Scientific.

Reichl, L. [1980]. *A Modern Course in Statistical Physics.* Austin: University of Texas Press.

Rowlinson, J. S., and Widom, B. [1982]. *Molecular Theory of Capillarity.* Oxford: Clarendon Press.

Rudnick, I. [1978]. *Phys. Rev. Lett.* **40:**1454.

Rushbrooke, G. S. [1963]. *J. Chem. Phys.* **39:**842.

Saito, Y., and Müller-Krumbhaar, H. [1984]. In *Applications of the Monte Carlo Method in Statistical Physics,* ed. K. Binder. Berlin: Springer-Verlag.

Schick, M., Walker, J. S., and Wortis, M. [1977]. *Phys. Rev.* **B16:**2205.

SCHRIEFFER, J. R. [1964]. *Superconductivity*. New York: Benjamin.

SCHULTZ, T., MATTIS, D., and LIEB, E. [1964]. *Rev. Mod. Phys.* **36**:856.

SHANNON, C. E. [1948]. *Bell Syst. Tech. J.* **27**:379, 623. (Reprinted in *The Mathematical Theory of Communication*. Urbana: University of Illinois Press.)

SILVERA, I. F., and WALRAVEN, J. T. M. [1986]. In *Progress in Low Temperature Physics*, Vol. 10, ed. D. Brewer. Amsterdam: North-Holland.

SINAI, IA. [1966]. In *Statistical Mechanics, Foundations and Applications*, ed. T. Bak. *IUPAP* meeting, Copenhagen, 1966.

SINAI, IA. [1970]. *Russ. Math. Surv.* **25**:137.

SOVEN, P. [1967]. *Phys. Rev.* **156**:809.

SKAL, A. S., and SHKLOVSKII, B. I. [1975]. *Sov. Phys. Semicond.* **8**:1029.

STANLEY, H. E. [1969]. *Phys. Rev.* **179**:501.

STANLEY, H. E. [1971]. *Introduction to Phase Transitions and Critical Phenomena*. Oxford: Oxford University Press.

STANLEY, H. E. [1974]. In *Phase Transitions and Critical Phenomena*, Vol. 3 ed. C. Domb and M. S. Green. New York: Academic Press.

STANLEY, H. E., and OSTROWSKY, N., eds. [1986]. *On Growth and Form*. Hingham, Mass.: Martinus Nijhoff.

STANLEY, H. E., REYNOLDS, P. J., REDNER, S., and FAMILY, F. [1982]. In *Real Space Renormalization*, ed. T. W. Burkhardt and J. M. J. van Leeuwen. Berlin: Springer-Verlag.

STAUFFER, D. [1979]. *Phys. Rep.* **54**:1.

STAUFFER, D. [1985]. *Introduction to Percolation Theory*. London: Taylor & Francis.

STEPHENS, M. J., and STRALEY, J. P. [1974]. *Rev. Mod. Phys.* **46**:617.

STINCHCOMBE, R. B. [1983]. In *Phase Transitions and Critical Phenomena*, Vol. 7, ed. C. Domb and J. L. Lebowitz. New York: Academic Press.

STRALEY, J. P. [1974]. *Phys. Rev.* **A10**:1881.

SWENDSEN, R. H. [1979]. *Phys. Rev. Lett.* **42**:461.

SWENDSEN, R. H. [1984a]. *Phys. Rev. Lett.* **52**:1165.

SWENDSEN, R. H. [1984b]. *J. Stat. Phys.* **34**:963.

SWENDSEN, R. H., and KRINSKY, S. [1979]. *Phys. Rev. Lett.* **43**:177.

SUZUKI, M. [1976]. *Prog. Theor. Phys.* **56**:1454.

SUZUKI, M. [1985]. *Phys. Rev.* **B31**:2957.

TEMPERLEY, H. N. V., ROWLINSON, J. S., and RUSHBROOKE, J. S. [1968]. *Physics of Simple Liquids*. Amsterdam: North-Holland.

THIELE, E. T. [1963]. *J. Chem. Phys.* **39**:474.

THOULESS, D. J. [1979]. In *Ill Condensed Matter*, ed. R. Balian, R. Maynard, and G. Toulouse. Amsterdam: North-Holland.

TOBOCHNIK, J., and CHESTER, G. V. [1979]. *Phys. Rev.* **B20**:3761.

TODA, M., KUBO, R., and SAITO, N. [1983]. *Statistical Physics I, Equilibrium Statistical Mechanics*. Springer Series in Solid-State Sciences, Vol. 30. Berlin: Springer-Verlag.

UHLENBECK, G. E., and FORD, G. W. [1963]. *Lectures in Statistical Mechanics*, Lectures in Applied Mathematics, Vol. 1. Providence, R. I.: American Mathematical Society.

VAN DER WAALS, J. D. [1893]. *Verh. K. Wet. Amsterdam*. (English translation in *J. Stat. Phys.* **20**:197, 1979.)

VASHISHTA, P., and SINGWI, K. S. [1972]. *Phys. Rev.* **B6**:875.

VELICKY, B., KIRKPATRICK, S., and EHRENREICH, H. [1968]. *Phys. Rev.* **175**:747.

VINCENTINI-MISSONI, M. S. [1972]. *Phase Transitions and Critical Phenomena*, Vol. 2, ed. C. Domb and M. S. Green. New York: Academic Press.

WAISMAN, E., and LEBOWITZ, J. [1970]. *J. Chem. Phys.* **52**:4707.

WAISMAN, E., and LEBOWITZ, J. [1972]. *J. Chem. Phys.* **56**:3086, 3093.

WALKER, L. R., and WALSTEDT, R. E. [1977]. *Phys. Rev. Lett.* **38**:514.

WANNIER, G. H. [1966]. *Statistical Physics*. New York: Wiley.

WEEKS, J. D., and GILMER, G. H. [1979]. *Adv. Chem. Phys.* **40**:157.

WEGNER, F. J. [1973]. *Phys. Rev.* **B5**:4529.

WERTHEIM, M. S. [1963]. *Phys. Rev. Lett.* **10**:321.

WERTHEIM, M. S. [1971]. *J. Chem. Phys.* **55**:4291.

WIDOM, B. [1965]. *J. Chem. Phys.* **43**:3898.

WILSON, K. G. [1971]. *Phys. Rev.* **B4**:3174, 3184.

WILSON, K. G., and FISHER, M. E. [1972]. *Phys. Rev. Lett.* **28**:240.

WILSON, K. G., and KOGUT, J. [1974]. *Phys. Rep.* **12**:75.

WISER, N. [1963]. *Phys. Rev.* **129**:62.

WOJTOWICZ, P. J., and SHENG, P. [1974]. *Phys. Lett.* **A48**:235.

YANG, C. N. [1952]. *Phys. Rev.* **85**:809.

YANG, C. N. [1962]. *Rev. Mod. Phys.* **34**:694.

ZIMAN, J. M. [1964]. *Principles of the Theory of Solids*. Oxford: Clarendon Press.

ZIMAN, J. M. [1979]. *Models of Disorder*. Cambridge: Cambridge University Press.

Index

DATE DUE

MAY 2 3 1992		
NOV 11 1992		
DEC 3 0 1992		
SEP 1 9 1993		
Plischke		
May 20 1997		
Keene P.L.		
MAY 3 1998		
		JUN 1 0 1991